Lecture Notes in Computer Science 14667

Founding Editors

Gerhard Goos
Juris Hartmanis

Editorial Board Members

Elisa Bertino, *Purdue University, West Lafayette, IN, USA*
Wen Gao, *Peking University, Beijing, China*
Bernhard Steffen, *TU Dortmund University, Dortmund, Germany*
Moti Yung, *Columbia University, New York, NY, USA*

The series Lecture Notes in Computer Science (LNCS), including its subseries Lecture Notes in Artificial Intelligence (LNAI) and Lecture Notes in Bioinformatics (LNBI), has established itself as a medium for the publication of new developments in computer science and information technology research, teaching, and education.

LNCS enjoys close cooperation with the computer science R & D community, the series counts many renowned academics among its volume editors and paper authors, and collaborates with prestigious societies. Its mission is to serve this international community by providing an invaluable service, mainly focused on the publication of conference and workshop proceedings and postproceedings. LNCS commenced publication in 1973.

Atsuyuki Morishima · Guoliang Li ·
Yoshiharu Ishikawa · Sihem Amer-Yahia ·
H. V. Jagadish · Kejing Lu
Editors

Database Systems for Advanced Applications

DASFAA 2024 International Workshops

BDMS, GDMA, BDQM and ERDSE
Gifu, Japan, July 2–5, 2024
Proceedings

Editors
Atsuyuki Morishima
University of Tsukuba
Tsukuba, Japan

Guoliang Li
Tsinghua University
Beijing, Beijing, China

Yoshiharu Ishikawa
Nagoya University
Nagoya, Japan

Sihem Amer-Yahia
University of Grenoble Alpes
Saint-Martin-d'Hères, France

H. V. Jagadish
University of Michigan
Ann Arbor, MI, USA

Kejing Lu
Nagoya University
Nagoya, Japan

ISSN 0302-9743 ISSN 1611-3349 (electronic)
Lecture Notes in Computer Science
ISBN 978-981-96-0913-0 ISBN 978-981-96-0914-7 (eBook)
https://doi.org/10.1007/978-981-96-0914-7

© The Editor(s) (if applicable) and The Author(s), under exclusive license to Springer Nature Singapore Pte Ltd. 2025

This work is subject to copyright. All rights are solely and exclusively licensed by the Publisher, whether the whole or part of the material is concerned, specifically the rights of translation, reprinting, reuse of illustrations, recitation, broadcasting, reproduction on microfilms or in any other physical way, and transmission or information storage and retrieval, electronic adaptation, computer software, or by similar or dissimilar methodology now known or hereafter developed.
The use of general descriptive names, registered names, trademarks, service marks, etc. in this publication does not imply, even in the absence of a specific statement, that such names are exempt from the relevant protective laws and regulations and therefore free for general use.
The publisher, the authors and the editors are safe to assume that the advice and information in this book are believed to be true and accurate at the date of publication. Neither the publisher nor the authors or the editors give a warranty, expressed or implied, with respect to the material contained herein or for any errors or omissions that may have been made. The publisher remains neutral with regard to jurisdictional claims in published maps and institutional affiliations.

This Springer imprint is published by the registered company Springer Nature Singapore Pte Ltd.
The registered company address is: 152 Beach Road, #21-01/04 Gateway East, Singapore 189721, Singapore

If disposing of this product, please recycle the paper.

Preface

Along with the main conference, the workshops of the Database Systems for Advanced Applications (DASFAA) conference continue to provide valuable forums for researchers and practitioners to explore focused problem domains in the database area. This year, we were pleased to host four successful workshops in conjunction with DASFAA 2024: – The 10th International Workshop on Big Data Management and Service (BDMS 2024) – The 9th International Workshop on Big Data Quality Management (BDQM 2024) – The DASFAA 2024 Workshop on Emerging Results in Data Science and Engineering (ERDSE 2024) – The 8th International Workshop on Graph Data Management and Analysis (GDMA 2024)

These workshops were selected through a rigorous Call-for-Proposals process and were organized by their respective Workshop Organizing Committees and Program Committees. Each workshop focused on a specific area that contributed to the main themes of DASFAA 2024. Following the acceptance of proposals, the workshops conducted their own publicity to solicit contributions and a thorough review of submissions for academic merit. As a result, in total twenty-six papers were accepted, including five for BDMS 2024, three for BDQM 2024, sixteen for ERDSE 2024 and two for GDMA 2024. We would like to express our sincere gratitude to all members of the Workshop Organizing Committees and Program Committees for their hard work and dedication in delivering such a great success to the DASFAA 2024 workshops. Our great gratitude also extends to the main conference organizers for their consistent support in making this year's workshops a valuable addition to the DASFAA conference series. Last but not least, we cannot thank enough the authors who contributed to the workshops: without them, the DASFAA 2024 workshops wouldn't be possible. We hope the workshops provided a stimulating and rewarding experience for all attendees, and we look forward to continued success and growth in the years to come.

July 2024

Atsuyuki Morishima
Guoliang Li
Yoshiharu Ishikawa
Sihem Amer-Yahia
H. V. Jagadish
Kejing Lu

Organization

BDMS 2024 Organization

Co-chairs

Yan Zhao	Aalborg University, Denmark
Xiaoling Wang	East China Normal University, China
Kai Zheng	University of Electronic Science and Technology of China, China

BDQM 2024 Organization

Co-chairs

Chengliang Chai	Beijing Institute of Technology, China
Yuyu Luo	Hong Kong University of Science and Technology, China
Xiaoou Ding	Harbin Institute of Technology, China

ERDSE 2024 Organization

Co-chairs

Satoshi Oyama	Nagoya City University, Japan
Jiyi Li	University of Yamanashi, Japan

Program Committee

Daisuke Kitayama	Kogakuin University, Japan
Dominik Köppl	University of Yamanashi, Japan
Hiroaki Shiokawa	University of Tsukuba, Japan
Hiroaki Ohshima	University of Hyogo, Japan
Makoto P. Kato	University of Tsukuba, Japan
Masafumi Oyamada	NEC, Japan

Jianwei Zhang	Iwate University, Japan
Kazutoshi Umemoto	University of Tokyo, Japan
Kento Sugiura	Nagoya University, Japan
Kei Wakabayashi	University of Tsukuba, Japan
Kejing Lu	Nagoya University, Japan
Shoko Wakamiya	Nara Institute of Science and Technology, Japan
Toshiyuki Shimizu	Kyushu University, Japan
Yang Cao	Tokyo Institute of Technology, Japan
Yasuhiko Kanemasa	Fujitsu, Japan
Yoshiyuki Shoji	Shizuoka University, Japan
Yousuke Watanabe	Nagoya University, Japan
Yuanyuan Wang	Yamaguchi University, Japan
Yusuke Yamamoto	Nagoya City University, Japan
Yuto Hayamizu	University of Tokyo, Japan
Yuyang Dong	NEC, Japan

GDMA 2024 Organization

General Chair

Lei Zou	Peking University, China

PC Chairs

Liang Hong	Wuhan University, China
Xiaowang Zhang	Tianjin University, China
Weiguo Zheng	Fudan University, China

Contents

BDMS

Analysis of Kinematics and Dynamics for Mobile Robots Under Nonholonomic Constraints 3
  Yongcun Shao

Exploring the Application of Nonlinear Partial Differential Equations in Computer Numerical Simulation Operating Systems 15
  Yongcun Shao

FAITH: A Fast, Accurate, and Lightweight Database-Agnostic Learned Cost Model 27
  Waner Li, Xu Chen, Zibo Liang, Runfan Ye, Ruyi Lai, Siying Yue, Qinyuan Su, Kai Zheng, Han Su, and Shaozhi Wu

Fast Approximate Temporal Butterfly Counting on Bipartite Graphs via Edge Sampling 42
  Jiaxi Pu, Yanhao Wang, Yuchen Li, and Xuan Zhou

Financial-ICS: Identifying Peer Firms via LongBERT from 10K Reports 58
  Jintao Huang

BDQM

Establishing a Decentralized Diamond Quality Management System: Advancing Towards Global Standardization 77
  P. H. T. Trung and L. K. Bang

Co-estimation of Data Types and Their Positional Distribution 94
  Shin-ya Sato

Enhancing Load Forecasting with VAE-GAN-Based Data Cleaning for Electric Vehicle Charging Loads 110
  Wensi Zhang, Shuya Lei, Yuqing Jiang, Tiechui Yao, Yishen Wang, and Zhiqing Sun

ERDSE

Audio-Guided Visual Knowledge Representation 129
   Fei Yu, Zhiguo Wan, and Yuehua Li

Boundary Point Detection Combining Gravity and Outlier Detection
Methods 147
   Vijdan Khalique, Hiroyuki Kitagawa, and Toshiyuki Amagasa

A Meta-learning Approach for Category-Aware Sequential
Recommendation on POIs 163
   Jia-Ling Koh, Po-Jen Wen, and Wei Lai

Automatic Post-editing of Speech Recognition System Output Using
Large Language Models 178
   Sheng Li, Jiyi Li, and Yang Cao

Comparative Analysis with Multiple Large-Scale Language Models
for Automatic Generation of Funny Dialogues 187
   Amon Shimozaki, Yousuke Tsuge, Tatsuya Kitamura,
   Tomohiro Umetani, and Akiyo Nadamoto

Effectiveness of the Programmed Visual Contents Comparison Method
for Two Phase Collaborative Learning in Computer Programming
Education: A Case Study 203
   Thanh Ha Nguyen, Yi Sun, Takeshi Nishida, Xiaonan Wang,
   Kazuhiro Ohtsuki, and Hidenari Kiyomitsu

Generating Achievement Relationship Graph Between Actions
for Alternative Solution Recommendation 211
   Tsukasa Hirano, Yoshiyuki Shoji, Takehiro Yamamoto, and Kouzou Ohara

Generating News Headline Containing Specific Person Name 220
   Taiga Sasaki, Takayuki Kuge, Yoshiyuki Shoji, Takehiro Yamamoto,
   and Hiroaki Ohshima

Investigating Evidence in Sentence Similarity Using MASK in BERT 228
   Kanako Nakai, Yuka Kawada, Takehiro Yamamoto, and Hiroaki Ohshima

Acceleration of Synopsis Construction for Bounded Approximate Query
Processing 236
   Tianjia Ni, Kento Sugiura, Yoshiharu Ishikawa, and Kejing Lu

Query Expansion in Food Review Search with Synonymous Phrase
Generation by LLM 252
    Arisa Ashizawa, Ryota Mibayashi, and Hiroaki Ohshima

Question Answer Summary Generation from Unstructured Texts by Using
LLMs 261
    Yuuki Tachioka

Real Estate Information Exploration in VR with LoD Control by Physical
Distance 269
    Yuki Nakayama, Yuya Tsuda, Yoshiyuki Shoji, and Hiroaki Ohshima

Voices of Asynchronous Learning Students: Revealing Learning
Characteristics Through Vocabulary Analysis of Notes Tagged in Videos 277
    *Xiaonan Wang, Yancong Su, Yi Sun, Takeshi Nishida, Kazuhiro Ohtsuki,
    and Hidenari Kiyomitsu*

Review Search Interface Based on Search Result Summarization Using
Large Language Model 293
    Marino Fujii, Yuka Kawada, Takehiro Yamamoto, and Takayuki Yumoto

Yes-No Flowchart Generation for Interactive Exploration of Personalized
Health Improvement Actions 302
    Naoya Oda, Yoshiyuki Shoji, Jinhyuk Kim, and Yusuke Yamamoto

GDMA

Enhancing Link Prediction Based on Simple Path Graphs 319
    Zhiren Li, Yuzheng Cai, and Hongwei Feng

Construction of EMU Fault Knowledge Graph Based on Large Language
Model 335
    Ziwei Han, Hui Wang, Yaxin Li, and Fangzhou Xu

Author Index 349

BDMS

Analysis of Kinematics and Dynamics for Mobile Robots Under Nonholonomic Constraints

Yongcun Shao(✉)

Suzhou City University, Suzhou, Jiangsu 215000, China
WZJ015@suda.edu.cn

Abstract. With the widespread application of mobile robots in industries, services, and exploration, their capabilities in motion planning and obstacle avoidance have become a focal point of research. Nonholonomic constraints, especially those of wheeled mobile robots, significantly impact the motion planning of robots. This paper focuses on the nonholonomic characteristics of Differential Wheeled Robots (DWRs), establishes their kinematic and dynamic models, and analyzes the influence of these constraints on robot motion planning. By introducing the Lagrange multiplier method, nonholonomic constraints are incorporated into the dynamic equations, and the unknown multipliers are eliminated through transformation, resulting in the dynamic equations for DWRs. Moreover, the paper analyzes the controllability of the system using the Chow-Rashevskii theorem, proving that the DWR system is locally controllable in the short term.

Keywords: Nonholonomic Constraints · Differential Wheeled Robots · Kinematic Model · Dynamic Model · Controllability Analysis

1 Introduction

With the rapid development of technology, mobile robots are playing an increasingly important role in various fields such as industrial production, logistics and distribution, home services, medical assistance, disaster relief, and space exploration. They are capable of performing tasks that are difficult or impossible for humans, greatly improving work efficiency and safety. One of the core technologies of mobile robots is autonomous navigation, which enables robots to conduct effective path planning and obstacle avoidance in unknown or dynamically changing- environments, thereby safely and efficiently reaching their destinations.

At the heart of autonomous navigation technology is motion planning, which involves designing the movement paths for robots from their current location to the target location while considering the robot's motion characteristics,- environmental constraints, and obstacle avoidance requirements. However, unlike fully actuated systems (such as cars and airplanes), many mobile robots, especially wheeled robotic vehicles (WRVs), are subject to nonholonomic constraints. These constraints mean that the robot's mobility is limited, for example, they cannot achieve lateral movement independently, which restricts the robot's maneuverability. This presents additional complexity and challenges in motion planning for robots.

A typical example of nonholonomic constraints is differential wheeled robots (DWRs), which have two parallel wheels that can be controlled independently. However, they cannot achieve lateral movement, which limits the robot's mobility. In practical applications, the existence of such nonholonomic constraints makes the motion planning problem more difficult for robots in complex environments. For instance, in narrow or crowded spaces, robots require precise motion planning to avoid collisions while ensuring e cient movement and energy conservation.

To address these issues, researchers have proposed various methods to solve motion planning for mobile robots under nonholonomic constraints. These methods include graph-based search methods, sampling-based methods, optimization-based methods, etc. However, these methods typically assume the robot to be a point mass, ignoring the robot's dynamic characteristics, which may lead to infeasible planned trajectories in high-speed or heavy-load situations.

In recent years, with the increasing application of dynamic models in motion planning, researchers have begun to try to incorporate the dynamic characteristics of robots into the planning process. The introduction of dynamic models can not only improve the feasibility of trajectories but also consider factors such as the robot's energy consumption and driving stability, thereby achieving more optimized motion planning.

This paper focuses on the dynamic modeling and motion planning problems of differential wheeled robots (DWRs), a typical type of nonholonomic mobile robot. Firstly, the paper establishes the kinematic and dynamic models of DWRs and analyzes the impact of nonholonomic constraints on robot motion planning. The paper utilizes the Chow-Rashevskii theorem to analyze the controllability of the system, proving that the DWR system is locally controllable in the short term. These research findings not only enrich the theoretical basis of motion planning for nonholonomic mobile robots but also provide strong support for the practical and industrial application of related technologies.

2 Review of Nonholonomic Mobile Robots

Motion planning for nonholonomic mobile robots is an intricate and dy- dynamically evolving field of study. Recent advancements in this area have tackled a multitude of complexities, encompassing methodologies for control, integration of sensory elements, establishment of communication protocols, navigation around obstacles, and optimization of energy consumption.

Motion planning for nonholonomic mobile robots is an intricate and dynamically evolving field of study. Recent advancements in this area have tackled a multitude of complexities, encompassing methodologies for control, integration of sensory elements, establishment of communication protocols, navigation around obstacles, and optimization of energy consumption.

Research contributions such as those by Li, Wen, and Chen [1] have expanded the theoretical framework of bearing-rigid systems to accommodate nonlinear robotic entities. They introduced a novel distributed control law that relies solely on bearing information for formation control. This groundbreaking approach ensures near-universal stability and adeptly addresses lip ambiguities within three-dimensional networks of robots, particularly those with uncertain system parameters.

Freudenthaler and Meurer [2] have contributed significantly with their innovative concept based on partial differential equations. Their work integrates methods from lateness-based motion planning and feedforward control with a backstepping-based boundary control mechanism. Their experimental validation, involving aswarm of mobile entities transitioning between various formation processes, has underscored the effectiveness of their approach.

Lin et al. [3] have proposed an adaptive image-based visual service strategy for leader-follower formation control under visibility constraints. Their approach ensures global stability through real-time estimation of key parameters such as the inverse height and optical center.

Miao et al. [4] have explored vision-based formation control, addressing challenges related to the field of view and uncertainties about feature depth. Their proposed image-based visual service schemes, designed for both adaptive and static controllers in uncertain systems, have been rigorously validated through numerical simulations and physical experiments.

In a similar vein, Chan, Jayawardhana, and de Marina [5] have investigated angle-constrained formation control for circular mobile entities. They introduced a gradient-based distributed control mechanism that leverages interior angle measurements, ensuring exponential convergence and collision avoidance among neighboring entities while utilizing cost-effective vision sensors.

Wang et al. [6] have made significant progress with an innovative algorithm that combines ant colony optimization (ACO) and the dynamic window approach (DWA) for path planning in uncharted and complex terrains. This integration ensures cooperative obstacle avoidance with high safety and global optimality, marking a pivotal development for applications in unfamiliar environments.

In parallel, Wang et al. [7]introduced a Udwadia-Kalaba approach-based distributed consensus algorithm for mobile entities dealing with communication delays. Their explicit equations of motion under directed spanning tree topology, within a second-order constraint paradigm, represent a significant step towards achieving consensus under challenging communication scenarios.

Gong et al. [8] presented a unique design and control paradigm for a cooperative transport system involving multiple mobile entities. This system features a novel mechanism with six degrees of freedom, capable of sensing various parameters, and effectively facilitates cooperative transport on uneven surfaces. The versatility of its design and control methods has been substantiated through meticulous experimental validation.

Liu et al. [9] proposed an advanced graph-based motion planner that accommodates both formation and obstacle constraints for multiple mobile entities. This innovative approach defines valid configurations that satisfy both constraints and employs a breadth-first search (BFS) method, ensuring optimal motions under formation constraints in obstacle-rich environments.

Furthermore, Chen et al. [10] conducted a comprehensive analysis of the kinematics of nonholonomic mobile entities, revealing a distributed mechanism for formation control under both fixed and switching topologies. Their work corporates stability estimation and globally uniform exponential convergence conditions, as verified through rigorous simulation results.

Liu et al. [11] addressed the complex interplay between distributed formation control and multi-objective avoidance. By proposing a motion planner skilled at balancing formation and obstacle avoidance through caging behavior and an improved artificial potential field algorithm, this work represents a significant advancement in handling real-world complexities.

Wang et al. [12] introduced a scheme for antidisturbance collision-free motion control based on nested constraints for multiple mobile entities. This innovative approach combines distributed cooperative control methods with artificial potential yield-based strategies, ensuring effective formation control with simultaneous collision avoidance.

Additionally, Wang et al. [13] developed a controller that ensures kinematic consistency for nonholonomic mobile entities by integrating kinematic and torque controllers for precise trajectory tracking. The reliability of their controller is showcased through meticulous simulation and experimentation on a mobile platform with two drive wheels.

Hameed et al. [14] provided a crucial perspective on enhancing energy efficiency in systems involving multiple mobile entities. They introduced an algorithm for optimizing power consumption for systems with automatic storage and retrieval using multi-mobile entities. Their algorithm, grounded in e cient motion planning, significantly improves energy consumption efficiency, as demonstrated through rigorous simulation results.

These innovative studies collectively contribute to the dynamic field of formation control in systems involving multiple mobile entities, offering profound insights, advanced strategies, and experimentally validated approaches that align with the highest academic standards.

3 Nonholonomic Constraints

3.1 Definition and Characteristics of Nonholonomic Constraints

Nonholonomic constraints are a class of kinematic constraints that restrict the motion of a mechanical system without directly affecting its energy. These constraints are inherent in the geometry of the system and cannot be integrated to yield a holonomic constraint, which is a constraint that can be expressed as a differential equation in the generalized coordinates. Nonholonomic constraints are typically represented in the form of differential relations that must be satisfied during the motion of the system. A prominent feature of nonholonomic systems is that they cannot be eliminated or transformed into an infinite number of algebraic equations, thus complicating the analysis and control of such systems.

3.2 The Impact of Nonholonomic Constraints on Mobile Robots

For mobile robots, nonholonomic constraints manifest in various forms, with the most common being the steering and mobility limitations of wheeled robots. For instance, a car or a differential wheeled robot (DWR) cannot change its direction instantaneously or move sideways due to the nonholonomic constraints imposed by its wheel structure. This limitation affects the robot's ability to navigate through narrow passages, make sharp turns, or move backward without turning in place.

The presence of nonholonomic constraints also affects the robot's motion planning and control strategies. Traditional motion planning algorithms often assume that robots can move freely in any direction, which is not the case for nonholonomic robots. Therefore, ignoring these constraints can lead to planned paths that are infeasible or unsafe for the robot to follow. Similarly, control strategies designed without considering nonholonomic constraints may not be able to accurately command the robot to reach desired poses or avoid obstacles.

3.3 Modeling Nonholonomic Constraints

Modeling nonholonomic constraints accurately is crucial for the development of effective motion planning and control algorithms. In the context of mobile robots, these constraints can be expressed as a set of differential equations that relate to the robot's velocity components. For example, in the case of a DWR, the nonholonomic constraint can be represented as the equation that relates the linear velocity and angular velocity of the robot, such as the condition that the linear velocity has no component perpendicular to the direction of the robot's heading.

3.4 Challenges in Handling Nonholonomic Constraints

Dealing with nonholonomic constraints presents several challenges. First, the presence of these constraints complicates the trajectory generation process, as the planner must ensure that all generated paths satisfy the nonholonomic conditions. Second, the nonholonomic nature of the robot's motion can lead to a large number of local minima in the cost function used for optimization, making it difficult to find the global optimum. Finally, the non-integrable nature of non-holonomic constraints means that they cannot be easily incorporated into the dynamic model of the robot, which is necessary for accurate control.

3.5 Overview of Methods to Address Nonholonomic Constraints

To overcome the challenges posed by nonholonomic constraints, researchers have developed various methods. Some approaches reformulate the motion planning problem to explicitly incorporate nonholonomic constraints, while others use approximation techniques to simplify the constraints. Advanced control strategies, such as model predictive control (MPC), have also been adapted to handle non-holonomic systems by incorporating the constraints directly into the control law. These methods aim to ensure that the robot's motion is both feasible and safe, taking into account the inherent limitations of nonholonomic systems.

4 Kinematic Modeling of Differential Wheeled Robots (DWRs)

4.1 Fundamentals of DWR Kinematics

Differential Wheeled Robots (DWRs) are a class of mobile robots characterized by two independently controlled wheels that allow for maneuverability and navigation in various environments. The kinematic model of a DWR is essential for understanding its motion capabilities and for designing control algorithms that enable efficient and safe navigation.

The kinematic model describes the relationship between the robot's body and its wheels, focusing on the robot's position, orientation, and velocity in the workspace. For a DWR, the kinematic model typically involves the following elements: - The position of the robot's center of mass (COM) in the global coordinate frame. - The orientation of the robot, often represented by the angle of the robot's heading with respect to a reference direction. - The linear velocity of the robot, which is the speed along the direction of the robot's heading. - The angular velocity, which represents the rate of change of the robot's orientation. Figure 1 shows Differential Wheeled Robots.

4.2 DWR Kinematic Equations

The kinematic equations for a DWR are derived from the conservation of mass and momentum principles. These equations relate the robot's wheel velocities to its linear and angular velocities. A common set of kinematic equations for a DWR is given by:

$$v = \frac{R(\theta) \cdot u}{L}$$

$$\omega = \frac{1}{L}(u_r - u_l)$$

where:
v is the linear velocity of the robot's COM.
ω is the angular velocity of the robot.

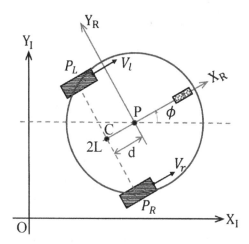

Fig. 1. Differential Wheeled Robots

$R(\theta)$ is a function that depends on the robot's heading angle θ and relates the wheel speeds to the robot's linear velocity.
u is the desired linear velocity command.
u_r and u_l are the velocities of the right and left wheels, respectively.
L is the distance between the wheels (wheelbase).

These equations assume that each wheel can operate independently and that robot's motion is influenced by the differential speeds of the two wheels.

4.3 Nonholonomic Constraints in DWR Kinematics

The nonholonomic constraints of a DWR arise from the robot's inherent physical limitations, which restrict certain motions even though they do not violate any holonomic constraints. For example, a DWR cannot change its orientation instantaneously or move sideways without turning.

These constraints are captured in the kinematic model through these structures of the kinematic equations. Specifically, the nonholonomic constraints ensure that the robot's velocity vector always lies in the plane tangent to the robot's path and that the angular velocity is related to the rate of change of the robot's heading.

4.4 Kinematic Modeling Challenges for DWRs

Developing an accurate kinematic model for a DWR involves several challenges. One of the primary challenges is accounting for the effects of wheel dynamics, such as ground friction and wheel compliance, which can affect the robot's motion. Additionally, the kinematic model must consider the robot's interactions with the environment, such as obstacles and uneven terrain, which can alter the robot's trajectory.

Another challenge is ensuring that the kinematic model is computationally e cient, as it often forms the basis for real-time motion planning and control algorithms. This requires the model to strike a balance between accuracy and computational complexity.

The kinematic model of a DWR is a critical component in the design of autonomous navigation systems. It provides a mathematical representation of the robot's motion capabilities and the constraints imposed by its physical structure. By understanding and accurately modeling the kinematics of DWRs, researchers, and engineers can develop effective motion planning and control strategies that enable these robots to navigate complex environments safely and efficiently.

5 Dynamic Modeling of Differential Wheeled Robots (DWRs)

5.1 Introduction to DWR Dynamics

The dynamic modeling of Differential Wheeled Robots (DWRs) is a complex process that involves understanding the forces and torques acting on the robot, as well as how these forces influence the robot's motion. A comprehensive dynamic model is crucial for the development of control strategies that ensure stability, efficiency, and safety in the robot's operation. This section will delve into the key aspects of DWR dynamics and how they are modeled.

5.2 Formulation of DWRs Dynamic Equations

The dynamics of a DWR can be described by a set of nonlinear differential equations that govern the relationship between the applied forces and torques and the resulting motion of the robot. These equations typically take into account the robot's mass, inertia, friction, and other physical properties. The general form of the dynamic equations for a DWR can be expressed as:

$$M(\dot{v},\dot{\omega}) \begin{bmatrix} F \\ T \end{bmatrix} + C(\dot{v},\dot{\omega},v,\omega) \begin{bmatrix} v \\ \omega \end{bmatrix} + K(\omega) = \begin{bmatrix} F_{\text{ext}} \\ T_{\text{ext}} \end{bmatrix} \quad (1)$$

where:

- $M(\dot{v},\dot{\omega})$ is the mass matrix that depends on the linear and angular accelerations.
- $C(\dot{v},\dot{\omega},v,\omega)$ is the Coriolis and centripetal matrix, accounting for inertial effects due to the robot's rotation.
- $K(\omega)$ represents the frictional forces and torques acting on the robot.
- F and T are the linear and angular velocities of the robot.
- F_{ext} and T_{ext} are the external forces and torques, such as those from the environment or other robots.

5.3 Incorporating Nonholonomic Constraints in Dynamics

The nonholonomic constraints of a DWR must also be considered in the dynamic model to accurately represent the robot's motion capabilities. These constraints are incorporated through the mass matrix and the Coriolis and centripetal matrix, which are functions of the robot's velocity and orientation. The nonholonomic nature of the DWRs's motion means that the dynamic model must satisfy certain differential relations that prevent the robot from performing certain maneuvers, such as sidestepping or instantaneous changes in direction.

5.4 Challenges in Dynamic Modeling

Creating an accurate dynamic model for a DWR presents several challenges. One of the primary difficulties is identifying and quantifying all the forces and torques that affect the robot's motion, including those due to the robot's own dynamics and interactions with the environment. Additionally, the dynamic model must account for variations in the robot's mass distribution, wheel characteristics, and ground conditions, which can significantly influence the robot's behavior.

Another challenge is ensuring that the dynamic model is computationally efficient and suitable for real-time control applications. The complexity of the dynamic equations can lead to high computational demands, which may not be feasible for time-sensitive control tasks. Therefore, it is often necessary to simplify the dynamic model while maintaining its accuracy and relevance for control purposes.

The dynamic modeling of DWRs is a critical aspect of designing effective control systems that ensure the robot's stable and efficient operation. By incorporating the nonholonomic constraints and accurately representing the forces and torques acting on the robot, the dynamic model provides a foundation for developing control algorithms that can adapt to various operating conditions and environmental interactions. Despite the challenges, a well-formulated dynamic model is essential for the advancement of DWR technology and its application in diverse settings.

6 Controllability Analysis

Controllability is a fundamental concept in control theory, used to assess whether there exists an input sequence that leads to a specific target state. A system described by $\dot{x} = f(x, u)$ is controllable if, for all $x_I, x_G \in X$, there exists a time $t > 0$ and an input sequence $\tilde{u} \in U$ such that the integral from x_I results in x_G. If for all $x_I, x_G \in X$, $x_G \in R(x_I, U)$, then the system is controllable, where R denotes the reachable set.

6.1 Controllability of Linear Systems

If a system is linear, its controllability can be judged by the rank of the controllability matrix. According to the Kalman controllability rank condition, if the controllability matrix is full rank, then the system is controllable.

6.2 Controllability of Nonlinear Systems

If the system is nonlinear, its controllability can be judged by linearizing the operating point and then determining the rank of the controllability matrix. The DWR system constructed earlier is controllable, but it may not be controllable if there are obstacles within the working range.

6.3 Short-Time Local Controllability (STLC)

Short-Time Local Controllability (STLC) can be defined as follows. If a system $\dot{x} = f(x, u)$ is STLC at x_I, then there exists a $t > 0$ such that for all $t' \in (0, t]$, $x_I \in \text{int}(R(x_I, U, t'))$, where int denotes the interior, and R denotes the short-time reachable set.

6.4 Affine Control Systems

An affine control system has the form $\dot{x} = f(x) + \sum_{i=1}^{m} g_i(x) u_i$, where the set $\{g_1, \ldots, g_m\}$ is called the system's vector field. Such systems are linear with respect to actions but nonlinear with respect to states.

6.5 DWR System

Taking the simplified DWRs model with the centroid and the center of rotation coinciding. Let $v = u_1$, $\omega = u_2$, rewrite the model as follows:

$$\begin{bmatrix} \dot{x} \\ \dot{y} \\ \dot{\phi} \end{bmatrix} = \begin{bmatrix} \cos\phi & 0 & 0 \\ \sin\phi & 0 & 0 \\ 0 & 0 & 1 \end{bmatrix} u_1 + \begin{bmatrix} 0 & 0 & 0 \\ 0 & 0 & 0 \\ 0 & 0 & 1 \end{bmatrix} u_2$$

6.6 Lie Algebra and Chow-Rashevskii Theorem

The controllability of an affine control system without drift can be described by the rank condition of the Lie algebra. According to the Chow-Rashevskii theorem: a system is STLC at $x \in X$ if and only if the Lie algebra dimension $\dim(L(\Delta)) = \dim(X) = n$, where n is the dimension of the space X.

6.7 Lie Algebra of the DWR System

The Lie algebra of the DWR system is:
$L(\Delta) = \text{span}\{[1, 0, 0]^T, [0, 1, 0]^T, [0, 0, 1]^T, [\sin\phi, -\cos\phi, 0]^T\}$. Thus, the Lie algebra of the differential mobile robot system is a three-dimensional Lie algebra, i.e., $\dim(L(\Delta)) = n$. According to the Chow-Rashevskii theorem, the system satisfies STLC and non-holonomicity, which means the DWR system is fully controllable.

7 Conclusion

The development of the DWRs' kinematic model, which incorporates nonholonomic constraints, ensures that any motion planning algorithm developed for such robots is feasible and adheres to the physical limitations inherent to their design. By accurately representing the robot's capabilities and restrictions, the kinematic model serves as a critical tool for navigating complex environments without collision or violation of the robot's nonholonomic nature.

The dynamic model of DWRs, on the other hand, offers insights into the forces and torques that govern the robot's motion. This model is essential for the design of control strategies that not only ensure the robot's stability and efficiency but also its safety in operation. By understanding the dynamics of DWRs, researchers and engineers can create algorithms that optimize energy consumption, improve response times, and maintain precise control over the robot's movements.

The paper's exploration of the challenges in handling nonholonomic constraints and the strategies to overcome them provides valuable guidance for future research. The emphasis on computational efficiency and the need for real-time applicability in modeling and control is particularly relevant in today's fast-paced technological landscape.

References

1. Li, X., Wen, C., Chen, C.: Adaptive formation control of networked robotic systems with bearing-only measurements. IEEE Trans. Cybern. **51**(1), 199–209 (2020)
2. Freudenthaler, G., Meurer, T.: Pde-based multi-agent formation control using flatness and backstepping: analysis, design, and robot experiments. Automatica **115**, 108897 (2020)
3. Lin, J., Miao, Z., Zhong, H., Peng, W., Wang, Y., Fierro, R.: Adaptive image-based leader-follower formation control of mobile robots with visibility constraints. IEEE Trans. Industr. Electron. **68**(7), 6010–6019 (2020)
4. Miao, Z., Zhong, H., Lin, J., Wang, Y., Chen, Y., Fierro, R.: Vision-based formation control of mobile robots with for constraints and unknown feature depth. IEEE Trans. Control Syst. Technol. **29**(5), 2231–2238 (2021)
5. Chan, N.P., Jayawardhana, B., de Marina, H.G.: Angle-constrained formation control for circular mobile robots. IEEE Control Syst. Lett. **5**(1), 109–114 (2020)
6. Wang, Q., Li, J., Yang, L., Yang, Z., Li, P., Xia, G.: Distributed multi-mobile robot path planning and obstacle avoidance based on ACO"CDWA in unknown complex terrain. Electronics **11**(14), 2144 (2022)
7. Wang, C., Ji, J., Miao, Z., Zhou, J.: Udwadia-Kaliba approach based distributed consensus control for multi-mobile robot systems with communication delays. J. Franklin Inst. **359**(14), 7283–7306 (2022)
8. Gong, Z., Nie, Z., Liu, Q., Liu, X.J.: Design and control of a multimobile-robot cooperative transport system based on a novel six-degree-of-freedom connector. ISA Trans. (2023)
9. Liu, W., Hu, J., Zhang, H., Wang, M.Y., Xiong, Z.: A novel graph-based motion planner of multi-mobile robot systems with formation and obstacle constraints. IEEE Trans. Robot. (2023)

10. Chen, N., Wang, Y., Jia, L.: Multimobile robot leader-follower formation distributed control under switching topology, pp. 2673–2678 (2020)
11. Liu, W., Zhang, H., Hu, J., Wang, M.Y., Xiong, Z.: A distributed formation controller with multi-obstacle avoidance for multi-mobile robot system, pp. 256–261 (2022)
12. Wang, G., Wang, X., Li, S., Niu, D.: A nested constraints based anti-disturbance collision-free formation control scheme for multi-mobile robot systems, pp. 4793–4798 (2022)
13. Wang, D., Zhou, J., He, C., Miao, Z.: Experimental verification of formation control in nonholonomic multi-mobile robots, pp. 5327–5331 (2021)
14. Hameed, H.M., Rashid, A.T., Rashid, M.T., Al Amry, K.A.: Constructing the optimal power algorithm for as/rs systems using multimobile robots. In: IOP Conference Series: Earth and Environmental Science, vol. 877, no. 1, p. 012014 (2021)

Exploring the Application of Nonlinear Partial Differential Equations in Computer Numerical Simulation Operating Systems

Yongcun Shao(✉)

Suzhou City University, Jiangsu 215000, Suzhou, China
wzj015@suda.edu.cn

Abstract. In this study, we introduce a novel method for enhancing diffusion images through a hybrid partial differential equation approach. This approach incorporates pulse filtering, an enhanced forward-backward diffusion filter, and the total variational diffusion method. Our algorithm effectively enhances small boundaries in images by utilizing an improved hybrid partial differential equation diffusion image filtering algorithm. By utilizing a pulse filter with specific parameter constraints, we enhance the processing capabilities for strong boundaries. Additionally, we employ the level set method to eliminate sawtooth artifacts at boundaries and enhance boundary smoothness. Furthermore, we propose a biorthogonal filtering method based on a degraded model that achieves biorthogonal mapping in the spatial domain, ensuring the invertibility of the degraded extended image concerning the degraded model. Simulation results demonstrate the superiority of our method compared to other techniques. The resulting amplified images exhibit excellent photorealism, with improved boundary details for both weak and medium brightness levels. This method effectively preserves image boundary details while effectively reducing noise.

Keywords: Image enhancement · image quality improvement · frequency domain processing · differential equation · image compression

1 Introduction

Since the 1980 s, computerized picture handling has been of broad concern and continues to evolve. Since the 1980 s, China has taken on DSP chips, from the underlying DSP chip advancement to the present, DSP has turned into the fundamental parts of correspondences, PCs, picture handling, purchaser hardware, and numerous different perspectives. As of now, the global television following and estimation are completely focused on continuous execution, for the most part accepting advanced signal handling as the fundamental means and fast computerized signal processor as the principal equipment. Through ceaseless advancement of the first calculation, further improvement of the current calculation, and nonstop overhauling of the computerized signal processor, the handling execution of the equipment is upgraded [1]. It makes continuous advanced picture handling conceivable and acquires higher accuracy and constant accuracy and ongoing. Picture intensification technique is the computerized picture after specific handling, to get a higher goal picture, that is to say, a solitary picture goal improvement reproduction, it is an extremely down-to-earth picture enhancement strategy. As per the steps, they are separated into two classifications: one is acquired from the high-goal unique picture by down examining to work with picture transmission and pressure; The subsequent stage is to utilize the light acceptance gadget to change the scene to get the picture. Another strategy is to involve the first scene as a high-goal picture, the utilization of different equipment to gather the picture, convert it into an electrical sign, and afterward disintegrate it into a computerized picture, in order to accomplish low-pass sifting of the picture [2]. For high-accuracy advanced picture approximating the first picture and point dispersion capability in view of energy variational hypothesis, it is difficult to get the high-accuracy point dissemination capability adjusting to the model limit by accepting the base energy blunder as the beginning stage. Picture examination and handling methods in light of nonlinear differential conditions have grown quickly, among which the most delegated ones are partial dispersion conditions, picture honing drive channel, positive and in reverse nonlinear dissemination condition, and so on, which are significant exploration contents in the fields of picture denoising and honing [3]. A few researchers have applied it to picture amplification. Focusing on the deficiencies of the current exploration, this task consolidates it with the customary wavelet change-based separating strategy, so it can successfully hone areas of strength for the zero intersection and upgrade the little edge subtleties.

Lately, with the constant advancement of computerized picture-handling innovation, picture amplification has turned into a significant examination subject in the field of picture handling. The conventional picture amplification technique for the most part utilizes the addition calculation to develop the picture, which will make the edge of the picture become obscured. To tackle this issue, a few researchers have proposed different picture amplification strategies in view of the hypothesis of variational math and partial differential conditions. These strategies can really save the edge data of the picture and stay away from the event of the picture being after amplification. Be that as it may, they actually have a few deficiencies during the time spent on picture amplification, for example, low amplification proportion, slow handling velocity, and restricted application situations. Accordingly, it is important to additionally investigate and explore the picture amplification innovation. In this task, we join the partial differential condition with the conventional wavelet change-based separating strategy to propose another picture amplification technique. This strategy can really safeguard the edge data of the picture and work on the visual nature of the picture after amplification. Simultaneously, it likewise has a high amplification proportion and quick handling speed and can be applied to various situations. Moreover, we additionally utilize mathematical recreation investigations to check the viability of this strategy. The trial results demonstrate the way that this technique can actually work on the visual nature of the picture after amplification and has great application possibilities in the field of computerized picture handling.

Inaccuracy and constant, the picture enhancement strategy is a computerized picture handling method that upgrades the goal of a solitary picture, recreating it to a higher goal. This functional technique is ordered into two primary stages. The primary stage includes downsampling a high-goal unique picture to work with transmission and pressure. The subsequent stage utilizes a light enlistment gadget to catch the scene and change it into a picture. On the other hand, the first scene can be caught as a high-goal picture utilizing different equipment parts. The picture is then changed over into an electrical sign and digitized for low-pass separating. To accomplish high-accuracy point dissemination works that precisely rough the first picture and conform to show limits, energy variational hypothesis can't be exclusively depended upon. Presently, nonlinear differential condition-based picture examination and handling strategies have quickly evolved. Outstandingly, fragmentary dispersion conditions, picture-hanging drive channels, positive and in reverse nonlinear dissemination conditions, and different procedures have arisen as key exploration regions in picture denoising and honing. A few researchers have applied these procedures to picture amplification. To address the constraints of existing exploration, this venture incorporates the customary wavelet change-based separating technique to successfully upgrade solid edge intersections and further develop little edge subtleties.

2 Implementation Plan for Multi-Target Tracking Based on DSP

In the multi-target tracking scheme based on DSP, it is necessary to use FPGA chips and DSP chips in conjunction to complete the tracking of target objects. The business flow of the relevant modules is shown in Fig. 2–1:

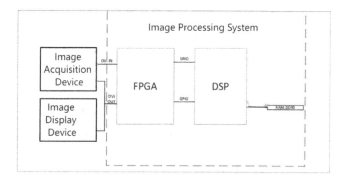

Fig. 1. DSP-based image processing process

From Fig. 2–1, the video signal from the device is transmitted to the FPGA module through the input port. The FPGA is responsible for processing the received video signal and encapsulating it into SRIO format data packets, which are sent to the DSP chip in real time. After receiving the video signal, the DSP chip performs target recognition and tracking processing, identifying the target's identification information and motion trajectory, and then returns the processing results to the FPGA through SRIO format data packets. The FPGA processes the data packets and restores them to video image format, which is transmitted to other external modules through the output port, forming a closed loop in business logic.

This venture expects to utilize two TMS320C6678 chips created by TI organization, enhanced by programmable CPLD and FPGA, joined with rapid video A/D, D/An, and other hardware, to construct a bunch of DSP chips in view of high accuracy, accuracy, low power and low power quick picture handling stage. The simple picture of CCD yield is examined and A/D change is performed at rapid speed [4]. The aim of this project is to construct a high-performance, low-power real-time digital image processing system based on dual DSP chips. To achieve this goal, we have selected two TMS320C6678 chips manufactured by Texas Instruments (TI), enhanced with programmable CPLD and FPGA, combined with rapid video A/D and D/A converters, as well as other necessary hardware components.

2.1 DSP Chip TMS320C6678

In the research of multi-target tracking technology based on DSP image processing systems, it is necessary to configure a DSP chip. The TMS320C66x series DSP chips released by TI Company are integrated with C66x fixed-point/floating-point and have an octal-core DSP. The typical speed is 1GHz, and according to the design standards, its computing power can reach 44.8 GMAC and 22.4 GFLOP, providing sufficient algorithm capabilities to meet the requirements of target tracking algorithms in a single physical scenario. Moreover, it ensures a good performance margin during long-term safe and stable operation. In this research project, we aim to find the best solution that balances image processing efficiency and performance consumption. Therefore, we choose the TMS320C6678 chip from the series to provide a research foundation for this project. The core technical parameters of the TMS320C6678 are as follows: CPU Core: 8 cores;
CPU Frequency: Single processing 1.25 GHz;
Onboard Memory: 4096KB*8;
External Memory: Supports 8 GB, x64, DDR3;
Data Communication Interface: Four-port SRIO communication module1, SGMII communication port2, two-port PCIe communication module1, four-port HyperLink communication module1;
Software Operating Environment: Supports VisualDSP++, CCS5.3;
According to the official product information released by TI Company, the sys-

tem framework diagram and description of the TMS320C6678 chip are shown in Figure.

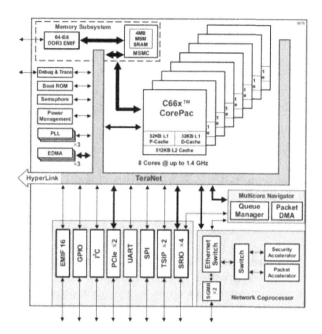

Fig. 2. TMS320C6678 chip frame diagram and description released by TI

2.2 FPGA Configuration Selection

According to the image processing flow shown in Fig. 4–1, this research requires the configuration of an FPGA chip. FPGA refers to the field-programmable array, which has the characteristics of high customization. Its technology has developed rapidly, continuously improving the degree of integration, and is often packaged with other chips to create multifunctional integration. Since the FPGA chip is not the core technology in this research, a mature and general FPGA chip is selected to construct the image processing system. According to the technical implementation requirements of multi-target tracking in this research, the FPGA chip selection needs to meet the following requirements:

According to the data [5], the mainstream manufacturers of the FPGA market include Xilinx, Microsemi, Altera, Lattice, etc. After the comparison of performance parameters, the Artix-7 XC7A100T-2FGG484-I chip produced by Celings is selected for the application of this topic. The Artix-7 XC7A100T-2FGG484-I chip seamlessly matches the previously selected DSP chip TMS320C6678 in terms of technical communication. Support to provide HDMI, DVI digital interface, and display model output to the display.

Fig. 3. requirements for selecting an FPGA chip

2.3 Software Development and Programming

According to the DSP chip option in the previous text, the TMS320C6678 chip supports the CCS5.3 version of the software operating environment [6]. The multi-target tracking software programming and CCS5.3 are carried out. The algorithm implementation logic is programmed in C language,and compiled,and an executable code file is generated, downloaded to the DSP chip, and run on the hardware platform. Part of the DSP algorithm is implemented as follows:

With the aforementioned hardware composition, our dual DSP real-time digital image processing system is capable of efficiently processing image data, providing a solid hardware foundation for image enhancement and other image processing tasks. The construction of this system not only improves the speed and quality of image processing but also opens up new possibilities for research and applications in related fields.

3 Innovative Dual-Directional Diffusion Technique for Image Enhancement

This section introduces a novel approach to image enlargement and quality improvement through an advanced hybrid bidirectional diffusion method. This technique is designed to harness the strengths of nonlinear partial differential equations (PDEs) and impulse filtering to achieve superior image enhancement results. The method is characterized by its ability to preserve fine details and boundaries while effectively reducing noise, ensuring that the amplified images exhibit high photorealism and clarity.

3.1 Pulse Filtering for Edge Preservation

A critical component of our hybrid method is the use of pulse filtering, which is tailored to enhance the processing capabilities for strong boundaries within the image. This step is essential for maintaining the integrity of the image's edges and contours. By applying specific parameter constraints to the pulse filter, we can significantly improve the way strong boundaries are processed, resulting in a more defined and crisp image output (Fig. 5).

```
typedef struct
{
int targetFR;
int targetID;
int detX;
int detY;
int detW;
int detH;
float score;
}TargetInfo;


typedef struct
{
TargetInfo targetInfo;
int FrTarNum;
}ResvInfo;


typedef struct
{
int targetFR;
int targetID;
float targetcost;
int* pNeiInds;
float* pNeiC;
int NeiNum;
float SPDist;
int preTarget;
int oriTarget;
int preTargetID;
int curTargetID;
}TrackInfo;
```

Fig. 4. Define structure

```
prvs = cv2.cvtColor(prvs, cv2.COLOR_BGR2GRAY)
next = cv2.cvtColor(next, cv2.COLOR_BGR2GRAY)
inst = cv2.optflow.createOptFlow_DIS(cv2.optflow.DISOPTICAL_FLOW_PRESET_MEDIUM)
inst.setUseSpatialPropagation(True)
flow = inst.calc(prvs, next, None)
```

Fig. 5. Image processing with optical flow

3.2 Enhanced Forward-Backward Diffusion Filtering

Our algorithm incorporates an advanced forward-backward diffusion filter that works synergistically with the pulse filter. This dual-action filter effectively amplifies subtle details and small boundaries in the image, ensuring that the enhanced image retains its original quality and sharpness. The forward-backward diffusion process is refined to intelligently adapt to the characteristics of the image, leading to a more natural and visually appealing result.

3.3 Total Variational Diffusion for Smoothness

A key innovation in our method is the application of the total variational (TV) diffusion approach. By employing the level set method, we eliminate sawtooth artifacts at boundaries and enhance boundary smoothness. This technique contributes significantly to the overall improvement in image quality, making the amplified images appear more natural and visually coherent.

3.4 Biorthogonal Filtering for Image Fidelity

We propose a biorthogonal filtering method based on a degraded model, which achieves biorthogonal mapping in the spatial domain. This ensures the invertibility of the degraded extended image with respect to the degraded model, a crucial step for maintaining the image's fidelity during the amplification process. This method is vital for preserving the image's original details and ensuring that the enhanced image closely represents the original scene.

By using the advantages of nonlinear differential equation diffusion and coupled impulse filter TV algorithm, a hybrid bidirectional filter is obtained

$$Q\{s(i,j,t)\} = s(i,j,t) - h * (\uparrow D(((h(i,j) * s(i,j,t)) \downarrow D) - v(i\prime, j\prime)))$$

The optimal diffusion time t_0 of the nonlinear differential equation filter is related to the magnification times and a definite diffusion time can be selected for different magnification times [7]. Simulation experiments show that the number of optimal nonlinear differential equations for different types of images mainly depends on the magnification of the image, and $h(i,j)$ also has an effect.

By integrating these advanced techniques, our hybrid bidirectional diffusion method stands out as a superior approach to traditional image enlargement techniques. It provides enhanced image quality and a more accurate representation of the original scene. The effectiveness of this method is demonstrated through extensive simulations, highlighting its potential for application across various fields of digital image processing.

4 System Simulation

In the realm of digital image processing, simulation plays a pivotal role in validating the effectiveness and practicality of newly developed techniques. To this end, we have conducted a series of system simulations to assess the performance of our proposed hybrid bidirectional diffusion image amplification method.

The simulation process is designed to emulate the practical application of our technique, taking into account various factors such as image magnification, noise levels, and the intricacies of different image types. The steps involved in our system simulation are as follows:

Model Discretization: We begin by discretizing the model to adapt it for computational processing. This involves converting the continuous equations into a form that can be implemented on a digital platform.

Image Interpolation: The original image is enlarged by interpolating the values between pixels. This step is crucial for simulating the process and involves using various interpolation methods to estimate the pixel values at the increased resolution.

Adaptive Nonlinear Filtering: Our algorithm applies an adaptive nonlinear filter to the interpolated image. This process takes into account the local gradient information at each pixel, allowing for a nuanced approach to detail preservation and noise reduction.

Boundary Preservation: Special attention is given to the edges and boundaries within the image. Our technique ensures that these critical areas are sharpened and preserved, avoiding the common issue of boundary blurring that can occur during image enlargement.

Noise Assessment: The impact of noise on the image is evaluated under different conditions. Our method is designed to minimize the amplification of noise while maximizing the enhancement of image details.

Performance Metrics: The results of the simulation are quantified using performance metrics such as mean square error (MSE) and visual quality assessments. These metrics provide an objective basis for comparing our method against other existing techniques.

Take the original image $T(m \times n)$ and enlarge the interpolation between the rows and rows twice to: $Z(M \times N, M = 2m - 1, N = 2n - 1)$ The four circle points are the data points of the original image and the remaining five cross points are the points that need to be interpolated. In the 8 neighborhoods of the target pixel, where α, n, δ, β is half point, $T_R = \{A, N, \Delta, B\}$ is defined. Let $c = (c^1, c^2) = \frac{\nabla s}{|\nabla s|^p}$, calculate divergence using central difference:

$$\nabla \cdot c = \frac{\partial c^1}{\partial x} + \frac{\partial c^2}{\partial y} \approx \frac{c_\alpha^1 - c_\delta^1}{f} + \frac{c_n^2 - c_\beta^2}{f}$$

f indicates that the mesh size is set to 1. The following uses the central difference to approximate c, taking half α as an example, other points are similar:

$$c_\alpha^1 = \frac{1}{|\nabla s_\alpha|^p} \frac{\partial s}{\partial x}\bigg|_\alpha \approx \frac{1}{|\nabla s_\alpha|^p} \frac{s_A - s_R}{f},$$
$$|\nabla s_\alpha|^2 \approx \frac{1}{f^2}((s_A - s_R)^2 + \frac{1}{16}(s_B + s_{BA} - s_N - s_{NA})^2)$$

Equation (4) is then discretized at pixel point as follows:

$$0 = \sum_{Q \in T_R} \frac{1}{|\nabla s_\alpha|^p}(s_R - s_Q) + \kappa(s_R - s_R^0)$$

If $Q = A$, then $q = \alpha$, other points are similar, and s_R^0 is the initial value of image Z at pixel point R.

$$\varphi_Q = \frac{1}{|\nabla s_Q|^p},$$
$$f_{RQ} = \frac{\varphi_Q}{\sum_{P \in T_R} \varphi_P + \kappa},$$
$$f_{RR} = \frac{\kappa}{\sum_{P \in T_R} \varphi_P + \kappa}, Q \in T_R$$

Then equation is reduced to:

$$s_R = \sum_{Q \in T_R} f_{RQ} s_Q + f_{RR} s_R^0$$

The equation is linearized and solved by iterative method:

$$s_R^{(n)} = \sum_{Q \in T_R} f_{RQ}^{(n-1)} s_Q^{(n-1)} + f_{RR}^{(n-1)} s_R^{(n-1)}$$

In order to accelerate the convergence of the algorithm, the initial value of Z is interpolated bilinear. In the actual operation, how to choose the appropriate parameters is of great significance to preserve the image boundary. Equation (4) applies an adaptive nonlinear filter process to an image where the value of each pixel point is obtained by interpolating the weights of four nearby pixel points. A representative one-dimensional inclined boundary is studied and it is found that it has a maximum value in the center of the boundary curve.

Since the gradient of the central pixel point c is the largest, the gradient of the pixel points a and e on both sides is relatively small, and the sum of weights $\sum f_{RQ} = 1$, therefore, the central pixel point has a lower weight, while the pixel points on both sides have a higher weight so that the effect of interpolation is that the image boundary is still clear. Figure 3 shows the interpolation results for $P = 4.8$.

The selection of parameter κ is related to the noise level of the image. If the image does not contain noise, $\kappa_0 = 0$ is selected. The more noise the image contains, the greater the value should be. At the same time, in order to avoid the divisor being zero, $|\nabla s|$ needs to be modified so that $|\nabla s|_a = \sqrt{a^2 + |\nabla s|^2}$, α is a smaller constant. First, reduce the original sampling by 2 times and 4 times, and then enlarge, the mean square error is defined as:

Specifically, in addition to information for content extraction (ie, place of origin, weight, item type, etc.), we also store information regarding the status of the package at "address received" (i.e. , "state" - default value is Null). Specifically, "state" changes to 1 if the corresponding package has been received and shipped by the shipping company (ie, "shipperID"); value 0 - pending (ie, waiting for shipper to pick up). Also, "unit" stores the number of orders (e.g., 10) as well as which "package" they are assigned to.

$$MSE = \frac{\sum_{i,j}^{M,N} (T(i,j) - Z(i,j))^2}{\sum_{i,j}^{M,N} (T(i,j))^2}$$

is the original image and enlarged image respectively. The numerical simulation of the Lena image without noise is carried out by using the EPTV interpolation method, and compared with quadratic cubic spline interpolation. Unfortunately, the EPTV interpolation method was not able to eliminate the noise in the image, resulting in a poorer quality of the interpolated image. The boundary of the interpolated image became wider and less sharp, and the image appeared to be blurred. In contrast, the quadratic cubic spline interpolation method performed better in terms of noise reduction and image clarity. The boundary of the interpolated image was narrower and sharper, and the image appeared to be more defined. However, the spline interpolation method caused overshooting at the image boundary, resulting in a jagged appearance. The numerical simulation of the interpolation methods was carried out using the Lena image without noise. The results showed that EPTV interpolation retained the clarity of the image and smoothed the image boundary, as compared to spline interpolation. The boundary became narrower and sharper after EPTV interpolation. When noise was introduced into the image, the interpolation methods performed differently. The effect of sample interpolation worsened with noise, and the boundary became more blurred. However, EPTV interpolation was able to not only eliminate noise but also preserve the clarity of the image. Table 1 compares the deviation after the secondary and fourth secondary correction of the Lena images using two methods: in the absence of noise and in the presence of noise. It can be seen that as the image magnification increases and noise becomes more severe, the algorithm studied in this paper performs better. The algorithms were programmed in detail using Matlab6.5, which allowed for further analysis and exploration of different interpolation methods. This study provides valuable insights into the use of interpolation methods for image processing and understanding their limitations in different scenarios.

5 Conclusion

The conclusion of this research marks the culmination of a comprehensive exploration into the application of nonlinear partial differential equations within the domain of computer numerical simulation operating systems, with a specific focus on image enhancement through a novel hybrid bidirectional diffusion method. The study has successfully demonstrated the efficacy of this technique in improving the quality of digital images, particularly in terms of boundary detail preservation and noise reduction.

Throughout the course of this research, we have introduced and elaborated on the various components that constitute our hybrid bidirectional diffusion image amplification method. These include pulse filtering, an enhanced forward-backward diffusion filter, and the total variational diffusion method, all of which contribute to the superior performance of our algorithm.

The simulations conducted have provided empirical evidence of the method's superiority over traditional techniques. The results have shown that our approach not only maintains the integrity of image boundaries but also effectively suppresses noise, leading to images with enhanced photorealism and clarity. This has been achieved through the adaptive nonlinear filtering process, which intelligently interpolates pixel values based on local gradient information, thereby optimizing the balance between detail enhancement and noise reduction.

Furthermore, the versatility of our method has been highlighted through its application to various image types and noise conditions. The ability to adapt to different magnification levels and to manage noise effectively is a testament to the robustness of our algorithm.

In summary, this research has made significant contributions to the field of digital image processing. The proposed hybrid bidirectional diffusion image amplification method stands out as an innovative solution that addresses many of the challenges associated with traditional image enlargement techniques. As we look to the future, the potential applications of this method are vast, ranging from high-resolution imaging in medical diagnostics to the enhancement of visual content in multimedia.

The findings of this study pave the way for further advancements and refinements in image processing technologies. We anticipate that the continued development and application of our method will yield even more impressive results, pushing the boundaries of what is possible in the realm of digital image enhancement.

References

1. Yapp, C., Schapiro, D., Sokolov, A.: A scalable, modular image-processing pipeline for multiplexed tissue imaging. Nat. Methods **19**, 311–315 (2022)
2. Hui, D.: Robot swarm motion control system design based on partial differential equation constraints. Comput. Measure. Control **31**, 130–135 (2022)
3. Ye Yong, W., Ze-Yan, Z.B.-J., et al.: Meshless numerical simulation of nonlinear heat transfer equation mlpg/rbf-fd. J. Eng. Thermophys. **43**, 98–103 (2022)
4. Xuanxuan, W.F.G.: Chebyshev assignment method for gaussian type radial basis function. J. Hunan City Univ. Nat. Sci. Edition **30**, 32–39 (2021)
5. Betz, V., Boutros, A.: Fpga architecture: principles and progression. IEEE Circuits Syst. Mag. **21**, 4–29 (2021)
6. Xu, W., Xing, F., Xu, D.: A modulation recognition algorithm of communication signals and implementation based on digital signal processor. In: IEEE MTT-S International Wireless Symposium, pp. 1–3 (2021)
7. Huan, D.J.X.: Numerical simulation of RC beam and column collapse resistance in strong wind environment. Comput. Simul. **39**, 319–323 (2022)

FAITH: A Fast, Accurate, and Lightweight Database-Agnostic Learned Cost Model

Waner Li[1], Xu Chen[1], Zibo Liang[1], Runfan Ye[1], Ruyi Lai[2], Siying Yue[3], Qinyuan Su[4], Kai Zheng[1,2(✉)], Han Su[1,2(✉)], and Shaozhi Wu[1,2]

[1] School of Computer Science and Engineering, Chengdu, China
{wanerli,xuchen,ziboliang,runfanye}@std.uestc.edu.cn
{zhengkai,hansu,wszfrank}@uestc.edu.cn
[2] Yangtze Delta Region Institute (Quzhou), University of Electronic Science and Technology of China, Quzhou, China
ruyilai@std.uestc.edu.cn
[3] Huaxin Consulting Co., Ltd., Cape Town, South Africa
yuesiying.hx@chinaccs.cn
[4] China Telecom Co., Ltd., Beijing, China
cdsuqinyuan@sctel.com.cn

Abstract. In recent years, there has been a notable rise in the application of machine learning to cost estimation for query optimization. Central to an effective cost model are the abilities of accuracy, efficiency, lightness, and generalizability. However, traditional cost models are based on heuristics thus lack of accuracy. On the other hand, the learned cost models frequently struggle to strike a balance between accuracy and efficiency, with many lacking broad applicability. To combat these challenges, we introduce FAITH, a fast, accurate, and database-agnostic learned cost model. FAITH harnesses data from multiple sources to learn cross-database meta-knowledge. It is then effectively refined, leveraging the unique data information from the target database via an Adapter we developed. Proven through various benchmarks, FAITH consistently showcases its prowess in delivering accurate and robust cost estimations.

Keywords: Cost estimation · Machine learning · Query optimization

1 Introduction

The query optimizer is a critical component for performance in every database and query engine. It is responsible for translating declarative queries into efficient execution plans. The cost model of a query optimizer guides the selection of plans from the search space. Therefore, for query optimizers, accurately predicting the cost or latency of execution plans is a significant task.

For an effective cost model, we argue that four requirements need to be satisfied: (1) Accuracy: The main task of the cost model is to estimate the costs of different query execution plans. Accurate cost estimation can assist the query optimizer in selecting the most optimal performance execution plan. Inaccurate estimation, on the other hand, can result in the selection of suboptimal plans, consequently diminishing query performance. (2) Efficiency: The query optimizer needs to frequently invoke the cost estimation model to evaluate different query execution plans. As a result, the cost model should swiftly analyze these plans and provide results. In other words, the cost model should demonstrate a high level of inference efficiency. (3) Lightweight: Database Management System(DBMS) typically run on limited hardware resources such as CPU and memory. Complex cost estimation models can consume a significant amount of system resources, leading to resource contention and performance degradation. Lightweight models can more effectively utilize these resources. (4) Robustness: Data statistics within a database, including factors such as data distribution, data volume, and the presence of indexes, are subject to change over time. Furthermore, the workload of a DBMS can also exhibit variations as time progresses. In light of these dynamic variables, it becomes essential for the cost model to demonstrate robustness, allowing it to adapt to these changes and consistently deliver efficient performance across diverse contexts. This adaptability is paramount in guaranteeing the reliability of the query optimizer. However, the previous works have not entirely met the aforementioned four requirements.

The design and improvement of traditional cost models usually require a lot of effort from human experts.PostgreSQL's cost model [5], for instance, consists of more than 4,000 lines of C code and takes into account various subtle considerations such as partially relevant index access, interesting order, tuple size, etc. Obviously, this would take months to complete. Additionally, traditional cost models often rely on fixed cost parameters, which may not adapt well to variations in different queries and hardware environments. As a result, human experts are required to manually adjust parameters and maintain the system. Moreover, some inaccurate assumptions underlie the traditional cost models [9]. For example, selectivities for each predicate are assumed to be individual, even though the underlying columns may be related, and the statistics are assumed to reflect the current state of the database, while the database characteristics may fluctuate. If these assumptions are invalid, the accuracy of cost estimation will be significantly reduced. So far, none of the traditional cost models have completely solved the aforementioned issues.

To overcome the shortcomings of traditional cost models, a series of works have utilized machine learning techniques to construct learned cost models. These models can avoid the aforementioned inaccuracies by self-tuning through training. Despite their potential, some limitations prevent them from being widely applied in practice. Bao's [6] cost model vectorizes query plans and then inputs the processed feature vectors into the model. However, it only encodes common scan operators and join operators, neglecting some unfrequent operators and the data characteristics of query plans. This would hamper the model's ability to fully extract plan features, thereby reducing the accuracy of cost estimation. Compared to Bao, the cost models of tree convolution [1,2,14] and

QueryFormer [16] not only encode operators but also retain and utilize more features related to query plans. Balsa incorporates data information from the workload into consideration, while QueryFormer encodes data information, positional information, and frequency distribution histograms of the tables in the plans. This approach can improve the accuracy of cost estimation, yet increases the complexity of the training process. And for cost models, handling more information can lead to a decrease in efficiency. Furthermore, these learned models are typically trained and tested on a single dataset, while the structures and features of query plans can vary across different databases, which implies that collecting training data by running queries must be repeated for each new database. Analogously, if the data characteristics of the database change over time, these models also need to undergo repeated training. Overall, these models need improvements in efficiency, robustness, and generalization to reduce repetitive training.

In summary, traditional cost models offer higher efficiency, while the learned models provide higher accuracy. Inspired by both types of models, we introduce a new approach in this paper. Our approach involves two stages: pre-training and fine-tuning. During pre-training, the model is trained on multiple datasets to acquire general knowledge of the databases. Specifically, it extracts common features across databases, achieving a zero-shot model's capability. In the fine-tuning phase, we further refine the model using local knowledge from the dataset, eliminating existing errors and enhancing the model's performance. After pre-training and fine-tuning, our cost model can meet the four requirements mentioned above:

(1) Accuracy: Our model has achieved the zero-shot model's performance, ensuring its accuracy when facing different databases. Just a few hours of training are needed for our model to achieve the same performance as the cost model of the open-source model (PostgreSQL). (2) Efficiency: Our model can perform rapid inference due to its lightweight structure. Additionally, local training has also improved its inference capabilities. (3) Lightweight: Our model has a small parameter size because it encodes only common knowledge across databases, making it lightweight. (4) Robustness: Our model has thoroughly learned the common features across databases, allowing it to robustly generalize to new databases through rapid fine-tuning.

In summary, we make the following contributions:

- We propose FAITH, a fast, accurate, and lightweight database-agnostic learned cost model.
- We designed an Adapter that allows the model to precisely fine-tune a subset of parameters based on the local knowledge of the target database, thereby eliminating the need for the model to be retrained.
- Our model is conducted on the benchmark datasets and achieves state-of-the-art accuracy and efficiency.

The remainder of the paper is structured as follows. Section 2 introduces some important notations and a formal definition of the cost estimation problem. Our approach's framework as well as its components are illustrated in Sect. 3.

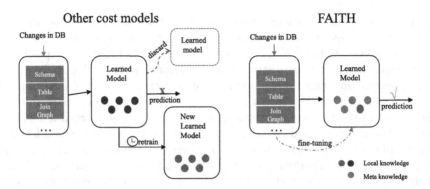

Fig. 1. FAITH VS other cost models. When facing changes in database characteristics or deployed in unseen databases, other cost models, having learned only old local knowledge (represented by the red dot), require retraining to adapt to new database-specific knowledge (represented by the green dot). In contrast, the FAITH model acquires cross-data meta-knowledge during the pre-training phase (represented by the blue dot). Therefore, it only needs to fine-tune with new local knowledge, eliminating the need for retraining. (Color figure online)

Section 4 presents our evaluation of experimental results and a case study. Lastly, the related work and conclusion are shown in Sect. 5 and Sect. 6.

2 Problem Definitions and Important Notations

Definition 1 (Physical Plan Tree). The physical plan tree in the context of query optimization is a structured representation that delineates how a query will be executed. Specifically, it translates the abstract operations specified in the logical plan into concrete strategies and methods that the DBMS can utilize to retrieve and process data. Each node in this tree represents a specific operation, such as a type of join or scan, and the method by which it will be carried out, like using a hash join or a nested loop join. The physical plan tree is typically generated by the query optimizer, which evaluates various possible execution strategies to determine the most efficient one based on estimated costs.

Definition 2 (Query Optimizition). Query Optimization, within the realm of database management, is a systematic process that seeks the most efficient manner to execute a query. Specifically, it evaluates the myriad ways a query can be executed and selects the optimal path that conservatively uses resources while ensuring rapid results. Each step in this process involves intricate decisions, such as choosing between different algorithms or determining the sequence of operations. The outcome of query optimization is often a physical plan tree. This process is driven by the query optimizer, which uses a combination of heuristics, statistics, and cost models to evaluate and choose the best strategy from the plethora of available options.

Problem Definition. The cost estimation problem can be defined as follows: Given a query plan p, the goal of cost estimation is to accurately estimate the potential resource consumption required to execute p, such as CPU cycles, memory usage, and I/O operations, thereby guiding the query optimizer to select the most optimal strategy for executing a given query.

3 Methodology

3.1 Framework

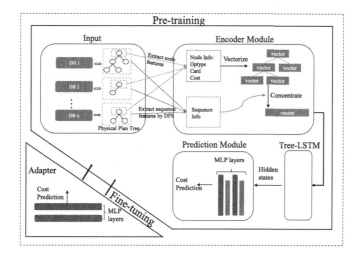

Fig. 2. Framework Overview of FAITH.

Our model's architecture is shown in Fig. 2. It consists of three basic components: (1) the encoder module, (2) tree-lstm [11], (3) the prediction module. In the initial stage, physical query plan nodes are encoded to obtain feature vector trees. Subsequently, these feature vectors are recursively fed into the Tree-LSTM module, resulting in hidden states. These hidden states are then input into the prediction module for cost estimation. Finally, the adapter is employed to finetune the model when dealing with the target database.

3.2 Encoder Module

The role of the encoder module is to extract features from the physical query plan tree and convert them into vectors. To obtain an effective representation of a physical query plan, the encoding of both individual node features and the tree structure is a critical task.

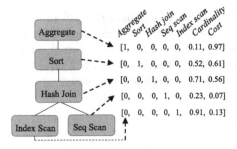

Fig. 3. Encoding of a query plan tree. Each node is subjected to one-hot encoding, and subsequently, they are concatenated following the Depth-First Search (DFS) order to construct the feature vector for the query plan tree.

Node Encoding. The encoding of a node is structured as Fig. 3. We perform one-hot encoding for operators such as index scan or merge join. Many learned cost models typically only encode common join and scan operators, whereas our model encodes all operators present in a dataset spanning 20 databases. This comprehensive approach enables it to effectively capture a wide range of plan features. Furthermore, we employ Min-max normalization for processing cost and cardinality, which assists in mitigating the influence of outliers within datasets. Subsequently, these normalized values are concatenated into the node's feature vector. In contrast to previous works, we do not encode database-specific information such as tables or columns. Our model does not consider the local knowledge of individual databases, focusing solely on fundamental features like node operators, costs, and cardinality. This approach facilitates the acquisition of cross-database meta-knowledge for future learning.

Plan Tree Encoding. The encoding of a plan tree should effectively capture the dependencies between parent and child nodes since the actions of a parent node are directly influenced by the actions of its child nodes. To address this challenge, we employ a Depth-First Search (DFS) algorithm that traverses the query plan tree effectively. The result is a comprehensive vector that amalgamates feature vectors and temporal information for all nodes throughout the tree. This vector, enriched with the hierarchical relationships between nodes, becomes a crucial input for the subsequent stages of processing, allowing us to achieve a more thorough and nuanced understanding of the plan tree's structure and dependencies.

3.3 Tree-LSTM

Modeling the extended paths of information flow within a query plan can be quite challenging. During query execution, significant long-range dependencies between nodes can arise. Consequently, when analyzing the cost of a node, it becomes necessary to consider information from all descendant nodes. The presence of these extended paths within a query plan poses challenges in devising effective representation models. To address these difficulties, we propose utilizing Tree-LSTM to learn the features of query plans. Tree-LSTM represents a generalization of LSTMs suitable for tree-structured network topologies. Unlike traditional LSTMs, Tree-LSTM can accommodate tree structures as both input and output, as opposed to just sequential data. It effectively captures the structural and semantic information of a tree by performing LSTM operations on the nodes within the tree.

To aggregate features of query plan trees, we define a dedicated temporal LSTM layer to transform the feature vectors obtained from the encoder module v_k^t to a hidden state h_k^t as follows:

$$h_k^t = LSTM(v_k^t, h_k^{t-1}; W) \tag{1}$$

where h_k^{t-1} is the hidden state of last node, W denotes temporal LSTM parameters. The temporal LSTM layer is executed recursively to obtain the final hidden state h_k^t of a query plan tree.

For example, in the plan query plan tree shown in Fig. 3, the root node "aggregate" is first input into the LSTM, resulting in the hidden state h_k^1. The child nodes "sort" and h_k^1 of the "aggregate" node are then input into the LSTM, obtaining the hidden state h_k^2. This process continues iteratively, ultimately yielding the information for the entire tree in h_k^n. The h_k^n is then input into the fully connected layers to predict the cost.

3.4 Prediction Module with Adapter

The structure of the prediction module is shown in Fig. 4. It operates differently during the pre-training phase and fine-tuning phase.

Pre-training. As depicted in Fig. 4, in the pre-training phase, the prediction module is composed of four linear layers. The Multi-Layer Perceptron (MLP) takes the hidden states produced by the Tree-LSTM as input. After passing through these linear layers for modeling and learning, it ultimately generates cost estimates. It's important to highlight that our model architecture incorporates the use of residual connections. Residual connections are a vital architectural element frequently employed in deep neural networks to mitigate the vanishing gradient problem and facilitate the training of exceedingly deep networks. The fundamental form of a residual block is as follows:

$$y = x + F(x, W) \tag{2}$$

The input x represents the feature maps or activations from the previous layer in the neural network. The output, denoted as y is the result of the operations performed within the residual block. $F(x, W)$ is the transformation operation within the residual block, typically composed of operations such as convolution and activation functions. Its objective is to learn to capture residual information in the input x, thereby enhancing the quality of feature representation.

Fine-tuning. During the fine-tuning stage, we developed Adapter designed to optimize the model for the target database. In the realm of deep learning, adapters are primarily employed to introduce and acquire new task-specific knowledge, all the while preserving the original parameters of pre-trained models. This approach affords flexibility for model transfer, multi-task learning, and knowledge integration, without encountering issues like catastrophic forgetting, thus enhancing parameter efficiency. Adapters function as external extensions that link to the output hidden states of intermediate layers in pre-trained models.

As illustrated in Fig. 4, in our methodology, the Adapter comprises linear layers represented in gray. Throughout the fine-tuning phase, we "freeze" the pre-trained Tree-LSTM model and the linear layers shown in blue, meaning that we do not alter their parameters from the pre-training phase. Adapter preserves the original parameters of pre-trained models without alteration, and it facilitates the ongoing infusion of knowledge. In essence, the introduction of new types of injected knowledge does not impact the parameters that have been previously acquired for existing knowledge. In contrast to conventional machine learning methods, our approach necessitates fine-tuning only a subset of parameters through the adapter, obviating the need for retraining the entire model.

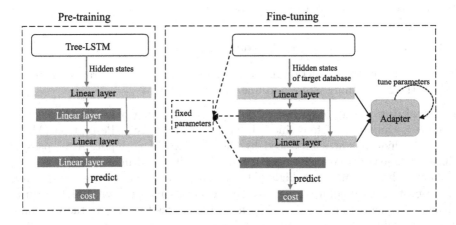

Fig. 4. The structure of prediction module

4 Experiments

4.1 Experimental Setup

In tasks related to query optimizer cost estimation, the commonly used benchmark is JOB-light [4], which is a workload derived from the IMDB dataset. JOB-light typically consists of a set of queries and associated workloads designed to emulate complex database queries and their performance requirements. However, given that our model requires training across various databases, and the IMDB benchmark primarily assesses cost estimators in a workload context on a single database, we have chosen to adopt a new testing benchmark proposed by Benjamin Hilprecht et al. [3] for zero-shot models. This benchmark incorporates well-established datasets such as SSB [8] and IMDB, along with publicly available datasets that possess distinct characteristics. This benchmark proves to be more advantageous for evaluating the performance of models trained on a diverse range of databases.

Datasets. The benchmark we employ comprises 20 databases, encompassing both publicly available real-world datasets and well-established benchmark datasets. For the purposes of cost estimation tasks, smaller databases will be scaled up to a larger size. These databases exhibit notable variations in features, including foreign key relationships, the number of columns, and the number of tables. Furthermore, the benchmark incorporates standard benchmarks such as TPC-H, ensuring its ability to encompass a broad spectrum of data characteristics.

Workloads. The benchmark offers a workload generator capable of generating different types of queries, and we opt for a workload that aligns with the zero-shot cost model. Additionally, the benchmark provides workload traces for the previously mentioned 20 databases. To elaborate, each of these databases involves executing 15,000 queries using the Postgres DBMS (version 12), thereby capturing information regarding query execution and runtimes. Queries that surpass a runtime of 30 s are excluded from the benchmark.

Implementation Details. LSTM with 64 units of hidden states in our model. Our model is trained with a batch size of 512 for 100 epochs using the Adam algorithm with a default setting. We employ 5% of the training data as validation data and set the learning rate to 1e–3.

Hardware and Platform. The hardware devices and platforms used in our experiments are as follows: the CPU I sIntel(R) Xeon(R) Silver 4214 CPU 2.20 GHz, the GPU is NVIDIA GeForce RTX 3080, the algorithms are implemented based on Python 3.9.18, and the deep learning framework is Pytorch 2.0.1.

4.2 Evaluation Baselines

- Zero-Shot cost model [3]: This model is a zero-shot cost estimation approach that enables the generation of a universal cost estimation model for unknown databases without the need for expensive training data collection.
- Scaled Optimizer Costs (Postgres): Since PostgreSQL's cost estimation returns abstract cost units, Scaled Optimizer [3] uses a simple linear model on top of PostgreSQL's database cost estimation to provide actual query execution times.
- MSCN: MSCN [4] is a multi-set convolutional network designed specifically for portraying relational query plans, utilizing set semantics to capture query characteristics and accurate cardinalities.

- QPPNet [7]: QPPNet is a plan-structured neural network. It can automatically discover complex performance models at both the operator and query plan levels, and this network architecture can match the structure of any optimizer-selected query execution plan and predict its latency.
- Tpool [10]: Tpool is an end-to-end cost estimation framework based on a tree-structured model, which embeds features of both queries and physical operations into the tree model and can simultaneously estimate cost and cardinality
- QueryFormer [16]: QueryFormer is a query plan representation model that employs a tree-structured Transformer architecture while integrating histograms obtained from the DBMS into the query plan encoding. It can be applied to various database optimization tasks.

4.3 Performance Evaluation

Accuracy. We trained our model using 19 datasets as training data and evaluated its accuracy on the remaining unseen database. In our evaluation, we employed the Median Q-Error as our primary metric. As shown in Fig. 5, our model performs better than scaled optimizer estimates on all datasets and outperforms the zero-shot model on 15 datasets, which demonstrates the high accuracy of our model. Even when confronted with databases featuring significantly different characteristics, our model exhibits robustness, as evidenced by a median error consistently remaining within the 1.61 threshold. This robust performance is instrumental for query optimization tasks across diverse database scenarios, ensuring reliable results.

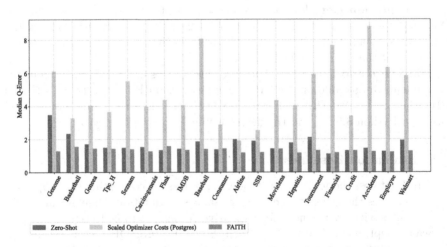

Fig. 5. The cost estimation errors of FAITH, Zero-shot, and scaled optimizer estimates across 20 datasets.

FAITH VS Workload-driven Baselines. In the first experiment, we conducted a comparative analysis of FAITH against the previously mentioned workload-driven models. Our evaluation was performed on well-established benchmarks, encompassing JOB-light, Synthetic, and Scale datasets.

Traditional workload-driven cost models face a limitation in that the representation of query plans is not transferable. Consequently, when deployed on previously unseen databases, it becomes imperative to retrain these models. In our approach, we employed MSCN [4] to generate 100,000 queries for training baseline models. However, for fine-tuning our model, we exclusively utilized a subset of 5,000 queries from this dataset.

We depict the Median Q-Error,95th Q-Error, and max Q-error in Fig. 6. Our model achieves and even surpasses the performance of these extensively trained models with minimal fine-tuning. Specifically, our model has the lowest max Q-Error on all three benchmarks, indicating strong performance across the entire dataset. This is attributed to the pre-training phase of our model, where it learns meta-knowledge about query plans sufficiently, ensuring its robustness when encountering unseen physical plans.

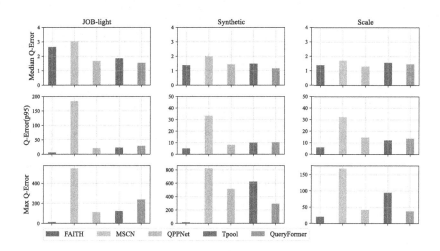

Fig. 6. The cost estimation errors of FAITH and workload-driven cost models

In our second experiment, we assessed the inference speed of FAITH and the baseline models, which measures the time required for the model to process a single query. The experimental outcomes are detailed in Fig. 1. The results clearly indicate that FAITH excels in both high accuracy and rapid inference speed. Notably, MSCN stands out with significantly faster inference speed in comparison to the other models. This swiftness is attributed to MSCN's focus on encoding solely the connections within the physical query plan. While this approach enhances both inference and training speed, it does come at the expense of a reduction in the model's accuracy. Zero-Shot's inference throughput is larger compared to FAITH. This is due to Zero-Shot's design of distinct MLP layers for different node types, leading to the model needing to switch between GPU and CPU during inference, thus hindering concurrent execution. Additionally, Zero-Shot employs a bottom-up approach for node inference, meaning a node has to wait for its child nodes to complete their inferences. For complex query plan trees, this undoubtedly substantially reduces the inference speed. Hence, when faced with diverse scenarios, it becomes imperative to strike a harmonious balance between the model's precision and efficiency.

Table 1. Inference speed of FAITH and baselines

Model	FAITH	MSCN	QPPNet	Tpool	QueryFormer	Zero-Shot
Inference speed(sec/query)	0.0016	0.0012	0.023	0.013	0.015	0.0086

Effectiveness of Adapter. The Adapter module holds a pivotal role within the prediction module of our model. In this section, we have designed experiments to showcase the effectiveness of the Adapter. Initially, we compared the time required for fine-tuning with the Adapter against the time needed for retraining the model for the target database. As illustrated in Fig. 7, the utilization of the Adapter results in a noticeable reduction in the training time.

Moreover, we recorded the Median Q-Error of FAITH both with and without the Adapter. The results are presented in the lower graph of Fig. 7. The graph reveals that the Adapter assists the model in maintaining a consistently high level of accuracy. This is attributed to the Adapter's capacity to achieve more precise model optimization by learning local knowledge specific to the target database. This experiment serves as a compelling testament to the Adapter's multifaceted capabilities, showcasing its proficiency not only in expediting the training process but also in elevating the model's accuracy and robustness to new heights.

5 Related Work

In this section, we will introduce some representative traditional cost models and learning-based models, as well as machine learning approaches commonly used in cost estimation tasks.

5.1 Traditional Cost Models

Open Source DBMS. PostgreSQL is a powerful, highly reliable, and highly scalable database management system suitable for applications of various scales. PostgreSQL uses various cost model parameters in query optimization to estimate the execution cost of queries. These parameters include page-based IO cost, tuple-based CPU cost, expression-based CPU cost, cost generated by parallel queries, and the impact of caching on costs. Each parameter has a default value, for example, the cost of processing one tuple is set to 0.1, and these values can be adjusted by users to suit specific environments and requirements.

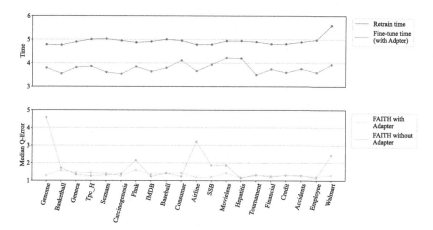

Fig. 7. A comparison of training time and estimation errors when using and not using an adapter.

Commercial DBMS. Oracle is a leading provider of relational database management systems known for its robust and scalable solutions widely used in enterprise applications. In Oracle, there are two cost models available: the IO cost model and the CPU cost model. The choice of which model to use is determined by the initialization parameter, which has four possible values: IO, CPU, FIXED, and CHOOSE. By default, the CHOOSE option is used, which selects the model based on available system statistics. In the cost models, all costs are converted into the number of single-block reads. Cost calculations involve both IO costs and CPU costs.

5.2 Learned Cost Models

Tree-LSTM Based Models. The Tree-LSTM family of approaches represents an improvement over the flattened method by employing hierarchical information aggregation following the tree structure. Notably, RTOS [15] and E2E-Cost [10] utilize the Tree-LSTM model [12] to aggregate node information from the bottom-up, starting from leaf nodes and culminating in the root node, ultimately using the final output as the representation of a physical plan. Meanwhile, Plan-Cost [7] introduces two distinct neural network modules tailored for leaf nodes and intermediate nodes, allowing them to capture parent-children dependencies and information flow paths. Nevertheless, these approaches, given their recursive nature, are susceptible to the forgetting problem and can prove challenging to train effectively when dealing with large query plans.

Tree-Transformer Based Models. Transformer [13] employs a mechanism known as the attention mechanism, allowing it to capture dependencies between different elements in the input sequence, rather than processing data sequentially. Balsa [14] and QueryFormer [16] utilize the Tree-Transformer to capture complex contextual information within queries, including keywords, operation sequences, and relationships between tables and fields. This contributes to a more accurate estimation of the plan's cost. Additionally, the Tree-Transformer architecture consists of multiple layers of self-attention and feedforward neural networks, enabling it to model complex relationships and capture hierarchical patterns in data. However, the computational complexity of Transformer models is very high, especially when dealing with large-scale data. This results in higher training and inference costs.

5.3 Related Machine Learning Approaches

Zero-shot Learning. It is a machine learning paradigm that extends beyond traditional supervised learning. In zero-shot learning, a model is trained to recognize and classify objects or concepts it has never seen during the training phase. This is achieved by providing the model with descriptions or attributes of these unseen classes, allowing it to generalize and make predictions based on the provided information. Zero-shot learning is particularly valuable when dealing with a vast number of possible classes or when new classes are constantly emerging, making it challenging to collect labeled training data for every category. This approach finds applications in various domains, including computer vision, natural language processing, and recommendation systems, where adaptability and versatility are essential.

6 Conclusion

In this paper, we propose FAITH, a fast, accurate, and Lightweight database-agnostic learned cost model. FAITH comprises three modules: the encoding module, Tree-LSTM, and the prediction module. Our approach consists of two components: pre-training and fine-tuning using Adapter. Through empirical experiments, we demonstrate that FAITH meets the four key requirements of a query optimizer's cost model, namely accuracy, efficiency, lightweight, and robustness. Additionally, we design experiments to validate the effectiveness of Adapter. Our model proves its ability to provide accurate cost estimates for query optimizers.

Acknowledgment. This work is supported by NSFC (No. 61802054, 61972069, 61836007, 61832017, 61532018), scientific research projects of Quzhou Science and Technology Bureau, Zhejiang Province (No. 2020D010, No. 2020D12) and Sichuan Science and Technology Program under Grant 2020JDTD0007.

References

1. Chen, X., Chen, H., Liang, Z., Liu, S., Wang, J., Zeng, K., Su, H., Zheng, K.: Leon: a new framework for ml-aided query optimization. Proceedings of the VLDB Endowment **16**(9), 2261–2273 (2023)
2. Chen, X., Wang, Z., Liu, S., Li, Y., Zeng, K., Ding, B., Zhou, J., Su, H., Zheng, K.: Base: Bridging the gap between cost and latency for query optimization. Proceedings of the VLDB Endowment **16**(8), 1958–1966 (2023)

3. Hilprecht, B., Binnig, C.: Zero-shot cost models for out-of-the-box learned cost prediction. arXiv preprint arXiv:2201.00561 (2022)
4. Kipf, A., Kipf, T., Radke, B., Leis, V., Boncz, P., Kemper, A.: Learned cardinalities: Estimating correlated joins with deep learning. arXiv preprint arXiv:1809.00677 (2018)
5. Leis, V., Gubichev, A., Mirchev, A., Boncz, P., Kemper, A., Neumann, T.: How good are query optimizers, really? Proceedings of the VLDB Endowment **9**(3), 204–215 (2015)
6. Marcus, R., Negi, P., Mao, H., Tatbul, N., Alizadeh, M., Kraska, T.: Bao: Making learned query optimization practical. In: Proceedings of the 2021 International Conference on Management of Data. pp. 1275–1288 (2021)
7. Marcus, R., Papaemmanouil, O.: Plan-structured deep neural network models for query performance prediction. arXiv preprint arXiv:1902.00132 (2019)
8. O'Neil, P., O'Neil, E., Chen, X., Revilak, S.: The Star Schema Benchmark and Augmented Fact Table Indexing. In: Nambiar, R., Poess, M. (eds.) TPCTC 2009. LNCS, vol. 5895, pp. 237–252. Springer, Heidelberg (2009). https://doi.org/10.1007/978-3-642-10424-4_17
9. Stillger, M., Lohman, G.M., Markl, V., Kandil, M.: Leo-db2's learning optimizer. In: VLDB. vol. 1, pp. 19–28 (2001)
10. Sun, J., Li, G.: An end-to-end learning-based cost estimator. arXiv preprint arXiv:1906.02560 (2019)
11. Tai, K.S., Socher, R., Manning, C.D.: Improved semantic representations from tree-structured long short-term memory networks. arXiv preprint arXiv:1503.00075 (2015)
12. Tai, K.S., Socher, R., Manning, C.D.: Improved semantic representations from tree-structured long short-term memory networks. arXiv preprint arXiv:1503.00075 (2015)
13. Vaswani, A., Shazeer, N., Parmar, N., Uszkoreit, J., Jones, L., Gomez, A.N., Kaiser, L., Polosukhin, I.: Attention is all you need. Advances in neural information processing systems **30** (2017)
14. Yang, Z., Chiang, W.L., Luan, S., Mittal, G., Luo, M., Stoica, I.: Balsa: Learning a query optimizer without expert demonstrations. In: Proceedings of the 2022 International Conference on Management of Data. pp. 931–944 (2022)
15. Yu, X., Li, G., Chai, C., Tang, N.: Reinforcement learning with tree-lstm for join order selection. In: 2020 IEEE 36th International Conference on Data Engineering (ICDE). pp. 1297–1308. IEEE (2020)
16. Zhao, Y., Cong, G., Shi, J., Miao, C.: Queryformer: a tree transformer model for query plan representation. Proceedings of the VLDB Endowment **15**(8), 1658–1670 (2022)

Fast Approximate Temporal Butterfly Counting on Bipartite Graphs via Edge Sampling

Jiaxi Pu[1], Yanhao Wang[1](✉), Yuchen Li[2], and Xuan Zhou[1]

[1] School of Data Science and Engineering, East China Normal University, Shanghai 200062, China
pujiaxi@stu.ecnu.edu.cn, {yhwang,xzhou}@dase.ecnu.edu.cn
[2] School of Computing and Information Systems, Singapore Management University, Singapore 178902, Singapore
yuchenli@smu.edu.sg

Abstract. Temporal bipartite graphs are widely used to represent time-evolving relationships between two disjoint sets of nodes, e.g., customer-product interactions in e-commerce and user-group memberships in social networks. Temporal butterflies, i.e., the complete bipartite subgraphs that occur between two nodes from each partition within a short period and in a prescribed order, are essential in modeling the structural and sequential patterns of such graphs. Counting the number of temporal butterflies is a fundamental task in temporal bipartite graph analysis. However, existing methods for butterfly counting on static bipartite graphs and motif counting on temporal unipartite graphs are inefficient for this purpose. Since exact counting can be time-consuming on large graphs, in this paper, we propose an edge sampling-based approach to approximating temporal butterfly counts accurately and efficiently. We provide an analytical bound on the number of edges to be sampled to obtain estimates with small relative errors and high probability. Finally, we evaluate our algorithm on six real-world temporal bipartite graphs to show its superior accuracy and efficiency compared to baselines.

Keywords: Butterfly counting · Bipartite graph · Temporal motif

1 Introduction

Temporal bipartite graphs are a natural data model that represents the interactions evolving over time between two distinct sets of nodes in various real-world applications [3,7,15]. For instance, the relationship between customers and products in e-commerce platforms, such as Amazon and Alibaba, can be modeled as temporal bipartite graphs. In this case, customers and products form two separate partitions of nodes, and purchase records, which occur over time, are represented by a sequence of temporal edges that connect customers to products, as illustrated in Fig. 1(a). Other examples of temporal bipartite graphs

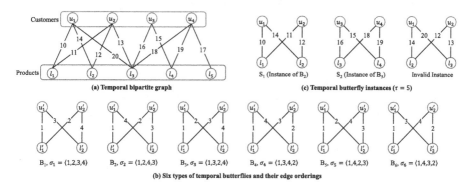

Fig. 1. Examples of temporal butterflies on a temporal bipartite graph.

include group memberships of users in social networks and project engagements of workers in crowdsourcing platforms.

In (static) bipartite graphs, *butterfly*, a complete bipartite subgraph with exactly two nodes from each partition, is one of the most fundamental structures that serve as building blocks of higher-order cohesive structures, including bitrusses [27] and communities [1]. Therefore, butterflies play an essential role in many bipartite network analysis problems, such as computing clustering coefficients [16], finding dense subgraphs [19], and learning node embeddings [6]. However, in a temporal setting where edges are associated with the timestamps of the events they represent, the above notions and methods are not applicable because they cannot capture the temporal dynamics of graphs. To fill this gap, we introduce the concept of temporal butterflies as $(2,2)$-bicliques whose edges occur in a prescribed order within a time interval $\tau \in \mathbb{R}^+$, which is in line with existing studies on motifs in temporal (unipartite) networks [5,11,13,20,21,25]. In Fig. 1(b), we list the six types of temporal butterflies with different edge orderings according to our definition. We also provide two instances of temporal butterflies in Fig. 1(c), which match B_2 and B_5 in Fig. 1(b), respectively. Additionally, we present an invalid instance that matches B_3 in terms of topology but does not satisfy the duration constraint $\tau = 5$ in Fig. 1(c).

In this paper, we investigate the problem of *temporal butterfly counting*, that is, to compute the number of occurrences of all six types of butterflies on a temporal bipartite graph. This problem is crucial for analyzing bipartite graphs with temporal information, such as characterizing collaborative patterns of users in a short period [7], understanding the dynamics of user-item interactions over time [15], and detecting bursting events [4]. Although there are numerous methods to count butterflies on static bipartite graphs [10,17,18,22,26,31], they are not efficient for the temporal setting. The primary issue is that they do not consider the edge orderings and timestamps. Consequently, they fail to distinguish different kinds of temporal butterflies and do not examine whether a butterfly instance is valid in the sense that all of its edges occur within the duration constraint. Furthermore, they are tailored for simple graphs, where only one butterfly instance exists between each set of four nodes at most. However, temporal bipartite graphs are multi-graphs that may have several butterfly instances of varying types between the same four nodes at different timestamps. Furthermore, the problem of motif counting has also been extensively studied on temporal unipartite graphs [5,8,11,13,14,20,21,25]. Unfortunately, they are still not efficient for temporal butterfly counting. First, most of them count only one motif at a time, which requires six separate runs to compute all butterfly counts. Second, the chronological edge-driven matching [12] used in these methods is inefficient in finding butterfly instances. Third, some of them are designed for other classes of motifs, such as triangles [14,25] and simple cycles [8], and cannot be trivially generalized to butterflies. To the best of our knowledge, the method in [2] is the only known method for temporal counting of butterflies, but it is specific to provide exact counts. We do not notice any prior work on sampling methods for approximate temporal butterfly counting.

Our Contributions. Counting the exact occurrences of butterflies can be time-consuming on large graphs [10,11,20,22]. Therefore, our aim is to propose efficient algorithms that accurately estimate the numbers of all six types of butterflies in a temporal bipartite graph. Our methods are supported by theoretical guarantees and achieve good empirical performance. Our main contributions include:

- We extend the concepts of butterflies on static graphs and motifs on temporal unipartite graphs to introduce the concept of *temporal butterfly* and formally define the temporal butterfly counting problem. (Section 3)
- We propose an edge sampling-based algorithm for approximate temporal butterfly counting on bipartite graphs. Specifically, it consists of (1) a sampling strategy to draw a set of edges randomly from the graph and (2) a wedge enumeration-based method to obtain the exact number of butterfly instances with respect to each sampled edge. Theoretically, our algorithm computes unbiased estimates of butterfly counts with bounded variances. We provide an analytical bound on the number of edge samples required to achieve a relative approximation error within $\varepsilon > 0$ with probability at least $1 - \delta$ ($0 < \delta < 1$). (Section 4)
- We experimentally evaluate the performance of our algorithm on six real-world datasets. The results demonstrate that it runs more than one order of magnitude faster than the state-of-the-art butterfly and temporal motif approximation methods while having comparable estimation errors. (Section 5)

2 Related Work

The butterfly counting problem was first addressed by Wang et al. [24], who proposed an exact algorithm based on matrix multiplication. Sanei-Mehri et al. [17] and Wang et al. [26] designed more efficient algorithms for exact butterfly counting based on layer-priority and node-priority wedge enumeration.

Sanei-Mehri et al. [17] also proposed sampling algorithms to count butterflies approximately with provable accuracy guarantees. Estimating the number of butterflies on streaming graphs was considered in [10,18,22]. Parallel and distributed algorithms for butterfly counting were proposed in [23,28,30]. Zhou et al. [31] studied the problem of approximating the expected number of butterflies on uncertain graphs. Several studies explored bitruss (or k-wing) and tip decomposition of bipartite graphs based on the butterfly structure [9,19,27,29]. However, all of these works are based on static graphs and are inefficient for our problem because they ignore temporal information.

Paranjape et al. [13] first introduced the definition of *temporal motifs* used in this work. They also provided a generic algorithm for exact temporal motif counting by enumerating all motif instances on static graphs and filtering them based on timestamps. Kumar and Calders [8] proposed an efficient algorithm to find all simple temporal cycles in a temporal graph. Mackey et al. [12] proposed a chronological edge-driven backtracking algorithm to enumerate all instances of a temporal query motif. Liu et al. [11] and Sarpe and Vardin [21] proposed time interval-based sampling algorithms for temporal motif counting. Wang et al. [25] proposed edge-centric sampling algorithms for temporal motif counting. But these algorithms [11,21,25] count only one motif at a time. Sarpe and Vardin [21] also studied the problem of counting all temporal motifs with the same static topology simultaneously. Pashanasangi and Seshadhri [14] designed fast algorithms to count all temporal triangles exactly. Gao et al. [5] proposed a distributed framework for exact temporal motif counting. However, these methods are designed for unipartite graphs and are inefficient for our problem.

Independently from this work, Cai et al. [2] proposed exact algorithms for butterfly counting and enumeration on temporal bipartite graphs. They adopted the same definition of *temporal butterfly* as ours but focused on effective pruning and indexing techniques for the exact counting problem. In the experiments, we will compare our sampling methods with the exact ones in [2] and show that our methods have higher efficiency at the expense of small estimation errors.

3 Preliminaries

A temporal bipartite graph is an undirected graph $T = (V, E)$ with $n = |V|$ nodes and $m = |E|$ edges. The node set V is partitioned into two disjoint subsets in the upper layer U and the lower layer L, respectively, where $U \cup L = V$ and $U \cap L = \emptyset$. The set of temporal edges is represented by $E \subseteq U \times L \times \mathbb{R}^+$, where each temporal edge $e \in E$ between two nodes $u \in U$ and $l \in L$ at time $t \in \mathbb{R}^+$ is denoted as a triple (u, l, t). Note that T is a kind of *multi-graphs*, where more than one edge may exist between the same nodes u and l at different timestamps. We use $d(u)$ to denote the degree of node u, i.e., the number of edges connected to u. By ignoring the timestamps of edges, we obtain a projected static bipartite graph $G(T) = (V, E_{static})$, where $E_{static} = \{(u,l)|(u,l,t) \in E\}$, from T.

A *butterfly*, or $(2,2)$-*biclique*, is one of the most fundamental motifs on (static) bipartite graphs. It is a subgraph of a bipartite graph consisting of four nodes

$u_x, u_y \in U$ and $l_x, l_y \in L$ such that the edges (u_x, l_x), (u_y, l_x), (u_x, l_y), and (u_y, l_y) all exist in E. Next, we follow the notion of temporal motifs proposed by Paranjape et al. [13] to extend the definition of *butterflies* to *temporal butterflies*.

Definition 1. (Temporal Butterfly) *A temporal butterfly $B = (V_B, E_B, \sigma)$ consists of a butterfly $G_B = (V_B, E_B)$ with a node set $V_B = \{u'_1, u'_2, l'_1, l'_2\}$ and an edge set $E_B = \{(u'_1, l'_1), (u'_2, l'_1), (u'_1, l'_2), (u'_2, l'_2)\}$ as well as an ordering σ of the four edges in E_B.*

Unlike the static case with only one type of butterfly, a temporal bipartite graph has six types of temporal butterflies, denoted as $B_1, \ldots, B_6$, with different edge orderings $\sigma_1, \ldots, \sigma_6$. Each type of butterfly can be represented by a sequence of four edges by fixing the first edge (u'_1, l'_1) and enumerating the permutations of (u'_2, l'_1), (u'_1, l'_2), and (u'_2, l'_2), as shown in Fig. 1(b).

Given a temporal butterfly as the template pattern, we aim to find all its occurrences in T. Furthermore, following a line of studies on temporal motif counting [5,8,11,13,14,20,21,25], we restrict our consideration to the instances where such patterns appear within a period $\tau \in \mathbb{R}^+$. In summary, we define the notion of *temporal butterfly τ-instance* as follows.

Definition 2. (Temporal Butterfly τ-Instance). *Given a temporal bipartite graph T and duration constraint $\tau \in \mathbb{R}^+$, a set S of four temporal edges $\{(u_x, l_x, t_1), (u_y, l_x, t_2), (u_x, l_y, t_3), (u_y, l_y, t_4)\}$ in T is a τ-instance of B_i ($i \in [1, \ldots, 6]$) if: (1) there exists a bijection f on the nodes such that $f(u_x) = u'_1$, $f(u_y) = u'_2$, $f(l_x) = l'_1$, and $f(l_y) = l'_2$; (2) the ordering of the four edges based on their timestamps $t_1, \ldots, t_4$ matches σ_i; and (3) all the edges occurs within τ timestamps, i.e., $\Delta(S) = \max_{j \in \{2,3,4\}} t_j - t_1 \leq \tau$.*

Without loss of generality, Definition 2 only considers the case where u_x and l_x are mapped to u'_1 and l'_1 in B_i. Other formulations with different node mappings are also allowed, but they are essentially identical to Definition 2.

A notion closely related to butterfly is the *wedge*, i.e., a triple $W = (u_x, l, u_y)$ where (u_x, l) and (u_y, l) both exist in E. We call l the center node and u_x, u_y the end nodes of W. In a temporal setting, $W = (u_x, l, u_y)$ is called a *wedge τ-instance* if $|t_y - t_x| \leq \tau$ for (u_x, l, t_x) and (u_y, l, t_y). Note that for simplicity, it is assumed that the center node of W is in the lower layer L. Any temporal butterfly τ-instance can be decomposed into two wedge τ-instances $W_{l_x} = (u_x, l_x, u_y)$ and $W_{l_y} = (u_x, l_y, u_y)$ with different center nodes l_x and l_y. This decomposition plays an important role in the design of butterfly counting methods.

Based on the above notions, we formally define the *temporal butterfly counting* (TBC) problem we study in this paper.

Definition 3. (Temporal Butterfly Counting). *Given a temporal bipartite graph T and duration threshold $\tau \in \mathbb{R}^+$, count the number of τ-instances C_i of each temporal butterfly B_i for $i = 1, 2, \ldots, 6$ in T.*

Counting the exact number of butterflies is often time-consuming on large static graphs [17], and the problem becomes even more challenging in temporal settings due to additional considerations for edge ordering and duration. Therefore, we turn our attention to accurately estimating each C_i. Specifically, for an error parameter $\varepsilon \in (0,1)$ and a confidence parameter $\delta \in (0,1)$, our goal is to compute an (ε, δ)-approximation $\widehat{C}_i$ of C_i such that $\Pr[|\widehat{C}_i - C_i| > \varepsilon C_i] < \delta$. Before presenting our framework for approximate temporal butterfly counting, we summarize the notation frequently used throughout this paper in Table 1.

Table 1. List of frequently used symbols.

Symbol	Description
$T = (V, E)$	Temporal bipartite network with node set V and edge set E
U, L	Nodes in the upper and lower layers of V
$B_1, \ldots, B_6$	Six types of temporal butterflies in Fig. 1(b)
τ	Duration constraint of temporal butterfly instances
$d(u)$	Degree of node u
S	Temporal butterfly τ-instance
$W = (u_x, l, u_y)$	Wedge (or wedge τ-instance) between u_x, l, and u_y
C_i	Number of τ-instances of B_i in T for $i = 1, \ldots, 6$
$\widehat{C}_i$	Approximate estimate of C_i
$\varepsilon, \delta \in (0,1)$	Error and confidence parameters in the approximation
$\widehat{E}$	Set of edge samples drawn uniformly at random from E
$C_i(e)$	Number of τ-instances of B_i starting with edge e
$\mathcal{N}_t(u)$	Set of nodes adjacent to u in the time interval $[t, t+\tau]$

4 Algorithm

In this section, we introduce a novel algorithm to approximately count temporal butterflies. Our algorithm is based on randomly sampling edges from the input graph T and efficiently and accurately counting the number of temporal butterfly instances for each edge sampled. The counting method is based on wedge enumeration [17,26] for butterfly counting on static graphs but is adapted to the temporal setting considering edge ordering and duration constraints. We first present the procedure of our algorithm and then provide a theoretical analysis of its unbiasedness, variance, approximation guarantee, and time complexity.

Algorithm Description. The procedure of our algorithm, TBC-E, is presented in Algorithm 1. After initializing all six counters $\widehat{C}_i$ for $i = 1, \ldots, 6$ to 0 (Line 1), it samples a set $\widehat{E}$ of edges from the input graph T (Line 3). Specifically, we adopt the edge sampling strategy that checks each edge $e \in E$ one by one and adds it to $\widehat{E}$ with a fixed probability $p \in (0,1)$. Then, for each edge $e = (u, l, t) \in \widehat{E}$, a wedge enumeration-based method is used to calculate the number $C_i(e)$ of τ-instances of B_i w.r.t. e for each $i = 1, \ldots, 6$ accordingly. Here, only instances where e is their first edge (u_x, l_x, t_1) with the smallest timestamp (or equivalently, u and l are mapped to u'_1 and l'_1) are considered to avoid repetition. As a first step, it initializes each counter $C_i(e)$ to 0. Then, it finds each node $u_y \neq u_x$ connected to l_x between t_1 and $t_1 + \tau$ (Line 7). As such, every wedge τ-instance W_{l_x} comprising two edges (u_x, l_x, t_1) and (u_y, l_x, t_2) with $t_2 > t_1$ is generated, and the remaining problem becomes to find another wedge τ-instance W_{l_y} that can form a temporal butterfly τ-instance together with W_{l_x}. Note that it also only considers wedge instances whose edges occur after t_1 since e should be the first edge in any instance. In particular, for each W_{l_x}, it finds two sets of nodes $\mathcal{N}_{t_1}(u_x)$ and $\mathcal{N}_{t_1}(u_y)$ adjacent to u_x and u_y within the time interval $(t_1, t_1 + \tau]$ (Lines 10–11) and computes the intersection of $\mathcal{N}_{t_1}(u_x)$ and $\mathcal{N}_{t_1}(u_y)$ as the candidate set L_y for the center nodes of W_{l_y} (Line 12). Subsequently, for each node $l_y \in L_y$, it enumerates the set $\mathcal{W}_{t_1, l_y}$ of all wedge τ-instances centered at l_y and ended at u_x, u_y within the interval $(t_1, t_1 + \tau]$, checks the order of edges in W_{l_x} and each $W_{l_y} \in \mathcal{W}_{t_1, l_y}$ to decide which type of temporal butterflies the instance comprising W_{l_x} and W_{l_y} belongs to, and updates the corresponding counters (Lines 13–17). Next, it updates each counter $\widehat{C}_i$ based on $C_i(e)$ (Line 18). To ensure that $\widehat{C}_i$ is an unbiased estimator, the weight of $C_i(e)$ should be the inverse of the probability that an edge $e \in E$ is included in $\widehat{E}$, that is, $\frac{C_i(e)}{p}$ (cf. Theorem 1). Finally, after processing all edges in $\widehat{E}$, each counter $\widehat{C}_i$ is returned as an approximation of C_i for $i = 1, \ldots, 6$ (Line 19).

Theoretical Analysis. Next, we provide thorough theoretical analyses of our TBC Algorithm. First, the following theorem indicates that $\widehat{C}_i$ returned by Algorithm 1 is unbiased and has a bounded variance.

Theorem 1. *For* TBC-E, $\mathbb{E}[\widehat{C}_i] = C_i$ *and* $\text{Var}[\widehat{C}_i] \leq \frac{1-p}{p} C_i^2$.

Proof. For TBC-E, let us define a random variable X_e for each $e \in E$ such that $X_e = 1$ if $e \in \widehat{E}$ and $X_e = 0$ if $e \notin \widehat{E}$. Intuitively, X_e is a Bernoulli variable with $\Pr[X_e = 1] = p$. For each $e \in E$, $C_i(e)$ is the number of τ-instances of each B_i ($i = 1, \ldots, 6$) with e serving as their first edge with the smallest timestamp, according to Algorithm 1. Also, note that every instance is counted exactly once when $e \in \widehat{E}$. Thus, we have $\mathbb{E}[\widehat{C}_i] = \frac{1}{p} \sum_{e \in E} C_i(e) \mathbb{E}[X_e] = \frac{1}{p} \sum_{e \in E} C_i(e) p = C_i$.

Algorithm 1: TBC-E

Input: Temporal bipartite graph T, temporal butterflies $B_1, \ldots, B_6$, duration constraint $\tau \in \mathbb{R}^+$, sampling probability $p \in (0,1)$
Output: Estimate $\widehat{C}_i$ of C_i for $i = 1, \ldots, 6$

1 Initialize each counter $\widehat{C}_i \leftarrow 0$ for $i = 1, \ldots, 6$ and $\widehat{E} \leftarrow \emptyset$;
2 **foreach** *edge* $e \in E$ **do**
3 $\quad \widehat{E} \leftarrow \widehat{E} \cup \{e\}$ with fixed probability p;
4 **foreach** *edge* $e \in \widehat{E}$ **do**
5 $\quad$ Fix $e = (u, l, t)$ as the first edge (u_x, l_x, t_1) in any instance;
6 $\quad$ Initialize $C_i(e) \leftarrow 0$ for $i = 1, \ldots, 6$;
7 $\quad \mathcal{N}_{t_1}(l_x) \leftarrow \{u_y \neq u_x \in U | (u_y, l_x, t_2) \in E \wedge t_2 \in (t_1, t_1 + \tau]\}$;
8 $\quad$ **foreach** *node* $u_y \in \mathcal{N}_{t_1}(l_x)$ **do**
9 $\quad\quad$ Find a wedge τ-instance W_{l_x} with (u_x, l_x, t_1) and (u_y, l_x, t_2);
10 $\quad\quad \mathcal{N}_{t_1}(u_x) \leftarrow \{l_y \neq l_x \in L | (u_x, l_y, t_3) \in E \wedge t_3 \in (t_1, t_1 + \tau]\}$;
11 $\quad\quad \mathcal{N}_{t_1}(u_y) \leftarrow \{l_y \neq l_x \in L | (u_y, l_y, t_4) \in E \wedge t_4 \in (t_1, t_1 + \tau]\}$;
12 $\quad\quad$ Obtain $L_y \leftarrow \mathcal{N}_{t_1}(u_x) \cap \mathcal{N}_{t_1}(u_y)$;
13 $\quad\quad$ **foreach** *node* $l_y \in L_y$ **do**
14 $\quad\quad\quad$ Find the set $\mathcal{W}_{t_1, l_y}$ of all wedge τ-instances centered at l_y and ended at u_x, u_y in the time interval $(t_1, t_1 + \tau]$;
15 $\quad\quad\quad$ **foreach** *wedge* $W_{l_y} \in \mathcal{W}_{t_1, l_y}$ **do**
16 $\quad\quad\quad\quad$ **if** *the order of edges in* W_{l_x} *and* W_{l_y} *matches* σ_i **then**
17 $\quad\quad\quad\quad\quad C_i(e) \leftarrow C_i(e) + 1$;
18 $\quad \widehat{C}_i \leftarrow \widehat{C}_i + \frac{C_i(e)}{p}$ for $i = 1, \ldots, 6$;
19 **return** $\widehat{C}_i$ *for* $i = 1, \ldots, 6$

Based on the definition of variance,

$$\begin{aligned}
\operatorname{Var}[\widehat{C}_i] &= \mathbb{E}[\widehat{C}_i^2] - \mathbb{E}^2[\widehat{C}_i] = \frac{1}{p^2} \mathbb{E}\left[\left(\sum_{e \in E} C_i(e) X_e\right)^2\right] - C_i^2 \\
&= \frac{1}{p^2}\left(\sum_{e \in E} C_i^2(e) \mathbb{E}[X_e^2] + 2 \sum_{e \in E} \sum_{e' \neq e \in E} C_i(e) C_i(e') \mathbb{E}[X_e X_{e'}]\right) - C_i^2 \\
&= \frac{1}{p} \sum_{e \in E} C_i^2(e) + 2 \sum_{e \in E} \sum_{e' \neq e \in E} C_i(e) C_i(e') - C_i^2 \\
&= \frac{1}{p} \sum_{e \in E} C_i^2(e) + \left(\sum_{e \in E} C_i(e)\right)^2 - \sum_{e \in E} C_i^2(e) - C_i^2 \\
&= \frac{1-p}{p} \sum_{e \in E} C_i^2(e) \leq \frac{1-p}{p} \left(\sum_{e \in E} C_i(e)\right)^2 = \frac{1-p}{p} C_i^2.
\end{aligned}$$

According to the above results, the proof is concluded. □

The following theorem indicates the minimum probability p in TBC-E to guarantee that $\widehat{C}_i$ is an (ε, δ)-approximation of C_i, i.e., $\Pr[|\widehat{C}_i - C_i| \geq \varepsilon C_i] \leq \delta$.

Theorem 2. *For any parameters $\varepsilon, \delta \in (0,1)$, TBC-E provide an (ε, δ)-estimator $\widehat{C}_i$ of C_i when $p \geq \frac{1}{1+\delta\varepsilon^2}$.*

Proof. We apply Chebyshev's inequality to obtain the lower bound of probability p, that is, $\Pr\left[|\widehat{C}_i - \mathbb{E}[\widehat{C}_i]| \geq \sqrt{\frac{\text{Var}[\widehat{C}_i]}{\delta}}\right] \leq \delta$. Since $\mathbb{E}[\widehat{C}_i] = C_i$ and $\text{Var}[\widehat{C}_i] \leq \frac{1-p}{p}C_i^2$ for TBC-E according to Theore 1, it requires $p \geq \frac{1}{1+\delta\varepsilon^2}$ for TBC-E to ensure that $|\widehat{C}_i - \mathbb{E}[\widehat{C}_i]| \leq \varepsilon C_i$ with probability at least $1 - \delta$. □

Time Complexity. Finally, we analyze the time complexity of TBC-E. First, the time complexity of edge sampling is $O(m)$ because each edge is considered at most once in the sampling process. Next, we consider how long it takes to compute $C_i(e)$ for any edge $e \in E$. First, it takes $O(d_\tau)$ time to obtain $\mathcal{N}_{t_1}(l_x)$, where d_τ is the maximum number of edges connected to one node within any time interval of length τ. Then, for each node in $\mathcal{N}_{t_1}(l_x)$, it also spends $O(d_\tau)$ time to obtain $\mathcal{N}_{t_1}(u_x)$ and $\mathcal{N}_{t_1}(u_y)$ and compute the union of two sets. Next, it only takes $O(1)$ time to decide the type of each instance. In summary, the time complexity of computing $C_i(e)$ is $O(d_\tau^2)$. Then, the expected number of edges in $\widehat{E}$ for TBC-E is mp and at least $p = \frac{1}{1+\delta\varepsilon^2}$ is required to achieve an (ε, δ)-approximation of C_i. Therefore, TBC-E provides an (ε, δ)-estimator of C_i for $i = 1, \ldots, 6$ in $O(\frac{md_\tau^2}{1+\delta\varepsilon^2})$ time.

5 Experiments

In this section, we evaluate the performance of our sampling method for approximate temporal butterfly counting on six real-world datasets. In our experiments, the following eight algorithms are compared.

- **ESampBFC** [17] is an approximation algorithm for butterfly counting on static bipartite graphs. To adapt it to temporal butterfly counting, we run it on the projected static graph $G(T)$ to sample a set of four-node quadruples with at least one butterfly among them and use the same procedure as Algorithm 1 to count the number of temporal butterflies on a subgraph induced by each four-node quadruple.
- **ES** [25] is an edge sampling algorithm to approximate the number of instances of a temporal motif. For temporal butterfly estimation, we invoke it six times using all butterflies in Fig. 1(b) as query motifs.
- **PRESTO** [21] is a time interval-based sampling algorithm for approximate temporal motif counting. It also requires six independent runs using all butterflies in Fig. 1(b) as query motifs for the temporal butterfly approximation. We do not compare with another time interval-based sampling algorithm in [11] since it is outperformed by PRESTO.
- **OdeN** [20] is an approximation algorithm to count the numbers of all kinds of temporal motifs with the same static topology (e.g., *butterfly* in this work).
- **ExactTBC** [2] is an exact algorithm for counting and enumerating temporal butterflies.

– TBC-E is our proposed edge sampling-based algorithm for approximate temporal butterfly counting in Sect. 4.

All algorithms were implemented in C++14 compiled by GCC v8.1 with -O3 optimizations and ran on a single thread in each experiment. For ES, PRESTO, OdeN, and ExactTBC, we used the implementations published by the original authors. All experiments were conducted on a server running Ubuntu 18.04 with an 8-core Intel Xeon Processor @3.0GHz and 64GB RAM. Our code and data are published at https://github.com/placido7/TBC.

We used six real-world datasets in the experiments. All datasets except *Reddit* were downloaded from KONECT[1]. The *Reddit* dataset was obtained from Kaggle[2]. In particular, *Linux* contains contributions of persons to various threads of the Linux kernel mailing list, where each edge (u, l, t) corresponds to a post of person u to thread l at timestamp t; *Twitter* denotes the hashtag-tweet relations in Twitter, where each edge (u, l, t) shows that a tweet u at timestamp t was associated with a hashtag l; *MovieLens* contains a sequence of movie ratings on the MovieLens website, where each edge (u, l, t) indicates that a user u has rated a movie l at timestamp t; *LastFM* represents the user-song listening habits from the Last.fm website, where each edge (u, l, t) connects a user u and a song l that u listened to at timestamp t; *Reddit* denotes the user-thread relations in one month on the Reddit website, where each edge (u, l, t) means that a user u posted under a thread l at timestamp t; *Wikipedia* is an edit network of the English Wikipedia, where each edge (u, l, t) represents an edit of user u on page l at timestamp t. Table 2 presents the basic statistics of the six datasets, where n is the number of nodes in T, m is the number of temporal edges in T, d_{max} is the maximum degree among all nodes in V, τ is the default duration constraint, *timespan* is the overall time range of the dataset, $\bowtie_G$ is the number of butterflies on $G(T)$, and $\bowtie_T$ is the number of temporal butterfly τ-instances on T. We computed $\bowtie_G$ and $\bowtie_T$ in each dataset with the exact butterfly counting algorithm in [2,17], respectively. The number of nodes and edges differs from those in the original datasets because isolated nodes and multiple edges with identical timestamps are removed in preprocessing. Figure 2 shows the percentages of six types of temporal butterflies (i.e., B_1–B_6) in the above six datasets.

We run every algorithm ten times with different (fixed) seeds. The accuracy of each algorithm is evaluated by the average of mean absolute percentage errors (MAPE) of six temporal butterfly estimates, that is, $\frac{1}{6}\sum_{i=1}^{6}|\frac{\widehat{C}_i-C_i}{C_i}|$, over ten runs. We also use the average relative error of each estimate, that is, $|\frac{\widehat{C}_i-C_i}{C_i}|$, over ten runs to analyze the performance of different algorithms on each kind of temporal butterfly individually. The efficiency of each algorithm is evaluated by the average CPU time over ten runs.

Accuracy vs. Efficiency. We first present the performance of each algorithm in terms of accuracy and efficiency in Fig. 3. On each dataset, we vary the sampling rate of each algorithm in the range $[0.125\%, 0.25\%, \ldots, 64\%]$. We illustrate the

---
[1] http://konect.cc/networks/.
[2] https://www.kaggle.com/general/31657.

Table 2. Statistics of datasets used in the experiments.

Dataset	n	$|U|$	$|L|$	m	d_{max}	τ	timespan	$\bowtie_G$	$\bowtie_T$
Linux	372K	41.8K	330.6K	996K	36.9K	24h	34.4yr	10.8M	2.1M
Twitter	520K	169.3K	350.8K	2.6M	62.4K	24h	3.2yr	69.5M	67.2M
MovieLens	80.5K	69.9K	10.6K	7.1M	25.8K	24h	14.0yr	484.9B	185.1M
LastFM	1.1M	992	1.1M	17.5M	164.8K	24h	8.6yr	28.2B	79.0M
Reddit	27.9M	2.5M	25.4M	49.9M	231.5K	24h	1mo	241.3B	8.6M
Wikipedia	28.9M	6.3M	22.7M	293.8M	2.8M	1h	16.5yr	5.0T	35.3M

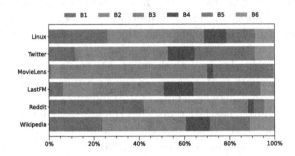

Fig. 2. Percentage of each type of temporal butterfly on the datasets in Table 2.

running time of ExactTBC with a vertical line in each plot. Note that we omit all results when an algorithm's MAPE exceeds 100% or its average running time exceeds 10,000 s.

First, we observe that TBC-E consistently achieves the best trade-off between accuracy and efficiency across the six datasets. It runs 1.8 to 25× faster than any other sampling algorithm when the MAPE is at a level of at most 10%. PRESTO underperforms TBC-E in almost all cases. This is mainly because it requires six independent runs to count all kinds of temporal butterflies. OdeN can compute all six butterfly counts simultaneously and thus outperforms PRESTO on most datasets. However, compared to TBC-E, OdeN still suffers from two limitations: (1) it incurs additional overhead to transform temporal graphs into their projected static graphs and (2) it does not exploit the property of bipartite graphs for butterfly enumeration. ES shows inferior performance to all the algorithms except ESampBFC on large datasets due to its low efficiency in per-edge butterfly counting. However, its performance is better on small datasets (e.g., MovieLens) because of the relatively low variances of estimates based on edge sampling. Another drawback of ES is huge memory consumption and does not provide any results on Wikipedia. ESampBFC is not comparable to any other algorithm on all datasets. From Table 2, we can see that the number of static butterflies is much larger than the number of temporal butterflies on each dataset. Since ESampBFC does not exploit temporal information for early pruning of quadruples without any valid instance, its efficiency is extremely low due to unnecessary computa-

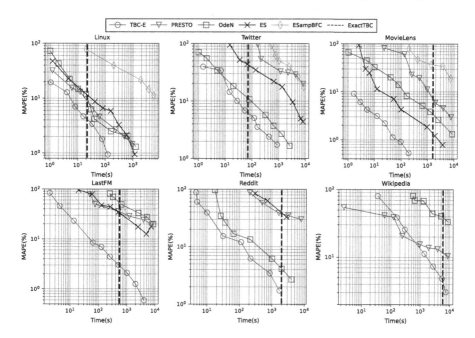

Fig. 3. Accuracy (MAPE) vs. time (s) of different algorithms by varying sampling rate.

tions on such quadruples. Finally, ExactTBC finishes within the time limit (i.e., 10,000 s) on all six datasets. The running time of ExactTBC is comparable to the time for TBC-E to achieve a MAPE of 1%~5% on five out of six datasets. However, on the MovieLens dataset, ExactTBC runs about 10 times slower than the time for TBC-E to achieve a MAPE of 1%.

Relative Errors of Different Temporal Butterflies. We illustrate the relative errors of the estimates for six types of temporal butterflies returned by different algorithms in Fig. 4. On each dataset, we set the time limit to that when TBC-E reaches a MAPE of at most 5% (e.g., 17 s on Linux and 5,500 s on Wikipedia). We then run all algorithms ten times within the time limit and compute their relative errors for each temporal butterfly. We omit the results of ESampBFC for its low efficiency and ES on Wikipedia due to out-of-memory. We used a box plot and six markers to show the mean (dashed line), median (solid line), and quartiles (box and whiskers) of relative errors for all estimates and average relative errors for six types of butterflies, respectively.

We observe that TBC-E and OdeN have mostly consistent relative errors between butterflies since they count all kinds of butterflies simultaneously. Their differences in relative errors are mainly attributed to sampling rates and the magnitudes of counts. However, for PRESTO and ES, the relative errors often vary among different types of temporal butterflies because they need to count each separately. In particular, they often exhibit inferior performance on B_4 and B_6 due to the deficiency of chronological edge-driven subgraph matching they use to

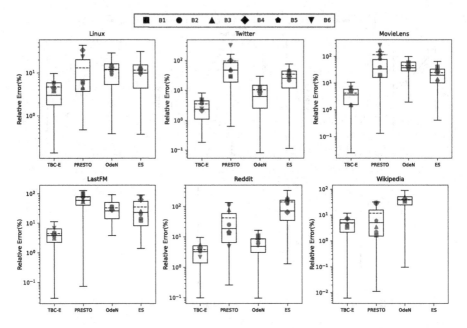

Fig. 4. Relative errors of estimates for six kinds of temporal butterflies (i.e., B_1–B_6) provided by different sampling algorithms.

count butterfly instances. As shown in Fig. 1, to enumerate all instances w.r.t. an edge $e = (u, l, t)$, after first mapping u and l with u'_1 and l'_1, the chronological edge-driven method for subgraph isomorphism must consider all edges between t and $t + \tau$ to be mapped with u'_2 and l'_2, leading to large sets of candidates and low efficiency. Given the same time limit, the sampling rates for B_4 and B_6 are much lower in PRESTO and ES, and thus the relative errors are naturally higher.

Effect of Duration Constraint τ. We first present the number of instances of each temporal butterfly by varying the duration constraint τ in $\{1, \ldots, 8, 24, \ldots, 168\}$ hours, that is, from one hour to one week, on the Linux and Twitter datasets in Fig. 5. The counts of all butterflies increase nearly linearly with τ. We then show the runtime of each algorithm by varying τ in the same range on the Linux and Twitter datasets in Fig. 6. We attempt different sampling rates for each algorithm and present the one when its MAPE is below 5% for each value of τ. In this way, we can fairly compare their time efficiency at the same level of accuracy. Generally, as the value of τ increases, the running time of each algorithm increases accordingly. In some cases, the running time of TBC-E and OdeN decreases with increasing τ because they require lower sampling rates to achieve a MAPE of 10% due to larger total counts and more uniform distributions of instances over time. TBC-E still always achieves the best performance for different τ values. We also notice that the running time of OdeN is less sensitive to τ because the transformation from temporal graphs to projected static graphs in OdeN is time-unaware. ExactTBC adopt a recursive merging approach

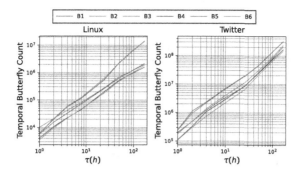

Fig. 5. Temporal butterfly counts by varying duration constraint τ.

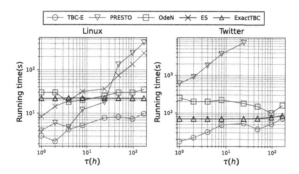

Fig. 6. Running time of different algorithms to achieve MAPEs of at most 10% by varying duration constraint τ.

to combine the wedge sets. This process reduces the computational overhead to determine whether an instance satisfies the duration constraints. Therefore, OdeN and ExactTBC can work better when the value of τ is greater.

6 Conclusion

In this paper, we propose an edge sampling-based algorithm for approximate temporal butterfly counting on bipartite graphs. We analyze the sample size and time complexity that our algorithm requires to provide an ε-approximation for every temporal butterfly count with probability at least $1 - \delta$ for any $\varepsilon > 0$ and $\delta \in (0, 1)$. Our experimental results show that our proposed method outperforms state-of-the-art methods on temporal butterfly counting. In future work, we will introduce more optimization techniques to improve the performance of our method and exploit its applications to downstream analysis tasks.

Acknowledgements. This work was supported by the National Natural Science Foundation of China (No. 62202169).

References

1. Aksoy, S., Kolda, T.G., Pinar, A.: Measuring and modeling bipartite graphs with community structure. J. Complex Netw. **5**(4), 581–603 (2017)
2. Cai, X., et al.: Efficient temporal butterfly counting and enumeration on temporal bipartite graphs. Proc. VLDB Endow. **17**(4), 657–670 (2023)
3. Chen, X., Wang, K., Lin, X., Zhang, W., Qin, L., Zhang, Y.: Efficiently answering reachability and path queries on temporal bipartite graphs. Proc. VLDB Endow. **14**(10), 1845–1858 (2021)
4. Chu, L., Zhang, Y., Yang, Y., Wang, L., Pei, J.: Online density bursting subgraph detection from temporal graphs. Proc. VLDB Endow. **12**(13), 2353–2365 (2019)
5. Gao, Z., Cheng, C., Yu, Y., Cao, L., Huang, C., Dong, J.: Scalable motif counting for large-scale temporal graphs. In: ICDE, pp. 2656–2668 (2022)
6. Huang, J., Shen, H., Cao, Q., Tao, S., Cheng, X.: Signed bipartite graph neural networks. In: CIKM, pp. 740–749 (2021)
7. Jurgens, D., Lu, T.C.: Temporal motifs reveal the dynamics of editor interactions in Wikipedia. Proc. Int. AAAI Conf. Web Soc. Media **6**(1), 162–169 (2012)
8. Kumar, R., Calders, T.: 2SCENT: an efficient algorithm to enumerate all simple temporal cycles. Proc. VLDB Endow. **11**(11), 1441–1453 (2018)
9. Lakhotia, K., Kannan, R., Prasanna, V.K., Rose, C.A.F.D.: RECEIPT: refine coarse-grained independent tasks for parallel tip decomposition of bipartite graphs. Proc. VLDB Endow. **14**(3), 404–417 (2020)
10. Li, R., et al.: Approximately counting butterflies in large bipartite graph streams. IEEE Trans. Knowl. Data Eng. **34**(12), 5621–5635 (2022)
11. Liu, P., Benson, A.R., Charikar, M.: Sampling methods for counting temporal motifs. In: WSDM, pp. 294–302 (2019)
12. Mackey, P., Porterfield, K., Fitzhenry, E., Choudhury, S., Jr., G.C.: A chronological edge-driven approach to temporal subgraph isomorphism. In: 2018 IEEE International Conference on Big Data (Big Data), pp. 3972–3979 (2018)
13. Paranjape, A., Benson, A.R., Leskovec, J.: Motifs in temporal networks. In: WSDM, pp. 601–610 (2017)
14. Pashanasangi, N., Seshadhri, C.: Faster and generalized temporal triangle counting, via degeneracy ordering. In: KDD, pp. 1319–1328 (2021)
15. Peters, L.J.J.M., Cai, J.J., Wang, H.: Characterizing temporal bipartite networks - sequential- versus cross-tasking. In: Complex Networks and Their Applications VII (Volume 2), pp. 28–39 (2019)
16. Robins, G., Alexander, M.: Small worlds among interlocking directors: network structure and distance in bipartite graphs. Comput. Math. Organ. Theor. **10**, 69–94 (2004)
17. Sanei-Mehri, S.V., Sariyuce, A.E., Tirthapura, S.: Butterfly counting in bipartite networks. In: KDD, pp. 2150–2159 (2018)
18. Sanei-Mehri, S.V., Zhang, Y., Sariyuce, A.E., Tirthapura, S.: FLEET: butterfly estimation from a bipartite graph stream. In: CIKM, pp. 1201–1210 (2019)
19. Sariyuce, A.E., Pinar, A.: Peeling bipartite networks for dense subgraph discovery. In: WSDM, pp. 504–512 (2018)
20. Sarpe, I., Vandin, F.: OdeN: Simultaneous approximation of multiple motif counts in large temporal networks. In: CIKM, pp. 1568–1577 (2021)
21. Sarpe, I., Vandin, F.: PRESTO: simple and scalable sampling techniques for the rigorous approximation of temporal motif counts. In: SDM, pp. 145–153 (2021)

22. Sheshbolouki, A., Özsu, M.T.: sGrapp: Butterfly approximation in streaming graphs. ACM Trans. Knowl. Discov. Data **16**(4), 76:1–76:43 (2022)
23. Shi, J., Shun, J.: Parallel algorithms for butterfly computations. In: APOCS, pp. 16–30 (2020)
24. Wang, J., Fu, A.W.C., Cheng, J.: Rectangle counting in large bipartite graphs. In: 2014 IEEE International Congress on Big Data, pp. 17–24 (2014)
25. Wang, J., Wang, Y., Jiang, W., Li, Y., Tan, K.L.: Efficient sampling algorithms for approximate temporal motif counting. In: CIKM, pp. 1505–1514 (2020)
26. Wang, K., Lin, X., Qin, L., Zhang, W., Zhang, Y.: Vertex priority based butterfly counting for large-scale bipartite networks. Proc. VLDB Endow. **12**(10), 1139–1152 (2019)
27. Wang, K., Lin, X., Qin, L., Zhang, W., Zhang, Y.: Efficient bitruss decomposition for large-scale bipartite graphs. In: ICDE, pp. 661–672 (2020)
28. Wang, K., Lin, X., Qin, L., Zhang, W., Zhang, Y.: Accelerated butterfly counting with vertex priority on bipartite graphs. VLDB J. **32**, 257–281 (2023)
29. Wang, Y., Xu, R., Jian, X., Zhou, A., Chen, L.: Towards distributed bitruss decomposition on bipartite graphs. Proc. VLDB Endow. **15**(9), 1889–1901 (2022)
30. Weng, T., Zhou, X., Li, K., Tan, K., Li, K.: Distributed approaches to butterfly analysis on large dynamic bipartite graphs. IEEE Trans. Parallel Distrib. Syst. **34**(2), 431–445 (2023)
31. Zhou, A., Wang, Y., Chen, L.: Butterfly counting on uncertain bipartite networks. Proc. VLDB Endow. **15**(2), 211–223 (2021)

Financial-ICS: Identifying Peer Firms via LongBERT from 10K Reports

Jintao Huang(✉)

University of Science and Technology of China, Hefei, China
huangjintao@mail.ustc.edu.cn

Abstract. We present Financial-ICS, a novel method that utilizes 10K reports as textual corpora to generate economic representations of firms, facilitating the identification of economically related peer firms based on cosine similarity. By integrating shifted block attention and global attention, we have reduced the computational and memory complexity of self-attention to $O(n)$, resulting in the LongBERT model. LongBERT is capable of processing texts with a maximum length of 131K tokens, enabling it to handle entire documents from 10K reports in one go. Our approach encompasses a comprehensive unsupervised training strategy, involving continual pre-training of LongBERT and resolving the anisotropy issue of the language model representation through contrastive learning. Additionally, We introduce prototype-based contrastive learning loss to increase the number of negative samples in each batch. Finally, we propose three evaluation metrics for a thorough assessment of Financial-ICS. The experimental results indicate that Financial-ICS outperforms a range of ICSs based on state-of-the-art algorithms across three metrics, as well as SIC, one of the most commonly used expert-designed ICSs. You can find our code at https://github.com/Jintao-Huang/financial-ics.

Keywords: Financial-ICS · LongBERT · Efficient Transformer · Contrastive Learning

1 Introduction

Industry classification systems (ICSs) play a crucial role in identifying economically related peer firms for a focal firm. Based on the identified peer firms, we are able to accurately assess market value, estimate competition and risks, determine executive compensation, and gain a deeper understanding of strategic behavior. The existing ICSs can be classified into two categories: expert-designed and algorithm-based. Expert-designed ICSs, such as SIC, NAICS, and GICS [2], establish a hierarchical industry structure to categorize companies into various industries, as illustrated in Fig. 1. However, there are several limitations to expert-designed ICSs.

Firstly, the process of updating and iterating this system demands a considerable investment of time and the expertise of knowledgeable professionals.

Fig. 1. The hierarchical industry structure in NAICS.

Secondly, the hierarchical industry structure often leads to a coarse-grained peer firm identification, lacking detailed ordering information, which is valuable for financial research. Lastly, when companies have diversified business portfolios, the hierarchical industry structure may struggle to effectively allocate them across industries.

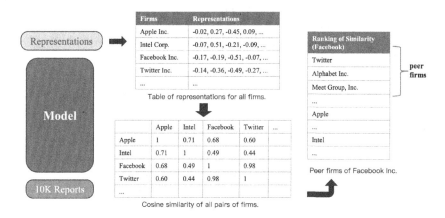

Fig. 2. Overview of algorithm-based ICSs.

To address these limitations and improve the accuracy and flexibility of industry classifications, a growing number of researchers are exploring algorithm-based ICSs, such as search-based peer firms [11] and TNIC [8]. As illustrated in Fig. 2, algorithm-based ICSs involve three primary steps: obtaining representations of all firms, calculating similarity scores between the focal firm and others, and identifying peer firms by sorting the scores.

TNIC, in particular, is an advanced algorithm-based ICS that is highly relevant to our study. It utilizes 10K reports as input, applies a bag-of-words method to produce economic representations of companies, and identifies peer firms by sorting them in reverse order according to cosine similarity. Nevertheless, the bag-of-words approach has limitations in terms of sparse representation and neglects contextual semantic information in 10K reports, resulting in a lack of expressive power.

The Transformer [22] has gained widespread popularity in the field of natural language processing due to its remarkable ability to tackle sequence problems, including language modeling [4,18], translation [12,19], and natural language inference [5,14]. The distinguishing feature of the Transformer model is its self-attention mechanism, which efficiently captures long-range dependencies with a operation path of $O(1)$. In contrast, models based on recurrent neural networks [9] have a path of $O(n)$. However, the computational and memory complexity of Transformer grows quadratically with the sequence length, restricting its capability to process long texts.

Several techniques and models [1,10,23,26] have been proposed to address the memory bottleneck of handling long texts, aiming to enhance the efficiency of self-attention. However, these models still have their limitations. Firstly, they struggle to process long texts of approximately 100K tokens, for example, Longformer can only handle a length of 4096 tokens. Additionally, models such as Linformer and Reformer do not utilize local attention to take advantage of the characteristic of natural language where related information tends to cluster together. Moreover, some models cannot efficiently implement their complex attention structures using naive torch. In this paper, we introduce LongBERT, a simple yet effective approach that leverages shifted block attention and global attention to reduce the complexity of self-attention to $O(n)$.

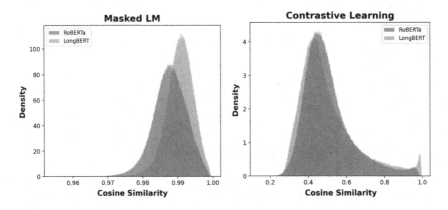

Fig. 3. Comparison of document representation similarity distributions between pretrained models and models after contrastive learning.

The pre-trained model with masked language model loss does not allow for direct utilization of the embeddings from its output layer. As depicted in Fig. 3, The document embeddings generated by RoBERTa and LongBERT from 10K reports have an anisotropy problem, where the representation vectors are concentrated within a narrow cone in the vector space [6], thus restricting the expressiveness of the representation. Contrastive learning aims to bring positive sample pairs closer and push negative sample pairs further apart. By employing contrastive learning, we can alleviate the anisotropy phenomenon and ensure that the output layer embeddings are uniformly distributed in the vector space [7].

In summary, the contributions of this paper are summarized as follows:

- We introduce Financial-ICS and provide a comprehensive workflow that encompasses the stages of pre-training, contrastive learning fine-tuning, and evaluation.
- we present LongBERT, which is specifically designed to handle long texts with a maximum length of 131K tokens. By incorporating shifted block attention and global attention, we significantly reduce the computational and memory complexity of self-attention from $O(n^2)$ to $O(n)$.
- We propose the prototype-based contrastive learning loss to increase the number of negative samples in each batch, mitigating the issue of small batch size caused by memory constraints.
- We propose three evaluation metrics for a thoroughly assessment of the economic relevance of peer firms discovered by Financial-ICS and the expressive capability of representations generated by LongBERT.

2 Related Work

2.1 ICSs

The current ICSs can be divided into two main categories: expert-designed and algorithm-based.

SIC, NAICS, and GICS are expert-designed ICSs. The Standard Industrial Classification (SIC) system, established in 1937 in the United States, is the oldest of these ICSs and utilizes four-digit codes to classify different industries. However, over time, SIC codes have been gradually replaced by the North American Industry Classification System (NAICS) codes. NAICS, which was jointly developed by the United States, Canada, and Mexico and released in 1997, provides a more unified and inclusive approach to industry classification. Additionally, the Global Industry Classification Standard (GICS) [2], developed by MSCI and Standard & Poor's in 1999, is widely accepted as an ICS for the global financial industry by financial professionals.

Search based peer firms and TNIC are algorithm-based ICSs. Search based peer firms [11] employs the "co-search" algorithm to analyze internet traffic on the SEC's EDGAR website in order to identify economically relevant peer firms. TNIC [8], utilizes product similarity as the foundation for industry classification, generating firm representations by analyzing the product descriptions in Item 1 and Item 1A sections of 10K reports. By leveraging information on product similarity between companies, TNIC can identify unique peer firms for each company.

2.2 Efficient Transformers

Efficient Transformers have been developed to address the $O(n^2)$ complexity of self-attention in traditional Transformer. Longformer [1] and BigBird [25] achieve a tradeoff between long-range dependency learning and complexity by integrating various attention mechanisms such as global attention, window attention, and even random attention. Reformer [10] utilizes LSH attention to assign q and k vectors to buckets, reducing complexity by only calculating attention for pairs within the same bucket that are highly likely to be similar. Memory compressed attention [13] and Linformer [23] reduce computation and memory consumption by compressing the length of the key and value sequences before computing attention, using 1D convolution and linear projection, respectively. Informer [26] divides the query into active and lazy parts, reducing complexity to $O(n \log n)$ by only calculating attention for active queries and using a uniform distribution for lazy queries.

2.3 Representation Learning

The BERT-like [5] pre-trained language model has demonstrated impressive performance across various downstream tasks. However, it has been observed that the vector representation of its output layer suffers from anisotropy [6]. In response to this issue, Sentence-BERT [20] fine-tunes pre-trained BERT using a siamese network architecture on natural language inference datasets to generate meaningful sentence embeddings. Other approaches such as ConSERT [24] and SimCSE [7] utilize contrastive learning methods to mitigate the anisotropy problem. ConSERT [24] introduces a generic framework for contrastive learning, including data augmentation, average pooling, and a contrastive loss layer. SimCSE [7] adopts the minimal data augmentation approach of dropout to encode the same sentence twice, creating positive sample pairs in an unsupervised manner.

3 Method

We introduce Financial-ICS, a method for identifying peer firms by utilizing the representations of 10K reports generated by LongBERT. Section 3.1 presents the unsupervised training framework for Financial-ICS, while Sect. 3.2 describes the efficient self-attention mechanism employed by LongBERT. In addition, Sect. 3.3 explains how the prototype-based contrastive learning loss are utilized to resolve the scarcity of negative samples within a single batch.

3.1 Training Framework

Financial-ICS leverages the 10K reports as input and utilizes the attention-optimized LongBERT to generate representation vectors. These vectors are then used to search for peer firms using cosine similarity. Our main challenge lies in generating representation vectors with rich contextual semantic information from the documents.

Financial-ICS employs a fully unsupervised strategy to train the model. We initialize the parameters of LongBERT using the pre-trained checkpoint of

RoBERTa. LongBERT, which is a modified version of RoBERTa with adjustments to the self-attention and positional encoding, allows us to reuse almost all parameters, thus significantly improving convergence speed. Subsequently, we conduct continual pre-training using the training set of 10K reports. This process enables the model to adapt to the new self-attention structure and positional encoding and enhance its understanding of financial data.

As depicted in Fig. 3, the pretrained LongBERT model exhibits anisotropy phenomenon. To address this issue, we fine-tuned the model using a contrastive learning loss. Our objective is to generate document representations that are uniformly distributed in the representation space, with economically related document representations displaying a higher cosine similarity. This aligns with the concept of contrastive learning.

3.2 Efficient Self-Attention

The attention formula in the traditional Transformer [22] is given by:

$$\text{Attention}(Q, K, V) = \text{softmax}\left(\frac{QK^T}{\sqrt{d_k}}\right) V = WV \quad (1)$$

In this equation, Q, K, and V are derived from X through linear projection. Here, d_k represents the dimension of K, and W represents the attention matrix. To understand self-attention, we can view it from a search perspective. We can think of Q as queries from the higher layer, and K and V as the information held by the lower layer, where K serves as indices and V serves as the associated information. If the dot product similarity between q_i and k_j is larger, then there is a greater weight of v_j flowing from position j to i, thereby achieving attention from position i to j.

However, traditional self-attention faces challenges in handling long sequences due to its computational and memory complexity, which grows at a rate of $O(n^2)$ with the sequence length. Considering that the maximum length of 10K reports exceeds 100K tokens, modifications to the self-attention mechanism and positional encoding of the model are necessary. To address this issue, we integrated shifted block attention and global attention into LongBERT's self-attention mechanism. Additionally, we have incorporated rotary position embedding [21] as the positional encoding because it has shown excellent performance in handling long-text tasks and possesses strong capabilities and flexibility to adapt to varying sequence lengths.

Shifted Block Attention As illustrated in Fig. 4, our analysis of RoBERTa's attention matrix revealed that q is more likely to exhibit higher attention scores with neighboring positions of k, showcasing the localized nature of information in natural language tasks. Consequently, we introduced block attention.

We define the length of the long text as n. By dividing the long sequence into blocks of length w, each block can only access the content within its boundaries, leading to a complexity of $O(wn)$. Since w is a constant, the complexity can be simplified to $O(n)$. In our experiments, we specifically set w to 512.

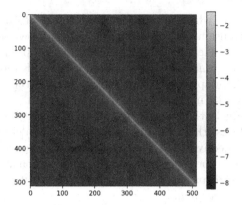

Fig. 4. Analysis of RoBERTa's attention matrix. we calculated the natural logarithm of the attention probabilities.

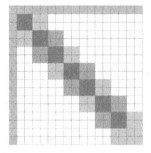

Fig. 5. The attention structure employed by LongBERT. The blue cells represent global attention, while the yellow and green cells correspond to block attention used by two different types of heads. The block in the diagram has a length of $w = 4$. (Color figure online)

However, this design does not facilitate the flow of information between blocks. Therefore, we introduced shifted block attention. As depicted in Fig. 5, we have shifted the attention position for half of the heads by $\frac{w}{2}$, while keeping the other half unchanged. Utilizing shifted block attention allows us to achieve a complexity of $O(n)$ while enhancing the model's attention span. For instance, in a LongBERT consisting of a total of l layers, the local attention span would be lw.

Global Attention. While the use of shifted block attention can greatly enhance the model's attention span, the maximum document length of the 10K reports exceeds 100K tokens, as indicated in Table 1. Despite employing shifted block attention, achieving a global perspective remains unfeasible. Consequently, we have introduced global attention to reduce the path length of long dependencies to $O(1)$.

We implement global attention using the global token, serving as a conduit for information exchange between blocks. As illustrated in Fig. 5, the global token has the capability to attend to all tokens, and likewise, all tokens are able to attend to it. In our experiments, we set the number of global tokens to 16. The self-attention structure of LongBERT can be efficiently implemented using native torch, without the requirement for custom CUDA kernels.

LongBERT strikes a balance between complexity and the ability to learn long dependencies by introducing block attention and global attention simultaneously. Block attention effectively leverages the principle of information locality in natural language, while global attention supplements the capability to learn long-range dependencies.

3.3 Contrastive Learning

LongBERT reduces the complexity of self-attention from $O(n^2)$ to $O(n)$, but still requires a large amount of GPU memory. This is not a problem during the pre-training phase, as we can use gradient accumulation to mitigate the impact of a low batch size. However, during the contrastive learning fine-tuning phase, a low batch size will result in fewer negative samples, affecting the training effectiveness. Therefore, we have designed a prototype-based contrastive learning loss.

We adopt a similar approach to SimCSE [7], utilizing minimal data augmentation technique of dropout to create positive sample pairs and employing global mean pooling to generate document representations. The design of our contrastive learning loss function is as follows:

$$\begin{aligned} p_i &= (h_{2i} + h_{2i+1})/2 \\ \text{label} &= [0, 0, 1, 1, ...] \\ \text{logits} &= \text{pairwise\_sim}(h, p)/\tau \\ \text{loss} &= \text{cross\_entropy}(\text{logits}, \text{label}) \end{aligned} \quad (2)$$

In the equations above, h_{2i} and h_{2i+1} represent positive sample pairs, while τ denotes the temperature.

The representations of positive sample pairs are generated from the same document using dropout. We calculate the mean of each pair of positive samples to create the corresponding document prototype p_i and assign them their respective labels. The number of prototypes can be interpreted as the number of categories, which is equivalent to the number of documents in a batch. The entire contrastive learning task can be seen as a classification task, in which we classify these samples into their corresponding prototypes, aiming to increase the cosine similarity between them while reducing the cosine similarity between the samples and other prototypes. This can be achieved by using cross-entropy loss.

Table 1. The statistics of datasets. We use RoBERTa's tokenizer for tokenization to generate token statistics.

	Documents	Firms	Industries (SIC)	Tokens	Max Tokens
Train Dataset	22675	3494	377	20.6k ±11.7k	114.3k
Val Dataset	1903	1898	340	23.3k±13.3k	113.1k
Demo Dataset	3345	3345	377	23.0k±12.7k	113.1k

In order to reduce memory consumption, we only compute the loss for the chosen s positive sample pairs and their prototypes, while the rest are computed without gradient context. In our experiments, due to GPU memory limitations, we set $s = 1$, but the total number of negative sample pairs is set to 7. By adopting this approach, we increase the number of negative samples for contrastive learning within a batch.

4 Experiment

4.1 Dataset

The Financial-ICS proposed in this study utilizes the annual 10K reports submitted by companies to the U.S. Securities and Exchange Commission (SEC) as the dataset. Each report consists of four parts and fifteen items. Previous research, TNIC [8], used Item 1 Business and Item 1A Risk Factors as the input for its algorithm. Item 1 provides a description of the company's main products and core business, while Item 1A discloses risk factors that could significantly impact the company's operations. We adopted the same sections of the 10K reports as TNIC to construct Financial-ICS.

We collected a total of 25K reports from 3514 companies spanning the period from January 1, 2010, to January 1, 2019. Each report contains the content of Item 1 and Item 1A. The dataset was divided into three parts: a training set, a validation set, and a demo set. The training set, consisting of data from 2010 to 2017, was used for pre-training and contrastive learning fine-tuning on LongBERT. The validation set, which included data in 2018, was employed to assess the effectiveness of Financial-ICS. The demo set consisted of the most recent 10K reports data from 2010 onwards, encompassing as many companies as possible to generate economic representations of companies and search for peer firms. The statistical data of the three subsets can be seen in Table 1.

Furthermore, for each company in the validation set, we obtained the corresponding SIC code, NAICS code, and stock market data for the year 2018. We collected 12-dimensional stock market data vectors to evaluate Financial-ICS, representing the data for each month of the year.

4.2 Metrics

We have devised three evaluation metrics to assess the economic relevance of the peer firms identified by Financial-ICS and the quality of the generated representations. In Sect. 4.1, we provide detailed information about our validation set, which includes the 10K reports, SIC code, NAICS code, and the 12-dimensional stock market data vectors over the year.

Firstly, peer firms with economic relevance tend to exhibit similar trends in the stock market, resulting in a tighter linear correlation closer to 1 compared to randomly sampled company pairs. To identify these peer firms, we set a similarity threshold and only considered company pairs with a similarity score above this threshold. The mean of the stock market linear correlation coefficients for peer firm pairs serves as an evaluation metric.

Secondly, the SIC code and NAICS code are widely used expert-designed ICSs, featuring hierarchical four-digit and six-digit codes, respectively. We assigned a similarity score to each company pair based on the longest common prefix of their SIC code and NAICS code. The label similarity score is calculated using the formula:

$$\text{score}_{\text{label}} = \max(0, \text{LCP}(c_1, c_2) - 1) \tag{3}$$

Here, LCP represents the longest common prefix function, and c_1 and c_2 represent the SIC code or NAICS code of two companies. The Spearman correlation coefficient between the cosine similarity of document representations and the label similarity score serves as the second evaluation metric.

Lastly, We appended a linear classifier to the representations generated by LongBERT and constructed a multi-class task for SIC3 and NAICS4, where SIC3 represents the first three digits of the SIC code, and NAICS4 follows a similar pattern. We validated whether the generated representations were linearly separable and contained rich information about company products and business. We excluded categories with fewer than 5 samples, utilized five-fold cross-validation, and evaluated using the F1 score.

4.3 Experimental Details

Baselines. In our study, we compared Financial-ICS with 11 different methods, which encompass 2 expert-designed ICSs (SIC and NAICS) and 9 algorithm-based ICSs. These algorithm-based ICSs employ various methods to generate company economic representations.

The random method randomly generates a 768-dimensional representation vector. For the bag of words approach [8], we followed the same settings as TNIC, extracting only nouns as tokens and removing common tokens with a frequency higher than 25%, resulting in a vocabulary size of 35K. We trained the average GloVe embeddings [17] on the 10K reports training set using $d = 768$. As for the RoBERTa CLS token, average RoBERTa embeddings [14], and average LongBERT embeddings, we continued pre-training them on the training set from RoBERTa's checkpoint and used either the mean of the CLS token or global mean pooling to generate the document's representation vector. In the case of SRoBERTa [20], SimCSE-RoBERTa, SimCSE-Longformer [1,7], and Financial-ICS, we performed fine-tuning after the pre-training. Since RoBERTa

Table 2. Comparison of the effectiveness of ICSs using different methods with Financial-ICS.

	Stock Pearson	Score Spearman		Logistic Regression F1	
		SIC	NAICS	SIC3	NAICS4
Random	0.2798	-0.0001	0.0004	0.0063	0.0115
Bag of words (TNIC)	0.4060	0.2913	0.2988	0.4936	0.5794
Avg. GloVe embeddings	0.3885	0.2904	0.2689	0.0601	0.0888
RoBERTa CLS token	0.3803	0.2802	0.2730	0.0579	0.0720
Avg. RoBERTa embeddings	0.3990	0.2859	0.2865	0.1280	0.1875
SRoBERTa	0.4177	0.3355	0.3559	0.5488	0.6333
SimCSE-RoBERTa	0.4197	0.3367	0.3441	0.6030	0.6730
SimCSE-Longformer	0.4235	0.3998	0.3808	0.6244	0.6982
SIC	0.4129	–	–	–	–
NAICS	**0.4451**	–	–	–	–
Avg. LongBERT embeddings	0.4154	0.3189	0.2939	0.2186	0.3403
Financial-ICS (Ours)	**0.4266**	**0.4145**	**0.4065**	**0.6294**	**0.7072**

and Longformer have token limitations (512 or 4096 tokens), we divided the documents into segments of the maximum length that the model can handle.

Training Hyper-Parameter. Our training consists of two main parts: pre-training and contrastive learning fine-tuning. We conducted experiments using 2 A100 GPUs. For both parts, we used the AdamW optimizer [15], with a weight decay of 0.01, enabled gradient checkpointing [3], and trained in a distributed manner. During pre-training, the learning rate was set to 1e-4, with warmup and linear decay, a batch size of 1, gradient accumulation of 16, and approximately 20K steps of training. During fine-tuning, the learning rate was set to 1e-6, using a constant learning rate strategy, a temperature of 0.1, and 200 steps of training.

4.4 Results

The evaluation results of ICSs are summarized in Table 2. Financial-ICS achieves the optimal performance among algorithm-based ICSs by directly processing the entire 10K reports document using LongBERT and surpasses SIC, one of the most popular ICS, in terms of the stock Pearson evaluation metric. When comparing expert-designed ICSs based on stock market correlation with peer firms, NAICS outperforms SIC. Interestingly, even for non-peer firms, the random method shows a non-negligible stock Pearson correlation of 0.28, suggesting that there is still stock price correlation within the year, aligning with intuitive expectations. Additionally, TNIC, utilizing the bag of words method, proves effective compared to the GloVe method and representations generated from pre-trained RoBERTa. However, TNIC's drawback lies in its embeddings being of tens of thousands of dimensions. On the other hand, both SRoBERTa and methods based on contrastive learning demonstrate promising performance, surpassing the TNIC method.

4.5 Inference Speed and Memory Consumption

We conducted experiments to compare the inference speed of RoBERTa, Longformer, and LongBERT at different batch sizes using the demo dataset, which had an average length of 23K tokens. For RoBERTa and Longformer, we segmented the text into lengths of 512 and 4096, respectively, before performing inference. From Table 3, we observe that due to LongBERT's $O(n)$ computation complexity and its concise implementation, its efficiency is 1.87 times faster than Longformer.

Table 3. Comparison of the inference speed (docs/s) among RoBERTa, Longformer, and LongBERT at various batch sizes.

Batch Size	RoBERTa	Longformer	LongBERT
1	2.99	3.03	7.61
4	7.74	3.74	7.49
16	9.83	4.07	–
64	10.37	4.07	–
256	10.73	–	–

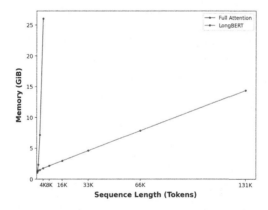

Fig. 6. The comparison of GPU memory usage between full attention and LongBERT for inferring texts of different lengths.

Additionally, we also compared the GPU memory consumption of full attention and LongBERT at different sequence lengths. As shown in Fig. 6, it is evident that the memory usage of LongBERT increases linearly with the sequence length. When inferring long texts of 131K tokens, only 14.4 GiB of GPU memory is required, which is generally acceptable.

(a) Apple Inc.

The Company designs, manufactures and markets mobile communication and media devices and personal computers, and sells a variety of related software, services, accessories and third-party digital content and applications. The Company"'s products and services include iPhone®, iPad®, Mac®, Apple Watch®, AirPods®, Apple TV®, HomePod™, a portfolio of consumer and professional software applications, iOS, macOS®, watchOS® and tvOS™ operating systems, iCloud®, Apple Pay® and a variety of other accessory, service and support offerings. The Company sells and delivers digital content and applications through the iTunes Store®, App Store®, Mac App Store, TV App Store, Book Store and Apple Music® (collectively ""Digital Content and Services""). The Company sells its products worldwide through its retail stores, online stores and direct sales force, as well as through third-party cellular network carriers, wholesalers, retailers and resellers. In addition, the Company sells a variety of third-party Apple-compatible products, including application software and various accessories, through its retail and online stores. The Company sells to consumers, small and mid-sized businesses and education, enterprise and government customers. The Company"s fiscal year is the 52- or 53-week period that ends on the last Saturday of September. The Company is a California corporation established in 1977.

(b) NVIDIA Corp.

Starting with a focus on PC graphics, NVIDIA invented the graphics processing unit, or GPU, to solve some of the most complex problems in computer science. We have extended our focus in recent years to the revolutionary field of artificial intelligence, or AI. Fueled by the sustained demand for better 3D graphics and the scale of the gaming market, NVIDIA has evolved the GPU into a computer brain at the intersection of virtual reality, or VR, high performance computing, or HPC, and AI.

The GPU was initially used to simulate human imagination, enabling the virtual worlds of video games and films. Today, it also simulates human intelligence, enabling a deeper understanding of the physical world. Its parallel processing capabilities, supported by up to thousands of computing cores, are essential to running deep learning algorithms. This form of AI, in which software writes itself by learning from data, can serve as the brain of computers, robots and self-driving cars that can perceive and understand the world. GPU-powered deep learning continues to be adopted by thousands of enterprises to deliver services and features that would have been impossible with traditional coding.

Fig. 7. Explanatory analysis of LongBERT's behavior. Only a partial content of the documents is showcased here.

(a) Microsoft Corp.

CIK	Firm Name	SIC	SIC Description	Cosine Similarity
789019	Microsoft Corp.	7372	Services-Prepackaged Software	1.0000
1660134	Okta, Inc.	7372	Services-Prepackaged Software	0.9622
1463172	Zendesk, Inc.	7374	Services-Computer Processing & Data Preparation	0.9605
1447669	Twilio	7372	Services-Prepackaged Software	0.9598
1104855	Support.com,Inc.	7374	Services-Computer Processing & Data Preparation	0.9594
1372612	Box, Inc.	7372	Services-Prepackaged Software	0.9591
1066194	eGain	7372	Services-Prepackaged Software	0.9587
1108524	Salesforce, Inc.	7372	Services-Prepackaged Software	0.9575
1470099	MobileIron Inc.	7372	Services-Prepackaged Software	0.9570
849399	Symantec Corporation	7372	Services-Prepackaged Software	0.9567
1448056	New Relic	7372	Services-Prepackaged Software	0.9556
...				

(b) Visa Inc.

CIK	Firm Name	SIC	SIC Description	Cosine Similarity
1403161	Visa Inc.	7389	Services-Business Services, NEC	1.0000
1141391	Mastercard Inc.	7389	Services-Business Services, NEC	0.9683
1123360	Global Payments Inc.	7389	Services-Business Services, NEC	0.9590
1029199	Euronet Worldwide	6099	Functions Related To Depository Banking, NEC	0.9564
1175454	Fleetcor Technologies, Inc.	7389	Services-Business Services, NEC	0.9524
721683	Total System Services, Inc.	7389	Services-Business Services, NEC	0.9507
1144354	Heartland Payment Systems, Inc.	7389	Services-Business Services, NEC	0.9442
1559865	Evertec, Inc.	7374	Services-Computer Processing & Data Preparation	0.9423
1671013	Cardtronics	7389	Services-Business Services, NEC	0.9357
1633917	PayPal Holdings, Inc.	7389	Services-Business Services, NEC	0.9321
1140184	The Western Union Company	7389	Services-Business Services, NEC	0.9281

Fig. 8. Peer firms identified by Financial-ICS for Microsoft Corp. and Visa Inc.

4.6 Explanatory Analysis

Financial-ICS utilizes LongBERT to generate document representations, but this process lacks transparency as it is considered a black box. However, explanations are of utmost importance in the financial domain. Therefore, we conducted an analysis to explain the model's behavior. We employed a customized version of the SHAP [16] architecture to suit the embedding model.

To analyze the model's behavior, we firstly introduced perturbations to the input of lengthy texts by replacing segments of tokens with "...". Then, we employed a concept based on prototype-based contrastive loss in Sect. 3.3 to convert it as a classification task. We randomly selected a sample of 100 documents, with the first document chosen for analysis. By perturbing the input and observing the corresponding changes in the model's output, we determined the contributions of each token to the output. The results of our explanations analysis for the 10K reports of Apple Inc. and NVIDIA Corp. were presented in Fig. 7. Notably, tokens such as "mobile" and "Apple" had a significant influence the output for Apple Inc., while tokens such as "NVIDIA", "GPU", and "AI" had a noticeable impact on the output for NVIDIA Corp.

4.7 Financial-ICS Demo

Figure 8 showcases the peer firms identified by Financial-ICS for Microsoft Corp. and Visa Inc. Remarkably, these peer firms generally share similar SIC codes with their respective focal firms. In contrast to expert-designed ICSs, we incorporate similarity scores and sequential information when determining these peer firms. Moreover, we have also observed cases where certain peer firms possess SIC codes entirely distinct from the focal firms, primarily due to their business diversification. Algorithm-based ICSs offer the ability to identify unique peer firms for each company. Lastly, in the event of introducing new 10K reports for a company or updating existing ones, only the corresponding company's embeddings need to be modified. Financial-ICS effectively addresses the limitations present in expert-designed ICSs while upholding comparable performance.

5 Conclusion

In this research, we introduce Financial-ICS, an advanced algorithm-based ICS that incorporates a comprehensive unsupervised training and evaluation process. By integrating shifted block attention and global attention, we have developed LongBERT, a high-performing model with a complexity of $O(n)$ that can directly handle documents containing more than 100K tokens. Additionally, we have implemented a prototype-based contrastive learning loss that expands the number of negative samples in a single batch. Furthermore, we have established three evaluation metrics to provide a thorough assessment of Financial-ICS.

In our experiments, LongBERT has demonstrated excellent time and space efficiency, along with a powerful capability to handle long texts. Additionally, Financial-ICS has surpassed a diverse range of ICSs that rely on cutting-edge algorithms or are expert-designed.

References

1. Beltagy, I., Peters, M.E., Cohan, A.: Longformer: the long-document transformer. arXiv preprint arXiv:2004.05150 (2020)
2. Bhojraj, S., Lee, C.M., Oler, D.K.: What's my line? a comparison of industry classification schemes for capital market research. J. Account. Res. **41**(5), 745–774 (2003)

3. Chen, T., Xu, B., Zhang, C., Guestrin, C.: Training deep nets with sublinear memory cost. arXiv preprint arXiv:1604.06174 (2016)
4. Dai, Z., Yang, Z., Yang, Y., Carbonell, J., Le, Q.V., Salakhutdinov, R.: Transformer-xl: attentive language models beyond a fixed-length context. arXiv preprint arXiv:1901.02860 (2019)
5. Devlin, J., Chang, M.W., Lee, K., Toutanova, K.: Bert: pre-training of deep bidirectional transformers for language understanding. arXiv preprint arXiv:1810.04805 (2018)
6. Ethayarajh, K.: How contextual are contextualized word representations? comparing the geometry of bert, elmo, and gpt-2 embeddings. arXiv preprint arXiv:1909.00512 (2019)
7. Gao, T., Yao, X., Chen, D.: Simcse: simple contrastive learning of sentence embeddings. arXiv preprint arXiv:2104.08821 (2021)
8. Hoberg, G., Phillips, G.: Text-based network industries and endogenous product differentiation. J. Polit. Econ. **124**(5), 1423–1465 (2016)
9. Hochreiter, S., Schmidhuber, J.: Long short-term memory. Neural Comput. **9**(8), 1735–1780 (1997)
10. Kitaev, N., Kaiser, L., Levskaya, A.: Reformer: the efficient transformer. arXiv preprint arXiv:2001.04451 (2020)
11. Lee, C.M., Ma, P., Wang, C.C.: Search-based peer firms: aggregating investor perceptions through internet co-searches. J. Financ. Econ. **116**(2), 410–431 (2015)
12. Lewis, M., et al.: Bart: denoising sequence-to-sequence pre-training for natural language generation, translation, and comprehension. arXiv preprint arXiv:1910.13461 (2019)
13. Liu, P.J., et al.: Generating wikipedia by summarizing long sequences. arXiv preprint arXiv:1801.10198 (2018)
14. Liu, Y., et al.: Roberta: a robustly optimized bert pretraining approach. arXiv preprint arXiv:1907.11692 (2019)
15. Loshchilov, I., Hutter, F.: Decoupled weight decay regularization. arXiv preprint arXiv:1711.05101 (2017)
16. Lundberg, S.M., Lee, S.I.: A unified approach to interpreting model predictions. In: Advances in Neural Information Processing Systems, vol. 30 (2017)
17. Pennington, J., Socher, R., Manning, C.D.: Glove: global vectors for word representation. In: Proceedings of the 2014 Conference on Empirical Methods in Natural Language Processing (EMNLP), pp. 1532–1543 (2014)
18. Radford, A., Wu, J., Child, R., Luan, D., Amodei, D., Sutskever, I., et al.: Language models are unsupervised multitask learners. OpenAI blog **1**(8), 9 (2019)
19. Raffel, C., et al.: Exploring the limits of transfer learning with a unified text-to-text transformer. J. Mach. Learn. Res. **21**(1), 5485–5551 (2020)
20. Reimers, N., Gurevych, I.: Sentence-bert: sentence embeddings using siamese bert-networks. arXiv preprint arXiv:1908.10084 (2019)
21. Su, J., Lu, Y., Pan, S., Murtadha, A., Wen, B., Liu, Y.: Roformer: enhanced transformer with rotary position embedding. arXiv preprint arXiv:2104.09864 (2021)
22. Vaswani, A., et al.: Attention is all you need. In: Advances in Neural Information Processing Systems, vol. 30 (2017)
23. Wang, S., Li, B.Z., Khabsa, M., Fang, H., Ma, H.: Linformer: self-attention with linear complexity. arXiv preprint arXiv:2006.04768 (2020)
24. Yan, Y., Li, R., Wang, S., Zhang, F., Wu, W., Xu, W.: Consert: a contrastive framework for self-supervised sentence representation transfer. arXiv preprint arXiv:2105.11741 (2021)

25. Zaheer, M., et al.: Big bird: Transformers for longer sequences. Adv. Neural. Inf. Process. Syst. **33**, 17283–17297 (2020)
26. Zhou, H., et al.: Informer: beyond efficient transformer for long sequence time-series forecasting. In: Proceedings of the AAAI Conference on Artificial Intelligence, vol. 35, pp. 11106–11115 (2021)

BDQM

Establishing a Decentralized Diamond Quality Management System: Advancing Towards Global Standardization

P. H. T. Trung and L. K. Bang[✉]

FPT University, Can Tho city, Vietnam
banglkce160155@fpt.edu.vn

Abstract. The global diamond industry currently lacks a unified standard for assessing diamond quality. Each diamond product is individually evaluated and certified by individuals or reputable organizations within the industry. However, when these certifying entities cease their operations, the certifications lose their value, resulting in a loss of value for the associated diamond products. Furthermore, varying evaluation criteria employed by different individuals and organizations lead to fluctuations in the value of diamonds when transitioning between entities or across different countries. In this paper, we propose an approach to address this issue by implementing a decentralized system for quality management and certification of diamond products using smart contracts, NFTs (Non-Fungible Tokens), IPFS (InterPlanetary File System), and distributed ledgers. Our model aims to ensure accurate assessment of diamond products according to predefined standards, providing individuals and organizations owning diamonds with confidence in the enduring value of their assets, irrespective of evaluations by different entities. This approach fosters consistency in the value of diamond products within the business sector and across nations, ultimately enhancing the global diamond industry.

Keywords: Diamond · Blockchain · Data Management · Non-Fungible Tokens (NFTs) · InterPlanetary File System (IPFS) · Decentralized Storage · Smart Contracts

1 Introduction

Traditional diamond quality management approaches, rooted in the "Four Cs" grading system, play a crucial role in the gemstone industry, with institutions like the GIA and IGI standardizing assessments and issuing globally recognized certifications. These methods, focusing on cut, color, clarity, and carat weight, have been instrumental in establishing market values, as seen in the $79 billion valuation of the diamond industry in 2019. The traditional systems also contribute significantly to consumer trust and ethical practices, as evidenced by the Kimberley Process Certification Scheme's success in ensuring over 99.8%

of diamonds were conflict-free by 2019. While maintaining the quality, value, and ethical sourcing of diamonds, these traditional practices have set a strong foundation for the industry's evolution amidst emerging technologies.

Traditional diamond quality management methods face critical limitations, particularly in transparency, which significantly impacts trust and market value. The lack of clear traceability from mine to market and potential ethical concerns, such as the infiltration of conflict diamonds, pose challenges to consumer confidence and security in the industry. Incidents like synthetic diamonds being mislabeled as natural highlight these shortcomings. Additionally, traditional grading does not fully encompass a diamond's history or ethical sourcing information, increasingly important to consumers. These transparency issues in an industry worth billions underscore the need for more open and traceable management practices to maintain market stability and consumer trust.

Integrating blockchain technology with traditional diamond management systems marks a pivotal shift towards enhancing transparency in the industry. Blockchain's decentralized and immutable ledger provides a secure, tamper-proof record of a diamond's entire journey, from mining to final sale. This integration allows for detailed tracking of each diamond's origin, processing, and ownership, significantly boosting consumer confidence in authenticity and ethical sourcing. For instance, a diamond can be assigned a unique blockchain identifier, linked to an NFT, detailing its complete history and quality certification, accessible to potential buyers. Such an approach offers a level of transparency and verifiability that traditional methods cannot match, revolutionizing how consumers and stakeholders interact with diamond certification and ownership data.

In our proposed diamond quality management model, blockchain technology serves as a foundational element, with Non-Fungible Tokens (NFTs) being used to uniquely represent and store crucial information about each diamond. We selected BNB, Fantom, Polygon, and Celo for deployment, chosen for their Ethereum Virtual Machine (EVM) compatibility and unique features such as transaction speed, scalability, cost-effectiveness, and mobile accessibility. Deploying our model on these platforms allowed us to leverage their specific strengths, enhancing the system's efficiency, and accessibility. This approach significantly augments the traditional diamond management system, introducing enhanced security, transparency, and a novel method of interaction with diamond data, ultimately revolutionizing the diamond industry's approach to quality management and certification.

In our analysis of blockchain-based diamond management, we evaluated three key functionalities - creating transactions, minting NFTs, and transferring NFTs - across BNB, Fantom, Polygon, and Celo platforms. Each platform displayed distinct strengths: BNB excelled in transaction speed and efficiency, Fantom stood out for its cost-effectiveness and speed, Polygon offered extremely low transaction and minting costs, and Celo shone with its mobile accessibility and moderate costs. The comparative analysis revealed a diversity in cost and performance, highlighting that each platform can be strategically chosen based on specific needs in the diamond industry, balancing factors like efficiency, cost-effectiveness, and scalability. This versatility underscores the practical adaptability of our blockchain-based model in enhancing diamond quality management.

2 Related Work

The integration of blockchain technology into the diamond industry has been a subject of growing interest among researchers and industry professionals. In the realm of luxury supply chains, especially in diamond authentication and certification, the role of blockchain technology has been increasingly recognized as a transformative factor. Choi et al. [1] explored different consumer utility-driven operations models, highlighting the value of blockchain-technology-supported platforms for diamond authentication and certification. Their work delves into both traditional retail network operations and blockchain technology-supported selling platforms, revealing the conditions under which blockchain enhances the diamond industry's operations.

Thakker et al. [7] surveyed the adoption of blockchain in the diamond industry, presenting both the opportunities and challenges of this integration. They noted that blockchain technology could bridge the gap between the diamond industry and financial markets, providing confidence in the legitimacy of diamonds and easing the recovery of stolen property.

Sumkin et al. [6] discussed blockchain's application in demonstrating the provenance of physical goods, specifically in the diamond supply chain. They argued that while blockchain could enable the traceability of responsibly produced goods, its implementation might also impact market segmentation strategies and sourcing from responsible suppliers.

Loebbecke et al. [3] focused on how blockchain technology impacts trust in transactions, particularly in trading high-value physical goods like diamonds. Their study analyzed the role of trust in trading diamonds with blockchain technology, concluding that blockchain simultaneously substitutes and complements trust in diamond trading.

Wright [8] examined the Kimberley Process Certification Scheme, a critical initiative in combating the trade in conflict diamonds. This scheme represented an early effort to bring transparency and ethical sourcing to the diamond industry, paving the way for blockchain's role in enhancing these aspects.

Maconachie [4] provided insights into the governance challenges in small-scale diamond mining communities, particularly in post-conflict Sierra Leone. The paper reviewed governance initiatives like the Kimberley Process and its impact on the diamond sector, offering a backdrop to the potential of blockchain in addressing similar challenges.

Furthermore, the white paper by CEDEX [2] addressed the transformation of diamonds into a new financial asset class through blockchain technology. It discussed the challenges in the diamond ecosystem and presented a blockchain-based approach for creating a diamond investment market.

Lastly, Santiago [5] explored the compliance and business practices in the diamond industry post the adoption of the Kimberley Process Certification Scheme, highlighting the gap between norm creation and stakeholder awareness.

These studies collectively underscore the potential of blockchain technology in revolutionizing the diamond industry, particularly in enhancing transparency, trust, and ethical sourcing. The blockchain's capability to provide a secure, immutable record of a diamond's journey from mine to market addresses many of the traditional challenges faced in diamond quality management.

3 Methodology

3.1 Traditional Model for Diamond Quality Management Process

Traditional Model's Components: The traditional diamond quality evaluation model is comprised of several crucial components, each playing a significant role in ensuring the authenticity and value of diamonds. This model is carefully structured to maintain the highest standards of diamond quality and integrity in the market.

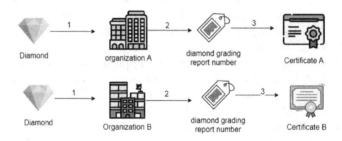

Fig. 1. Classic model for pediatric healthcare

Diamond. The central element of this model. Diamonds are the subject of evaluation, and the model aims to standardize their management and quality assessment. Each diamond is unique and requires careful inspection to determine its value and characteristics.

Organization. A reputable and authoritative organization is responsible for the assessment of diamond quality. This organization utilizes various criteria such as brightness and cut to evaluate each diamond. The credibility of the organization is vital in ensuring the trustworthiness of the evaluation.

Diamond Grading Report Number. Diamond grading report number sequence engraved on each diamond for identification. This diamond grading report number distinguishes each diamond and is crucial for tracking and documenting its attributes and history. It maintains the integrity and traceability of the diamond.

Quality Certification. A certificate that records the detailed information of the evaluated diamond, including its diamond grading report number and attributes. This certification serves as a formal validation of the diamond's qualities such as cut, clarity, color, and carat weight, providing essential information and assurance about the diamond's quality and authenticity.

Traditional Model's Steps: In the intricate world of diamond trading, the assessment and certification of diamonds are paramount to maintaining their value and trustworthiness in the market. This meticulous process is divided into several key stages, each playing a vital role in determining the quality and authenticity of these precious stones. Let's explore these stages to understand better how diamonds are evaluated and certified in the traditional diamond industry, refer to Fig. 1.
Step 1: Initial Assessment. When a diamond is first produced, it is sent to a specialized quality evaluation organization. At this stage, the diamond undergoes a detailed inspection. The examination covers various attributes such as size, shape, color, and clarity. This step is essential to determine the fundamental characteristics of the diamond, which are critical for its valuation and categorization.
Step 2: Assigning a Diamond Grading Report Number. After the initial evaluation, each diamond is given a unique grading report number. This code is crucial for several reasons. It allows for the identification and tracking of the diamond throughout its lifecycle. Additionally, the code is linked to detailed information about the diamond, encompassing all aspects of its evaluation. This system ensures transparency in the diamond trade and facilitates buyers in accessing comprehensive information about the diamond they are considering.
Step 3: Quality Certification. The final step in the process is the issuance of a quality certificate by the evaluating organization. This certificate includes in-depth information about the diamond, such as the number of facets, the level of transparency, cut patterns, and other specifics that correspond to the unique grading report number. This certification serves as an authoritative proof of the diamond's quality and characteristics. It plays a vital role in building buyer trust and is an indispensable tool in transactions, providing essential information for both current and future dealings.

Traditional Model's Limitation: The traditional model for diamond quality evaluation and management, while being a cornerstone in the diamond industry for many years, presents several limitations that can impact its effectiveness and reliability. The traditional diamond quality evaluation model faces several key challenges. First, there's an inconsistency in quality standards due to different organizations having their own evaluation criteria, leading to varied valuations for similar diamonds. This creates market confusion. Second, the validity of quality certifications is dependent on the issuing organization's operational status, risking loss of certification validity if the organization ceases operations.

Third, the reliance on physical certificates poses risks as they can be lost or damaged, making it difficult to reissue and prove a diamond's quality and history. Finally, the system's vulnerability to fraud is heightened due to the lack of standardization, allowing counterfeit certificates and misrepresented diamonds to undermine the industry's trust and integrity. Given these limitations, there is a growing need for a new approach to diamond management and certification. This is where modern technologies like blockchain, smart contracts, NFTs (Non-Fungible Tokens), and IPFS (InterPlanetary File System) come into play.

3.2 Innovative Blockchain Model for Diamond Quality Management Process

In the realm of diamond quality management, the integration of blockchain technology heralds a transformative approach, characterized by a suite of advanced components each playing a pivotal role in redefining industry standards. This modern model includes:

Unified Evaluation Standard: This standard, established by reputable organizations worldwide, applies to all diamond products. It ensures consistency in diamond quality assessment, providing a reliable benchmark for evaluation.

User Interface (UI): The UI is a crucial point of interaction for manufacturers, evaluators, and owners. It allows for the assessment and reporting of diamond quality, and enables owners to access comprehensive information about a diamond they own or intend to trade. This interface facilitates transparency and ease of access to diamond data.

Distributed Ledger (Blockchain): This ledger serves as a transparent and clear repository of information regarding the quality of diamonds, indexed by their grading report numbers. It ensures that all data related to a diamond's quality and history are securely and immutably recorded, providing a single source of truth.

Smart Contracts: These are used to execute new transactions, queries, and data updates for a diamond. Smart contracts automate the enforcement of agreements and conditions, reducing the need for intermediaries and enhancing the efficiency and security of transactions.

Non-Fungible Tokens (NFTs): Each diamond's data and certification can be represented as a unique NFT, preventing the falsification of information. NFTs provide a distinctive and tamper-proof digital identity for each diamond, ensuring its authenticity.

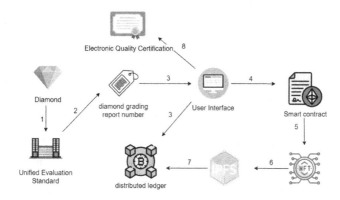

Fig. 2. Innovative Blockchain Model for Child Care

InterPlanetary File System (IPFS): IPFS offers a fast and efficient way to transfer data. It's used to store and share data in a distributed manner, enhancing the accessibility and durability of diamond-related information.

Electronic Quality Certification: This is the digital certification issued by the evaluating authority according to the unified standard and stored within the distributed ledger. It represents a secure and easily verifiable record of a diamond's quality, replacing traditional paper certificates (Fig. 2).

The process of evaluating diamond quality using blockchain technology follows a comprehensive and systematic approach, integrating several high-tech steps to ensure accuracy, transparency, and security. This process can be described as follows:

Step 1 Initial Evaluation Using Unified Evaluation Standard: The process begins with the meticulous evaluation of the diamond against the Unified Evaluation Standard. This comprehensive standard ensures that every aspect of the diamond, such as cut, color, clarity, and carat, is rigorously assessed. This initial step establishes a baseline of quality, crucial for the subsequent blockchain integration.

Step 2 Assignment of Grading Report Number: Upon passing the quality assessment, each diamond is assigned a unique grading report number. This identifier is crucial as it links the physical diamond to its digital records, forming the basis for its blockchain identity. This step is pivotal in ensuring that each diamond can be individually tracked and its history can be accurately recorded.

Step 3 Updating Information in the System and Distributed Ledger: Next, the diamond's data, including its quality metrics and grading report number, is updated in the blockchain system. This involves synchronizing the information with the distributed ledger, a process that ensures the data is not only secure but also immutable. It provides a permanent, unalterable record of the diamond's quality and characteristics.

Step 4 Creation of Smart Contracts: In this step, smart contracts specific to each diamond are generated within the blockchain. These contracts are programmed to automate the secure and efficient updating of information related to the diamond, such as ownership transfers or changes in certification. Smart contracts play a crucial role in maintaining the integrity and accuracy of the diamond data over time.

Step 5 Generation of Unique NFT for Each Diamond: Here, the transformative role of NFTs comes into play. Each diamond is represented as a unique NFT, encapsulating its identity, ownership, and quality information. This NFT, being a one-of-a-kind digital asset, provides a powerful tool for proving the authenticity and ownership of the diamond. It ensures that the information about each diamond is distinct, secure, and easily verifiable, greatly reducing the risk of fraud.

Step 6 Data Storage on IPFS: The diamond's detailed information, along with its transaction and certification history, is stored on the IPFS. The use of IPFS ensures decentralized storage, making data highly resilient to tampering and loss. This system also facilitates faster and more reliable access to diamond information from anywhere in the world.

Step 7 Data Synchronization with the Distributed Ledger: The information on the IPFS is synchronized back to the blockchain's distributed ledger. This ensures that every piece of data related to the diamond is consistently updated and securely stored across both platforms, reinforcing the integrity and reliability of the information.

Step 8 Issuance of Digital Quality Certification: Finally, a digital quality certificate, securely stored within the blockchain and linked to the specific NFT, is issued for the diamond. This certificate provides a digitally verifiable proof of the diamond's quality, leveraging the blockchain's security and transparency.

This blockchain-based process revolutionizes the traditional diamond evaluation method, bringing in unmatched levels of transparency, security, and efficiency. It not only streamlines the process of diamond certification but also builds a robust framework for verifying the authenticity and quality of diamonds in the global market.

4 Implementation

In the blockchain-based diamond quality management system, the processes of initiation, querying, and updating play crucial roles in managing a diamond's data lifecycle. The initiation phase involves the precise recording of a diamond's characteristics based on a unified standard, establishing its digital identity. Querying allows stakeholders to efficiently access and analyze detailed information about the diamond's quality, history, and ownership, ensuring transparency and authenticity. The updating process, often automated by smart contracts, is essential for maintaining current and accurate data, reflecting changes in ownership or diamond characteristics. These integrated processes significantly enhance the system's security, efficiency, and transparency, marking a significant advancement over traditional diamond management methods (Fig. 3).

4.1 Data and NFT Initialization

The first step in the blockchain-based diamond quality management process involves meticulous collection of each diamond's data, including detailed identification information (grading report number, source, unique characteristics) and a thorough record of its quality attributes based on the Unified Evaluation Standard. The system also logs details about the professionals involved in the diamond's evaluation and certification, recording their qualifications and roles in the process. A distributed ledger system is utilized to manage this sensitive and crucial data, enabling secure, real-time data sharing and access across various entities in the diamond industry. This system minimizes latency and ensures immediate access to updated, accurate information for all relevant parties, thereby enhancing the efficiency, transparency, and security of the diamond management process.

A crucial aspect of data collection is the specific details about each diamond. This includes its type (e.g., rough, polished), quality characteristics (such as cut, color, clarity, carat), and any specific attributes that might impact its valuation and suitability for various uses. Given the different values and uses of diamonds based on these characteristics, this precise categorization and documentation are essential. For example, while a high-grade diamond might be sought after for luxury jewelry, others might be more suitable for industrial applications. Each diamond's unique information is meticulously recorded and managed within the blockchain system to ensure accurate tracking and valuation throughout its lifecycle.

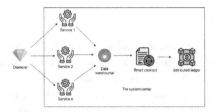

Fig. 3. Process of Data and NFT Initialization

4.2 Data Retrieval

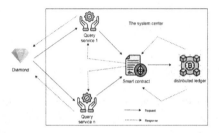

Fig. 4. Data Retrieval Process

Figure 4 depicts the steps involved in data retrieval within the blockchain-based diamond quality management system. Leveraging a distributed model, this system enables various stakeholders in the diamond industry, such as jewelers, evaluators, and buyers, to access data concurrently. For instance, jewelers and buyers might use the system to gather information about the provenance of diamonds, their certification details, or upcoming auctions and sales events. For professionals involved in diamond evaluation and certification, the system offers the capability to access in-depth data about individual diamonds, which is vital for planning assessments or organizing diamond grading sessions. This system ensures that all parties involved in the diamond industry are well-informed and can effectively coordinate their activities.

4.3 Data Update Mechanism

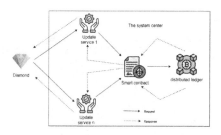

Fig. 5. Procedure for Data Updates

As illustrated in Fig. 5, the process of updating data in the blockchain-based diamond quality management system is initiated following rigorous data verification. If certain data is not available on the blockchain, the system automatically notifies the user of its absence. In instances where updates are necessary, such as alterations in a diamond's certification status or ownership changes, the system first retrieves the existing data and then modifies the relevant details. For significant alterations in diamond records, mirroring the concept of NFTs (Non-Fungible Tokens), any major update results in the creation of a new entry within the system. This method ensures that the historical records of each diamond's evaluation, certification, and transaction history are preserved intact, maintaining a transparent and accurate chronicle of the diamond's journey through the industry.

5 Comprehensive Assessment Methodology

In the evolving landscape of blockchain applications for diamond quality management, various platforms compatible with the Ethereum Virtual Machine (EVM) play a pivotal role. For our in-depth analysis, we intend to deploy our bespoke smart contracts designed for diamond certification and tracking on four eminent EVM-compatible platforms: *Binance Smart Chain (BNB Smart Chain)*[1], *Polygon*[2], *Fantom*[3], and *Celo*[4]. Each of these platforms offers unique features and advantages, and our implementation aims to assess and leverage the specific capabilities of each. Moreover, in line with the rising trend of decentralized storage solutions and the growing interest in Non-Fungible Tokens (NFTs), we plan to integrate the storage of detailed diamond certification data on the InterPlanetary File System (IPFS) using the Pinata platform, a developer API for IPFS. This approach will enhance the transparency and security of diamond data management within the blockchain ecosystem.

---

[1] https://github.com/bnb-chain/whitepaper/blob/master/WHITEPAPER.md.
[2] https://polygon.technology/lightpaper-polygon.pdf.
[3] https://whitepaper.io/document/438/fantom-whitepaper.
[4] https://celo.org/papers/whitepaper.

5.1 Environment Simulation

Fig. 6. The sample of 20 nodes (i.e., first 5 nodes) with the public key and private key

In a decentralized blockchain setup for diamond quality management, nodes are fundamental, ensuring data integrity, redundancy, and security. Our simulation environment has been meticulously crafted to mirror the behavior of an actual decentralized blockchain network. Figure 6 illustrates a scenario with 20 nodes used in our simulation, each uniquely identified by its corresponding public and private key pairs.

In our model, each node represents a potential stakeholder in the diamond industry, such as jewelers, certification agencies, diamond miners, or buyers. The public key serves as an identifier, enabling nodes to recognize and interact with each other within the blockchain, while the private key ensures secure, encrypted transactions and access to proprietary data. This key system guarantees that although any node with the required permissions can view a record, only the node with the corresponding private key can alter it.

Our deployment utilized a mix of virtual and physical servers to establish a diverse testing environment, closely emulating real-world scenarios in diamond management. This variety in the setup allowed us to test the resilience, latency, and throughput of our system under different conditions. Additionally, the use of both server types helped identify any potential issues or vulnerabilities that might be more evident in one environment than the other.

By utilizing this simulated environment, we could conduct transactions, create and transfer NFTs representing diamond certification records, and assess the overall performance of our blockchain system. Continuous testing and refinement enabled us to fine-tune our approach, ensuring it effectively meets the challenges and dynamics of a decentralized network in diamond quality management and tracking.

5.2 Utilize the IPFS for the Secure Storage of Data About Diamond Quality Management Process

In today's era of digital innovation, secure storage, effortless accessibility, and swift retrieval of data are crucial, particularly in industries like diamond quality management. The InterPlanetary File System (IPFS), with its decentralized architecture, offers a pioneering solution that enhances data security and is robust against data loss. This is highly relevant for sensitive information in the diamond industry, where storing certification and transaction records on IPFS can be extremely advantageous. The use of IPFS ensures that data about a diamond's provenance, quality assessments, and ownership history is stored in a secure, distributed manner, facilitating easy and fast access while maintaining the integrity of the data.

As illustrated in Fig. 7, a typical diamond certification record is displayed, containing essential information such as quality metrics and the diamond's provenance crucial to the valuation process. These records are converted into a unique digital hash to maintain the authenticity and integrity of the data within the blockchain system. This process ensures that every detail pertaining to the diamond, from its grading to its history, is securely stored and verifiable, enhancing the transparency and trust in diamond transactions.

Figure 8 demonstrates the method of creating a unique hash link for a diamond certification record on the InterPlanetary File System (IPFS). This hash serves as a distinctive identifier for the stored data. Employing the Pinata platform, a well-known IPFS developer API, diamond certification records are securely stored on IPFS. Pinata provides advantages such as user-friendly interfaces and advanced encryption techniques, thereby ensuring the confidentiality of data and managing access control effectively. This approach enhances the security and reliability of diamond data storage and retrieval in the blockchain-based management system.

The decentralized nature of IPFS is instrumental in protecting against single-point failures in the network, introducing a critical layer of redundancy essential for the constant availability of diamond certification records. Figure 9 showcases how a diamond record identifier is represented on the Pinata platform, illustrating the smooth integration between IPFS and Pinata in the context of diamond data management. This setup ensures that diamond certification and transaction records are securely and perpetually accessible, thereby enhancing the reliability and efficiency of the blockchain-based diamond management system.

As depicted in Fig. 10, when querying the specific identifier associated with a diamond certification record, the complete information regarding the diamond

```
const body = {
    diamondType: "diamond Type 1",
    weight: "n carat",
    shape: "oval",
    testingPlace: "testing Place 1",
    productPlace: "product place",
    id: "Diamond grading report number",
    testerName: "tester Name 1",
    time: "test date",
    color: "color 1",
};
const options = {
    pinataMetadata: {
        name: "diamondManagement.json",
    },
```

Fig. 7. Example information of diamond Record

```
Lock
  Deployment
CID QmWGsm38xWU1Y7RPZ4tam8s8gjxjXgABDRAr7FHLnA7URh
    ✓ Should set the right unlockTime (4355ms)


1 passing (4s)
```

Fig. 8. Generation of Unique Hash Link for Information of Diamond on IPFS

Fig. 9. Information of diamond Identifier on Pinata Interface

can be accessed. The integrated use of IPFS and Pinata in our methodology ensures to all parties involved in the diamond industry - from miners and jewelers to certifiers and buyers - that the diamond data remains secure, transparent, and easily accessible. This combination not only streamlines the process of retrieving detailed diamond information but also reinforces the trustworthiness and efficiency of the blockchain-based diamond management system.

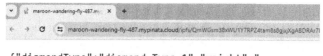

```
{"diamondType":"diamond Type 1","weight":"n
carat","shape":"oval","testingPlace":"testing Place
1","productPlace":"product place","id":"Diamond
grading report number","testerName":"tester Name
1","time":"test date","color":"color 1"}
```

Fig. 10. Retrieval of information of diamond Using Identifier

5.3 Transaction Fee Analysis

Table 1. Transaction fee

	Contract Creation	Create NFT	Transfer NFT
BNB Smart Chain	0.02731136 BNB ($8.37)	0.00109162 BNB ($0.33)	0.00057003 BNB ($0.18)
Fantom	0.009576826 FTM ($0.001860)	0.000405167 FTM ($0.000079)	0.0002380105 FTM ($0.000046)
Polygon	0.006840590024626124 MATIC($0.01)	0.00028940500115762 MATIC($0.00)	0.000170007500612027 MATIC($0.00)
Celo	0.0070973136 CELO ($0.004)	0.0002840812 CELO ($0.000)	0.0001554878 CELO ($0.000)

The information in Table 1 demonstrates the varied transaction fees for different operations across several notable blockchain platforms: BNB Smart Chain, Fantom, Polygon, and Celo. All these platforms are compatible with the Ethereum Virtual Machine (EVM) and are compared based on the costs associated with Contract Creation, NFT Creation, and NFT Transfer. This comparison is crucial for those in the diamond industry using blockchain technology, highlighting the cost implications of choosing a particular platform for specific blockchain operations.

Contract Creation Costs: The fees for creating a smart contract, essential for initiating blockchain-based diamond tracking and certification processes, differ across platforms. BNB Smart Chain, designed for cost-efficiency, charges approximately $8.37 per contract. Fantom, known for its rapid contract deployment, has a fee of about $0.001860, while Polygon, a Layer-2 solution, costs around $0.01. Celo, which focuses on mobile-friendly solutions, charges approximately $0.004.

NFT Minting Fees: As NFTs become increasingly popular for representing diamond ownership and certification, understanding minting costs is essential. Binance Smart Chain's minting fee stands at around $0.33, whereas Fantom offers a significantly lower fee of about $0.000079. Polygon has a nearly nominal fiat fee, and Celo's minting fee is also minimal, at approximately $0.000.

NFT Transfer Costs: The fees associated with transferring NFTs, representing diamonds, vary by platform. BNB Smart Chain has a fee of about $0.18 for this operation, Fantom charges approximately $0.000046, Polygon's fee is minimal in fiat terms, and Celo's transfer fee is around $0.000.

Overall Analysis: Table 1 provides a detailed analysis of transaction fees for essential blockchain operations on different platforms within the context of diamond industry applications. Stakeholders should consider these costs along with other platform-specific factors like security, scalability, and community support when selecting the most suitable blockchain platform for their diamond management needs.

The variability in fee structures across these platforms underscores the importance of choosing the right blockchain platform based on economic considerations, especially for those in the diamond industry where high-volume transactions are common.

6 Conclusion

In this paper, we have introduced a groundbreaking framework based on blockchain technology, meticulously tailored for the intricate domain of diamond quality management. Our approach transcends the limitations of conventional diamond tracking systems, incorporating heightened security, accessibility, and transparency through the integration of Non-Fungible Tokens (NFTs) and the InterPlanetary File System (IPFS).

We have demonstrated how our framework effectively eliminates the risks inherent in centralized data storage and management in diamond tracking, thereby enhancing the integrity and security of diamond records. By employing NFTs, each record of a diamond becomes a unique, immutable asset on the blockchain, offering greater control and assurance to jewelers, certification bodies, and buyers. The deployment of IPFS for decentralized data storage guarantees continuous accessibility to these crucial records, substantially improving the system's resilience and reliability.

Our research extends beyond theoretical concepts; we have successfully implemented our framework across diverse blockchain platforms, verifying its versatility and practical applicability in real-world scenarios. Additionally, by utilizing the Pinata platform, an IPFS-based data storage solution, we have fortified the system's defense against single points of failure, ensuring a more secure and efficient process in diamond quality management.

References

1. Choi, T.M.: Blockchain-technology-supported platforms for diamond authentication and certification in luxury supply chains. Transp. Res. Part E Logistics Transp. Rev. **128**, 17–29 (2019)
2. Exchange, G.D.C.: White paper transforming diamonds into a new financial asset class (2018)
3. Loebbecke, C., Lueneborg, L., Niederle, D.: Blockchain technology impacting the role of trust in transactions: reflections in the case of trading diamonds. In: ECIS Proceedings (2018)

4. Maconachie, R.: Diamonds, governance and 'local' development in post-conflict sierra leone: lessons for artisanal and small-scale mining in sub-saharan africa? Resour. Policy **34**(1–2), 71–79 (2009)
5. Santiago, A.P.: Guaranteeing conflict free diamonds: from compliance to norm expansion under the kimberley process certification scheme. S. Afr. J. Int. Affairs **21**(3), 413–429 (2014)
6. Sumkin, D., Hasija, S., Netessine, S.: Does blockchain facilitate responsible sourcing? an application to the diamond supply chain (2021)
7. Thakker, U., Patel, R., Tanwar, S., Kumar, N., Song, H.: Blockchain for diamond industry: opportunities and challenges. IEEE Internet Things J. **8**(11), 8747–8773 (2020)
8. Wright, C.: Tackling conflict diamonds: the kimberley process certification scheme. Int. Peacekeeping **11**(4), 697–708 (2004)

Co-estimation of Data Types and Their Positional Distribution

Shin-ya Sato

Nippon Institute of Technology, 4-1 Gakuen-Dai, Saitama, Japan
shin-ya.sato@acm.org

Abstract. Tables in documents serve as valuable sources of information, and there has been a growing interest in research on information extraction from them. This paper delves into the methods for information extraction from text data we refer to as *catalogs*, bearing resemblance to tables. A catalog is a collection of text data where data of specific types (categories) are arranged roughly in a predefined order, such as a product list composed of product name, manufacturer, and production date. By tokenizing the text data, catalogs can be transformed into tables. However, catalogs are not always created with explicit table-like structures in mind, which may lead to potential discrepancies in data alignment in the resulting tables. Moreover, catalogs lack information like table headers, which have been commonly used for interpreting tables. In this paper, we propose a method to extract data of specific type from datasets comprising multiple catalogs. Distinctively different from conventional table-based information extraction methods, our method does not rely on external knowledge or labeled data, but exclusively on the estimated distribution of types of data within the dataset. We have validated the efficacy of our approach using synthetically generated data and product data from e-commerce sites.

Keywords: table understanding · co-occurrence analysis · Shannon entropy

1 Introduction

The increase in the amount of data that can be processed by computers has significantly contributed to the development of technologies such as machine learning and has supported our intellectual activities. There has been a concerted effort to augment the volume of data by aggregating it from diverse data sources and subsequently integrating these collections. Furthermore, the extraction of information such as named entities from this data has been actively pursued [12], setting the stage for advanced knowledge derivation. Tables in documents are increasingly being recognized as valuable data sources, prompting active research into their interpretation [11,15,20]. While tables are, by design, intended to organize data according to predefined rules, the reality often deviates from this ideal. Consequently, techniques for extracting information from tables that contain such discrepancies are also actively being studied [2–4,10].

In our endeavors to analyze and leverage data generated from e-commerce sites, we frequently come across text data that resemble tables, wherein terms of specific types are arranged in roughly a predefined order. Throughout this study, we refer to these collections of texts as *catalogs*. A quintessential example of a catalog is a product list. To facilitate understanding, here we provide a specific example: a product list from a wine store consisting of wine names, makers, and years of production (vintage), as shown below.

```
"Hermitage" "Guigal" "1998"
"Clos Vougeot" "Rene Engel" "2002"
"Were Dreams" "Jermann" "2001"
```

A catalog can be transformed into a table by tokenizing the text. In the resulting table, ideally, data of the same type should align in their respective column positions, as exemplified where the first column represents wine names. However, occasional deviations arise due to the intrusion of unexpected data or errors in tokenization. While this mirrors situations with erroneous table data, it diverges from genuine table data in that there are no supplementary details like table headers.

This paper presents a method to extract specific entities, such as product names, from a collection of catalogs. Previous studies on extracting entities from tables have employed supervised learning by leveraging table headers and external knowledge. In contrast, our method adopts an unsupervised learning approach to extract entities without the need for table headers, external knowledge, or labeled data. Instead of relying on other data sources, our approach utilizes token distribution information across multiple catalogs to estimate the likelihood of a token being the target entity. The main contribution of this paper is to detail this method[1]/ and demonstrates its effectiveness.

Our discussion is structured as follows: Sect. 2 presents the problem setting. Section 3 reviews related work, while Sect. 4 provides an in-depth look at our proposed method. Evaluation experiments and their results are detailed in Sect. 5. Subsequently, Sect. 6 points out the limitations of the proposed method.

2 Problem Setting

As mentioned in Sect. 1, a catalog can be converted into a table without headers using tokenization. This research delineates the challenge of extracting entities (e.g., wine names) from a catalog (e.g., inventory data from a wine store) into two processes: tokenization and entity extraction from the derived table. The focus of this paper is on the latter. Specifically, we address challenges such as extracting product names from the data presented in Fig. 1. Each $\mathscr{C}_i$ in Fig. 1

---

[1] The code of this study will be available at https://github.com/satoshinya/co-estimation.

is a catalog of products (wines) sold at the respective store. Each row of the catalog is a sequence of tokens representing attributions of the product, such as its name, producer, and year of production. Of these, wine names are in boldface type in the figure to make them easily recognizable. $\mathcal{C}_3$ is an example where data of a type not found in other catalogs, namely "Free shipping", is mixed in the catalog. $\mathcal{C}_4$ represents an instance where tokenization failed.

$\mathcal{C}_1$
Hermitage	Guigal	1998
Clos Vougeot	Rene Engel	2002
Were Dreams	Jermann	2001

$\mathcal{C}_2$
2010	Clos Rougeard	Saumur Champigny
2002	Rene Engel	Echezeaux
1990	Penfolds	Grange

$\mathcal{C}_3$
Free shipping	Grange	Penfolds	1997
Echezeaux	Rene Engel	2001	
Hermitage	J.L. Chave	2010	

$\mathcal{C}_4$
Penfolds	Grange	2001		
Rene	Engel	Echezeaux	2001	
Philippe	Charlopin	Parizot	Echezeaux	1997

Fig. 1. Sample dataset of inventory catalogs from four wine stores.

In the dataset illustrated in Fig. 1, the same entity is shared across multiple catalogs. This characteristic is commonly observed in many real-world datasets. If the data arrangement is consistent within each catalog, one can start with a single token and collect the same type of tokens by following the *column co-occurrence relationship*, the relationship between two tokens that belong to the same column. Bearing this in mind, this paper reinterprets the entity extraction problem as a problem analogous to the set expansion [19], as follows:

Token Identification in Catalogs (TIC)
Given a set of catalogs and a token appearing within them, identify tokens in each catalog that represent entities of the same type as the provided token.

When tackling this problem using unsupervised learning without external knowledge, the column co-occurrence relationship becomes crucial information. However, due to the potential misalignment in column positions, one cannot rely solely on this information. Such situations can be visualized through a co-occurrence network, treating tokens as nodes and their co-occurrence relationships as links. Figure 2 presents the co-occurrence network derived from the data in Fig. 1 where nodes that signify wine names are marked in red. It is evident that the marked nodes are inadvertently connected with unmarked nodes due to token misplacement. The challenge is extracting only the wine names under this circumstance.

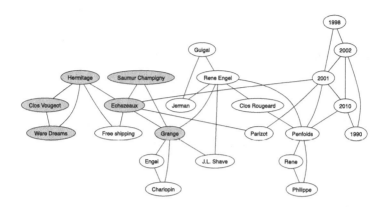

Fig. 2. Co-occurrence network of tokens in Fig. 1.

In actual datasets, it has also been observed that the degree of data misalignment varies among catalogs. This means that within the dataset, there can be catalogs with well-arranged data, as well as those like free-format. We leverage this fact in addressing TIC. The details of the method will be described in Sect. 4.

For the subsequent discussion, we introduce several notations here. Let the token specified as an input in TIC be referred to as a *seed token*, and tokens to be identified as *target tokens*. Let $c(\mathscr{C}, k)$ denote the k-th column of catalog $\mathscr{C}$, that is, the set of the k-th token in each row in $\mathscr{C}$. To give an example, $c(\mathscr{C}_1, 3)$ in Fig. 1 is {"1998", "2002", "2001"}.

3 Related Work

3.1 Table Understanding

Tables offer a simple yet powerful means to structure information within documents. The inverse process of representing information in tables—extracting tables from documents and interpreting their structure and content—is an emerging research domain referred to as table understanding [11,15,20]. A plethora of studies focus on leveraging the structure of tables for tasks such as relationship extraction between entities. On the other hand, it is not uncommon to encounter real-world tables where inconsistencies abound, such as different types of data within the same column. Research efforts are also being directed towards addressing these anomalies [2–4,10]. One prevalent strategy to mitigate these inconsistencies involves estimating the types or categories of cell values and columns. External knowledge bases are commonly employed in these estimations. For instance, in [10], both table headers and cell values are queried against Freebase [1] to generate potential estimates, which are then iteratively refined to enhance accuracy. On the other hand, [2] prepares multiple sets of prior knowledge for determining cell or column types and uses the outcomes of these determinations as features for further training, ultimately drawing conclusions.

Preceding studies have explicitly estimated the types of cell values or columns by harnessing both table-specific contexts, such as headers, and external knowledge or labeled data for training. Unlike these approaches, our research relies solely on table data, targeting the extraction of entities of the same type as the specified entity.

3.2 Co-Occurrence Analysis

As mentioned in Sect. 2, following the column co-occurrence relationships can be considered as a method for collecting tokens of the same type. However, when there are misalignments in the data sequence, not all co-occurrences necessarily indicate a same-type relationship. Thus, it is essential to discern between co-occurrences that genuinely represent the focused relationship (i.e., same-type) and those that occur coincidentally. This is not just a concern for the discussion at hand but is a general issue that needs to be addressed in any co-occurrence analysis.

Co-Occurrence Measures. One basic approach to address the aforementioned issue is to quantify the degree of co-occurrence, thereby enabling the selection of co-occurrence relationships. Such measures include log-likelihood [6], pointwise mutual information [5], and Jaccard index [9]. In Sect. 5.2, a solution for TIC using one such measure is presented, and its performance is compared with the proposed method.

Network Community Detection. Co-occurrence network analysis is a method suitable for capturing the overall picture of co-occurrence relationships and for gaining a comprehensive understanding of them. Figure 2, introduced in Sect. 2, is the co-occurrence network for the column co-occurrence relationships derived from the dataset presented in Fig. 1, in which nodes representing product (wine) names are marked in red. We can see that the marked nodes are distributed close to each other on the network, constituting a densely connected subnetwork. Subnetworks of this type are called communities, and the procedure of finding communities in a network is called community detection [7,8,16]. Community detection algorithms include random walk [14], spectral clustering [13], modularity optimization [8].

Real-world networks can be very large[2], and community detection is computationally expensive in general. To address this issue, Leiden algorithm [18]

---

[2] The network used in our evaluation experiment has 332,053 nodes and 534,790,063 links.

was developed to improve the computational complexity and the detection performance of the previous modularity optimization algorithms. In this paper, Sect. 5.2 presents a solution for TIC using the Leiden algorithm, and its performance is compared with the proposed method.

3.3 Label Spreading

Although this is outside of the current problem setting, it would be worth mentioning that semi-supervised learning can be applied for problems where seed tokens are provided for each data type. For example, one option would be to map each data type to a label and apply the label spreading method [21] to the co-occurrence network of tokens.

4 Proposed Method

4.1 Co-estimation Process

As repeatedly pointed out in previous sections, the fact that two tokens co-occur in a column can be valuable information for inferring types of tokens, and for solving TIC. However, the value of the information decreases if the catalog in which the tokens co-occur is poorly formatted. Suppose that we have an index, which should be referred to as *degree of tidiness* or *DoT* for short, that measures how well a catalog is formatted. Then, we can take a strategy of prioritizing the use of high-value information based on DoT.

Note that there is an interdependence between types of tokens and DoT: DoT contributes to the estimation of the types of tokens, and conversely, DoT is determined based on the distribution of the types of tokens. The proposed method takes advantage of this interdependence. Specifically, in the method, hypotheses about token types and DoT are made, and one is successively updated on the basis of the other.

For defining the co-estimation process, we introduce quantities that correspond to the hypothesis about token types. Under the situation of TIC where the type of each token is unclear, let $\mathcal{E}(T)$ be the estimated degree to which T is of the same type as the seed token. By definition, $\mathcal{E}(T)$ is the degree to which T is what we are trying to find. Therefore, assuming that there is only one token to find in each row, the target token can be identified in row R using $\mathcal{E}(T)$ as follows:

$$\arg\max_{T \in R} \mathcal{E}(T). \tag{1}$$

To illustrate the co-estimation process, we prepared Fig. 3 schematically depicting an example dataset and quantities used in the process. It shows three catalogs, with the bottom half of each showing how the catalog and rows are composed of tokens, similar to Fig. 1. Additionally, the upper half of each shows a bar chart of quantities calculated for each column. In the figure, $C(\mathscr{C}_i, k)$ is abbreviated as C_{ik}, and T_1 is marked for later explanation. We also prepared Table 1 that lists some symbols and notations used in the discussion.

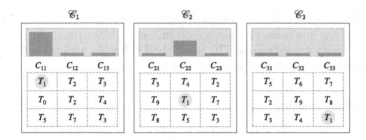

Fig. 3. Schematical representation of a dataset and quantities used in the co-estimation process.

Table 1. Notations.

Symbols	Description
$\mathscr{C}, \mathscr{C}_i$	Catalogs
R	Row in a catalog
T, T_i	Tokens (T_0 is the seed token)
$C(\mathscr{C}_i, k), C_{ik}$	k-th column of catalog $\mathscr{C}_i$
$\mathcal{E}(T)$	Estimated degree to which T is the target token
$\mathcal{O}(T)$	Set of columns containing token T
$\mathcal{W}(\mathscr{C})$	Degree of tidiness of catalog $\mathscr{C}$

4.2 Steps in Co-estimation Process

The co-estimation process consists of three steps. Each step is explained below using Fig. 3.

Step 1. For each column C, the average of $\mathcal{E}(T)$ is computed to estimate how intensively target tokens appears in C. The specific formula for calculating the average, denoted as $s(C)$, is as follows:

$$s(C) = \frac{1}{|C|} \sum_{T \in C} \mathcal{E}(T). \tag{2}$$

The graphs in Fig. 3 represent example distributions of $s(C)$, from which it can be inferred that the occurrence of target tokens is concentrated in C_{11} and C_{22}.

Step 2. DoT is calculated based on the distributions of $s(C)$ in this step. We define DoT of catalog $\mathscr{C}$, denoted as $\mathcal{W}(\mathscr{C})$, as follows:

$$\mathcal{W}(\mathscr{C}) = e^{-H(\mathscr{C})}, \tag{3}$$

where

$$H(\mathscr{C}) = -\sum_{C \in \mathscr{C}} \frac{s(C)}{S(\mathscr{C})} \log_2 \frac{s(C)}{S(\mathscr{C})}, \quad S(\mathscr{C}) = \sum_{C \in \mathscr{C}} s(C).$$

Note that $H(\mathscr{C})$ is the Shannon entropy of the (normalized) distribution of $s(C)$ in $\mathscr{C}$. Therefore, $\mathcal{W}(\mathscr{C})$ gets larger when target tokens appear in fewer columns (e.g., $\mathscr{C}_1$) and smaller when they appear evenly in more columns (e.g., $\mathscr{C}_3$).

Step 3. This step updates $\mathcal{E}(T)$ to the value estimated from the results of the above calculations. Remark that $\mathcal{E}(T_i) = \mathcal{E}(T_j)$ holds if T_i and T_j co-occur in column $C(\mathscr{C}, k)$ of well formatted catalog $\mathscr{C}$. Therefore, $\mathcal{E}(T_i) = s(C)$ also holds under the same condition. In other words, the value of $s(C(\mathscr{C}, k))$ can be an estimate of $\mathcal{E}(T)$, depending on $\mathcal{W}(\mathscr{C})$. To take a more specific example, of the possible reference values $s(C_{11})$, $s(C_{22})$, and $s(C_{33})$ for determining the value of $\mathcal{E}(T_1)$, the credibility of the value of $s(C_{33})$ is lower than others because $\mathcal{W}(\mathscr{C}_3)$ is lower. Based on the discussion up to this point, one can choose the weighted average of $s(C(\mathscr{C}, k))$ with corresponding weight $\mathcal{W}(\mathscr{C})$ as an estimate of $\mathcal{E}(T)$. In this case, the specific formula for the estimated value, denoted as $\mathcal{E}'(T)$, is as follows:

$$\mathcal{E}'(T) = \sum_{C(\mathscr{C},k) \in \mathscr{O}(T)} \frac{\mathcal{W}(\mathscr{C})}{\sum_{C(\mathscr{C},k) \in \mathscr{O}(T)} \mathcal{W}(\mathscr{C})} s(C(\mathscr{C}, k)), \quad (4)$$

where $\mathscr{O}(T)$ is the list that is compiled by enumerating each column every time token T appears in it.

Another estimate can be obtained by modifying the weights in Eq. (4). For instance, the following equation uses exponential functions to emphasize the magnitude of $\mathcal{W}(\mathscr{C})$:

$$\mathcal{E}'(T) = \sum_{C(\mathscr{C},k) \in \mathscr{O}(T)} \frac{e^{\mathcal{W}(\mathscr{C})}}{\sum_{C(\mathscr{C},k) \in \mathscr{O}(T)} e^{\mathcal{W}(\mathscr{C})}} s(C(\mathscr{C}, k)), \quad (5)$$

The proposed method employs Eq. (5) and uses learning rate α to update the value as follows:

$$\mathcal{E}(T) \leftarrow (1-\alpha)\mathcal{E}(T) + \alpha\mathcal{E}'(T) \quad (6)$$

4.3 Iterations of Co-estimation

In the proposed method, each value of $\mathcal{E}(T)$ is gradually adjusted by iterating the update procedure described above. Let $\mathcal{E}(T)^{(n)}$ denote the value obtained as a result of n iterations.

Prior to the iterations, $\mathcal{E}(T)^{(0)}$ is initialized to 1 if T is the seed token (i.e., T_0), and 0 otherwise. Then, $\mathcal{E}(T_0)^{(0)}$ is distributed to other tokens through the co-occurring relationships in the iterations. In other words, each token earns its score from T_0 depending on its relationship with T_0. The score earned is redistributed to tokens with lower scores in the subsequent iterations.

Figure 4 shows how $\mathcal{E}(T)^{(n)}$ actually changes as the iterations progress, taking T_1 in Fig. 3 as an example. It has been found empirically from experiments that the peak value of $\mathcal{E}(T)^{(n)}$ is a reasonable choice for the value that $\mathcal{E}(T)$ should take. However, it is not clear yet whether all $\mathcal{E}(T)^{(n)}$ exhibit similar unimodal patterns, so the proposed method selects the value of $\mathcal{E}(T)$ based on the predetermined parameter M as follows:

$$\mathcal{E}(T) = \max_{n \leq M} \mathcal{E}(T)^{(n)}. \tag{7}$$

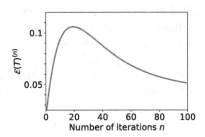

Fig. 4. Variation of $\mathcal{E}(T)^{(n)}$ with the increase of n at $T = T_1$.

5 Evaluation

To evaluate the proposed method, we conducted two experiments. One used synthetic data designed to mimic real-world situations, the other employed actual real-world data. In this section, we discuss each experiment in order.

5.1 Experiment with Synthetic Data

In the first experiment, we created multiple catalogs with misaligned data and investigated the performance of the proposed method on them.

Dataset. The dataset used in this experiment emulates a collection of store product lists, similar to the data exemplified in Sect. 2. Each row of a catalog in the dataset contains three items: the product name, its producer, and the year of production. The catalog where all rows consistently list these items in the same order is deemed a *properly aligned catalog*.

To mirror real-world data, we generated properly aligned catalogs using the following procedure. First, we selected products and producers and defined the relationships indicating which products are manufactured by which producers. This relationship can be represented as a bipartite graph, as illustrated in Fig. 5. For instance, the graph signifies that product P_1 is produced by two manufacturers, M_1 and M_2.

Subsequently, a weight is assigned to each store. This weight acts as a parameter adjusting the number of products the store deals with, or, in other words, the volume of data inscribed in the catalog of the store. Similar weights are also

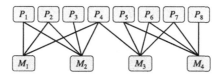

Fig. 5. An example of a bipartite graph representing the relationships among products and product manufacturers. P_i and M_i respectively represent a product and a product manufacturer.

assigned to products, producers, and production years. Furthermore, for each store s, its product selection, denoted as $L(s)$, is probabilistically determined based on these weights.

Under the aforementioned setup, rows of the catalogs are generated by following the pseudocode shown below, where K represents the total number of rows to be generated.

```
for i in [1, K]
{
    Select store s;
    Choose product p from L(s);
    Based on the product-producer mapping, select producer m;
    Choose production year y;
    Add (p, m, y) to s's catalog;
}
```

In the above procedure, each store, product, and production year are chosen randomly with probabilities proportional to their respective weights. For this experiment, we employed a normal distribution to represent the distribution of weights for entities such as stores.

To produce catalogs with misaligned data, modifications are made to some rows of a properly aligned catalog by altering the order of the items. The steps are as follows. Firstly, choose catalogs for which data alignment will be modified. To do this, select a value $0 < Q < 1$. Construct the set S_Q of catalogs to be modified such that the proportion of S_Q in all catalogs is equal to Q. This is achieved through random sampling. Then, for each row in a catalog chosen from set S_Q, shuffle the order of items with a probability P. These steps are illustrated in Fig. 6.

Evaluation Procedure. In this experiment, we chose four values for Q, specifically $\{0.1, 0.2, 0.3, 0.4\}$, and ten values for P, ranging from 0.0 to 0.9 in increments of 0.1. For each of the 40 parameter combinations (Q, P), we generated 100 datasets.

We applied the proposed method to solve TIC for these datasets and assessed its performance. Specifically, each dataset and a product name randomly selected from that dataset were used as the input. The challenge was to extract all product names from the dataset, and the results were evaluated using the F-measure.

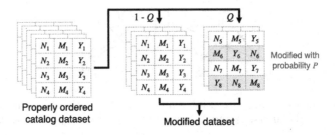

Fig. 6. Procedure of generating a dataset containing partially corrupted catalogs.

Thus, for each parameter combination (Q, P), 100 F-measures were obtained. We calculated their average to determine the final evaluation metric. The extraction of product names from the dataset was conducted based on Eq. (7).

Now, if there were no data misalignments in all catalogs within a dataset, it is expected that by following the relationship of "being in the same column," one could solely gather product names without inadvertently incorporating producers or production years. Therefore, in this experiment, we also compared against this approach as a baseline. We used the datasets with $Q = 0.1$ for the baseline.

Results. Figure 7 depicts the variation in the F-measure with the increase of P for each Q. Under the condition $P = 0$, which corresponds to the case where no data modification is performed, both the proposed method and the baseline extract almost all product names correctly. However, for $P > 0$, the performance of the baseline degrades significantly. This degradation can be attributed to the baseline's inability to distinguish between product names and producers as well as production years when they coexist in the same column. In contrast, the proposed method does not show such a dramatic drop in performance, indicating its capability to mitigate the effects of misalignment.

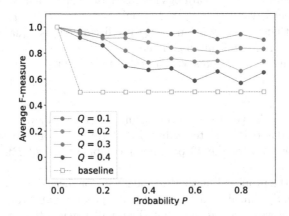

Fig. 7. The performance of the proposed method on the synthetic data.

Figure 8 presents a scatter plot illustrating the relationship between QP (the product of Q and P) and the F-measure of the proposed method. QP signifies the proportion of misalignment present within the dataset. Consider two pairs of parameters (Q_1, P_1) and (Q_2, P_2) where $Q_1 < Q_2$ and $Q_1 P_1 = Q_2 P_2$ hold true. Under these conditions, the amount of misalignment in the datasets generated by these parameters should be equivalent. Yet, the graph shows better performance for the Q_1 case. When comparing Q_1 to Q_2, misalignment occurs more locally in Q_1. It is considered that the performance improves because this situation in Q_1 is conducive to the strategy of the proposed method, which aims to estimate the degree of data misalignment per catalog and more preferentially adopt estimation obtained from those with a lower degree of misalignment.

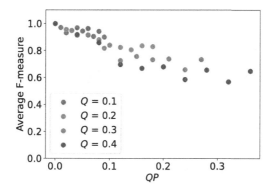

Fig. 8. Impact of the proportion of the amount of misalignment on the performance.

5.2 Experiment with Real-World Data

We conducted the second experiment to verify the effectiveness of the proposed method on real-world data.

Datasets. We prepared another dataset similar to the one shown in Fig. 1 by extracting data on wines from Rakuten Ichiba Dataset [17], which is the dataset on products sold in EC stores in the Rakuten Ichiba marketplace in Japan. Let the dataset be denoted as DS. Texts in DS, written in Japanese, were first normalized, strings enclosed by parentheses or quotation marks were treated as a single word, and tokenized by splitting at whitespaces. Note that, in Japanese, there are usually no whitespaces between words, and a whitespace explicitly indicates a semantic or structural break.

The amount of data in DS is so large that it is difficult to label all of them for evaluation. Therefore, we also created a labeled dataset DS-L using the following steps. First, rows were randomly selected from DS. At this step, we excluded rows where tokenization failed, such as when a wine name was split into multiple tokens. In the next step, each token was assigned a label of manually identified data type. Basic statistics of DS and DS-L are summarized in Table 2.

Table 2. Statistics of the real-world datasets.

	DS	DS-L
Catalogs	1,018	104
Rows	478,424	199
Tokens	332,053	1,300

Baselines. We compared the proposed method with two other methods for evaluation, one using a co-occurrence measure and the other using a community detection algorithm, both discussed in Sect. 3.2.

For the co-occurrence measure, we used the Jaccard index

$$J(T_1, T_2) = \frac{|\mathcal{O}_*(T_1) \cap \mathcal{O}_*(T_2)|}{|\mathcal{O}_*(T_1) \cup \mathcal{O}_*(T_2)|} \tag{8}$$

where $\mathcal{O}_*(T)$ is the set of unique elements in $\mathcal{O}(T)$. The target token in row R was obtained using this index as follows:

$$\arg\max_{T \in R} J(T_0, T). \tag{9}$$

For detecting communities, we used the Leiden algorithm. We divided the co-occurrence network of tokens in DS into subnetworks N_i using the algorithm. Then, we obtained target token(s) in row R by

$$R \cap N_{i_0} \tag{10}$$

where i_0 is the index of the subnetwork to which the seed token belongs, i.e., $T_0 \in N_{i_0}$. Since the Leiden algorithm is probabilistic, the method was evaluated by averaging the results of multiple (specifically, 10) trials. We used leidenalg package[3] for the actual calculations.

Evaluation Procedure. The following procedure was used for the evaluation. (1) Choose a seed token. (2) Solve TIC. That is to find the target token for each row in DS. (3) Evaluate the answers against the correct answers in DS-L.

As the evaluation index, we used the F-measure. As for seed tokens, the names of five items were selected so that the number of columns in which each name appeared (i.e., $|\mathcal{O}_*(T_0)|$) was not close to the others. The learning rate α was set to 0.1, and the number of iterations M was set to 100.

---
[3] https://github.com/vtraag/leidenalg.

Results. The evaluation results are summarized in Table 3. The baseline methods using the Jaccard index and the Leiden algorithm, and the proposed method are respectively denoted as Jaccard, Leiden, and Co-E. It shows that Co-E significantly outperforms the other methods for all seed tokens. Methods using Jaccard, Leinden, and the proposed method have in common that they are all based on the co-occurrence information. On the other hand, the proposed method differs in that it adjusts the weight of each piece of information according to the reliability of its source. The fact that the proposed method clearly outperforms the others indicates the effectiveness of this adjustment on the real-world data.

Table 3. Comparison of the performance of each method on real-world data.

T_0	$|\mathcal{O}_*(T_0)|$	Jaccard	Leiden	Co-E
Ben Rye	2	0.29	0.44	**0.69**
Nerojbleo	20	0.53	0.41	**0.81**
Romanaee-Conti	45	0.48	0.45	**0.78**
Clos Vougeot	85	0.48	0.48	**0.83**
Chianti Classico	108	0.42	0.48	**0.80**

6 Limitations

The proposed method has a drawback that it cannot estimate scores for tokens that do not co-occur with other tokens. This is a natural consequence of the fact that the method is based on co-occurrence information. This drawback can be overcome to some extent by increasing the amount of data since an increase in the amount of data generally leads to an increase in the chances of co-occurrence.

$\mathcal{E}(T)$ is calculated based on Eq. 7, which is derived from empirical observations. Additionally, the parameter α, used for computing $\mathcal{E}(T)$, is arbitrarily selected. In these respects, the proposed method lacks a firm theoretical foundation. Overcoming these shortcomings could enhance the robustness of the proposed method and its performance in solving the TIC problem.

7 Conclusion

The paper has described the method for finding data of a specified type in a dataset by co-estimating data types and their positional distribution. The effectiveness of the method has been verified through evaluation experiments using both synthetic and real-world data.

We considered the case where the distribution of each type of token is concentrated in a particular *column*, for the sake of clarity in explaining our method. The idea of our method should be applicable to more generalized cases where tokens are distributed around a specific *region* for each of their types.

Acknowledgements. In this paper, we used "Rakuten Dataset" (https://rit.rakuten.com/data_release/) provided by Rakuten Group, Inc. via IDR Dataset Service of National Institute of Informatics.

References

1. Bollacker, K., Evans, C., Paritosh, P., Sturge, T., Taylor, J.: Freebase: a collaboratively created graph database for structuring human knowledge. In: Proceedings of the 2008 ACM SIGMOD International Conference on Management of Data, pp. 1247–1250. SIGMOD '08, Association for Computing Machinery (2008). https://doi.org/10.1145/1376616.1376746
2. Bonfitto, S., Cappelletti, L., Trovato, F., Valentini, G., Mesiti, M.: Semi-automatic column type inference for CSV table understanding. In: Bureš, T., Dondi, R., Gamper, J., Guerrini, G., Jurdziński, T., Pahl, C., Sikora, F., Wong, P.W. (eds.) SOFSEM 2021: Theory and Practice of Computer Science, pp. 535–549. Springer International Publishing (2021).https://doi.org/10.1007/978-3-030-67731-2_39
3. van den Burg, G.J.J., Nazábal, A., Sutton, C.: Wrangling messy csv files by detecting row and type patterns. Data Min. Knowl. Discov. **33**(6), 1799–1820 (2019). https://doi.org/10.1007/s10618-019-00646-y
4. Cheng, K., et al.: Tab-cleaner: weakly supervised tabular data cleaning via pre-training for E-commerce catalog. In: Proceedings of the 61st Annual Meeting of the Association for Computational Linguistics (Volume 5: Industry Track), pp. 172–185. Association for Computational Linguistics, Toronto, Canada (2023). https://doi.org/10.18653/v1/2023.acl-industry.18
5. Church, K.W., Hanks, P.: Word association norms, mutual information, and lexicography. Computat. Linguist. **16**(1), 22–29 (1990). https://aclanthology.org/J90-1003
6. Dunning, T.: Accurate methods for the statistics of surprise and coincidence. Comput. Linguist. **19**(1), 61–74 (1993)
7. Fortunato, S.: Community detection in graphs. Phys. Rep. **486**(3), 75–174 (2010). https://doi.org/10.1016/j.physrep.2009.11.002
8. Girvan, M., Newman, M.E.J.: Community structure in social and biological networks. Proc. Natl. Acad. Sci. **99**(12), 7821–7826 (2002). https://doi.org/10.1073/pnas.122653799
9. Grefenstette, G.: Explorations in Automatic Thesaurus Discovery. Kluwer Academic Publishers, USA (1994)
10. Hitzler, P., Cruz, I., Zhang, Z.: Effective and efficient semantic table interpretation using tableminer+. Semantic Web **8**(6), 921–957 (2017). https://doi.org/10.3233/SW-160242
11. Liu, J., Chabot, Y., Troncy, R., Huynh, V.P., Labbé, T., Monnin, P.: From tabular data to knowledge graphs: a survey of semantic table interpretation tasks and methods. Web Semantics **76**(C) (2023).https://doi.org/10.1016/j.websem.2022.100761
12. Nasar, Z., Jaffry, S.W., Malik, M.K.: Named entity recognition and relation extraction: state-of-the-art. ACM Comput. Surv. **54**(1) (2021). https://doi.org/10.1145/3445965
13. Newman, M.E.J.: Spectral methods for community detection and graph partitioning. Phys. Rev. E **88**(4), 042822 (2013). https://doi.org/10.1103/physreve.88.042822

14. Pons, P., Latapy, M.: Computing communities in large networks using random walks. In: Yolum, p., Güngör, T., Gürgen, F., Özturan, C. (eds.) Computer and Information Sciences - ISCIS 2005, pp. 284–293. Springer Berlin Heidelberg, Berlin, Heidelberg (2005)
15. Pujara, J., Szekely, P., Sun, H., Chen, M.: From tables to knowledge: recent advances in table understanding. In: Proceedings of the 27th ACM SIGKDD Conference on Knowledge Discovery & Data Mining, pp. 4060–4061. Association for Computing Machinery (2021). https://doi.org/10.1145/3447548.3470809
16. Radicchi, F., Castellano, C., Cecconi, F., Loreto, V., Parisi, D.: Defining and identifying communities in networks. Proc. Natl. Acad. Sci. **101**(9), 2658–2663 (2004). https://doi.org/10.1073/pnas.0400054101
17. Rakuten Group, Inc.: Rakuten Ichiba data. Informatics Research Data Repository, National Institute of Informatics. (dataset) (2020). https://doi.org/10.32130/idr.2.1
18. Traag, V.A., Waltman, L., van Eck, N.J.: From Louvain to Leiden: guaranteeing well-connected communities. Sci. Rep. **9**(1), 5233 (2019). https://doi.org/10.1038/s41598-019-41695-z
19. Wang, R.C., Cohen, W.W.: Language-independent set expansion of named entities using the web. In: Seventh IEEE International Conference on Data Mining (ICDM 2007), pp. 342–350 (2007). https://doi.org/10.1109/ICDM.2007.104
20. Zhang, S., Balog, K.: Web table extraction, retrieval, and augmentation: a survey. ACM Trans. Intell. Syst. Technol. **11**(2), 13:1–13:35 (2020). https://doi.org/10.1145/3372117
21. Zhou, D., Bousquet, O., Lal, T.N., Weston, J., Schölkopf, B.: Learning with local and global consistency. In: Proceedings of the 16th International Conference on Neural Information Processing Systems, pp. 321–328. NIPS'03, MIT Press, Cambridge, MA, USA (2003)

Enhancing Load Forecasting with VAE-GAN-Based Data Cleaning for Electric Vehicle Charging Loads

Wensi Zhang[1], Shuya Lei[1], Yuqing Jiang[1(✉)], Tiechui Yao[1], Yishen Wang[1], and Zhiqing Sun[2]

[1] State Grid Smart Grid Research Institute Co., Ltd., Beijing, China
3220231399@bit.edu.cn, yaotiechui17@mails.ucas.ac.cn
[2] State Grid Hangzhou Power Supply Company, Hangzhou, China

Abstract. With the popularization of environmental protection ideas, people are increasingly valuing low-carbon lifestyles and the economy. Electric vehicles play a crucial role in this transformation to reduce carbon emissions. However, integrating electric vehicles into the power grid poses challenges, especially the possibility of destructive load peaks, which may endanger the stability and safety of the power grid. Accurately predicting the load of electric vehicles and managing grid scheduling are crucial for solving this problem. The current solutions are mainly divided into two categories: statistics-based methods and machine learning-based methods. Statistical methods require a large amount of long-term data modeling, making data collection a significant challenge. Similarly, machine learning-based methods have good long-term prediction performance on high-quality data, but they do not perform well in terms of short-term prediction accuracy. To overcome these obstacles, a comprehensive electric vehicle charging load prediction framework is proposed, which utilizes an innovative variational autoencoder to generate adversarial networks (VAE-GAN) for data processing, Principal Component Analysis (PCA) for feature extraction, and an improved CNN-GRU model for prediction. The experimental results show that the accuracy of short-term power load prediction is significantly improved, which verifies the effectiveness of the framework in processing small sample load data and provides advanced tools for intelligent management of electric vehicle charging stations.

Keywords: load forecasting · electric vehicles · data cleaning · variational autoencoder · generate adversarial networks · gated recurrent network

1 Introduction

With the rise of new forms of carbon neutrality, the world is increasingly advocating low-carbon life and a low-carbon economy. This trend has given rise to many new industries dedicated to reducing carbon emissions, with Electric Vehicles (EVs) becoming a prominent representative and seen as one of the key industries in the transition to a low-carbon economy. According to the International Energy Agency, electric vehicle sales will account for more than 20% of total global vehicle sales by 2023. This indicates that as the low-carbon economy advances, a large number of EVs will be integrated into the electricity system.

However, a large number of EVs may be connected to the grid at a certain time to meet the charging demand, resulting in unacceptable load peaks in the distribution network. This not only interferes with the normal operation of the distribution network, but also may cause security and power quality problems such as voltage instability, line overload, and large-scale power outages. To solve this problem, it is essential to forecast the load of EVs and realize the reasonable load scheduling of the power grid.

Existing Solutions. In the current literature research, the specific methods of load forecasting are mainly divided into two aspects: statistics-based and neural networks-based methods.

Statistics-based methods usually need to obtain long-term charging data of EVs for modeling. For example, Yuan et al. [2] adapt Monte Carlo simulation to simulate the charging behavior of vehicles, and then accumulate the charging curves of each vehicle to form the overall charging load curve, to achieve prediction. The advantage of this method is to realize the comprehensive analysis and simulation of the problem through random sampling and comprehensive consideration of many factors. On the other hand, to improve the accuracy of short-term load forecasting, Sun et al. [11] use multi-label and KNN to determine the weight of each cluster to forecast points and establish a load forecasting model through Multi-LR to achieve high-precision forecasting and reduce the forecasting time. Madhukumar et al. [8] establish 19 load prediction algorithms based on the data of two electricity meters on campus, and find that the Gaussian Process Regression (GPR) model has the best effect. However, the traditional statistics-based method has some limitations in predicting short-term charging behavior, and its accuracy is not high. This makes the optimization of the model an important problem to be solved by statistics-based methods.

With the rise of artificial intelligence, neural network-based methods have become a research direction with their high accuracy. For example, Zhu et al. [16] use three deep learning methods to predict the charging load of electric vehicles from the perspective of charging stations. By comparing the experimental results of RNN, LSTM and GRU, they conclude that GRU has the best performance, which provides a beneficial direction for the optimization of GRU. Su et al. [10] reduce the error of the original data through data cleaning and outlier screening, and adopt wavelet noise reduction to reduce the data noise, and finally use the LSTM to predict the short-term load of charging stations. However, a single neural network makes it difficult to meet the demand of load forecasting, thereby the multi-network model gradually emerges. Meng et al. [9] propose a short-term prediction method that combines Empirical Pattern

Decomposition (EMD), Bidirectional Long Short-Term Memory (BiLSTM), and attention mechanisms. Compared to traditional machine learning models, deep learning model shows better prediction accuracy. In addition, Yan et al. [14] use GMM (Gaussian Mixture Model) to cluster user samples, extract typical user behavior characteristics, and build an EV load prediction model based on CNN-GAN (Convolutional Neural Network - Generative Adversarial Network). Through semi-supervised regression, they successfully obtain a CNN-GAN prediction model with high accuracy and also provide more possibilities for CNNs to process text data. However, due to the fact that so since neural network-based prediction forecasting methods often only focus on the prediction model itself and, but ignore the data quality issues of model training, without processing existing data such as importance of data cleaning and, feature extraction, the data quality is not ideal and other links in the whole prediction process.

Challenges. The realization of high-accuracy electric vehicle load forecasting still faces complex and severe challenges. The root causes of this challenge can be divided into two important aspects: external user charging behavior and internal battery status. In terms of external user charging behavior, there are many factors, including the living habits of individual owners, temperature, weather conditions, holiday factors, the distribution of charging facilities, and the psychological status of owners. These factors not only affect the charging decision of the owner to varying degrees but also directly shape the mode and law of the charging load. Due to the complexity of charging behavior, the prediction model needs to be carefully and comprehensively considered in multiple dimensions and scenarios to capture the trend of charging behavior more accurately.

On the other hand, when considering the internal state of the battery status, factors such as charging power, charging time, battery characteristics, and the starting battery will directly affect the formation of the charging load. The complexity of the battery state is that it covers many key parameters in the charging process, which produces dynamic and nonlinear changes under the influence of these parameters. Effectively predicting these parameters is critical to understanding the actual energy requirements of EV charging stations. Therefore, in the process of charging load forecasting, it is necessary to deeply study and solve these influencing factors to ensure that the model can provide highly accurate forecasting results in various practical situations.

Contributions. To solve the above problems, we propose a comprehensive EV charging load forecasting framework, emphasizing the whole forecasting process. An innovative VAE-GAN is proposed to handle the cleaning of load data, then the principal component analysis method of PCA is used to extract the main features from many influential factors. Finally, the improved CNN-GRU model is used for prediction. Our main contributions include:

(1) We propose a new framework that covers three key components: data cleaning, feature extraction, and predictive models. The noise level in the data is successfully reduced, and data quality is significantly improved, thus making

an innovative breakthrough in the accuracy of short-term load prediction. It provides a more reliable and advanced tool for the intelligent management and optimization of electric vehicle charging piles.

(2) To improve the quality of the data, we propose an innovative VAE-GAN to process the load data, which successfully reduces the noise level in the data and significantly improves the quality of the data, thus achieving an innovative breakthrough in the accuracy of load prediction.

(3) To achieve the short-term load prediction, we adopt a composite network structure consisting of a stacked CNN convolutional layer and GRU layer. By leveraging the complementary strengths of the convolutional and GRU layers, the model can effectively capture intricate patterns in load fluctuations. This approach enhances the accuracy of short-term load prediction significantly, thereby offering robust data support for the planning and operation of EV charging stations.

(4) Experimental results show that the proposed method significantly improves the accuracy of short-term power load prediction. In addition, through the verification of charging pile meter information in the smart grid, the effectiveness of the framework in predicting small sample load data is successfully verified.

2 Related Work

Statistics-Based Methods. Based on mathematical statistics, Long et al. [7] propose a support vector regression load prediction method for charging stations through in-depth analysis of load characteristics of EV charging stations, combined with actual charging power and weather forecast data of charging piles. This method improves the scalability and robustness of prediction. On the other hand, Hu et al. [4] extend the existing radial basis function method and proposed a new hybrid radial basis function network. Compared with the traditional radial basis function, this method has a higher accuracy. Additionally, Xu [13] and other researchers have effectively predicted the erratic charging patterns of electric vehicles and constructed daily load curves based on Monte Carlo simulations. Their prediction model, validated through experiments, accounts for electric vehicle battery characteristics and user travel habits, with high accuracy and effectiveness. However, statistics-based forecasting methods are more suitable for long-term forecasting, and their accuracy is often unsatisfactory.

Machine Learning Based Methods. In machine learning algorithms, decision trees and neural networks are widely used. Decision trees are commonly used for classification and regression tasks and are widely applied in load forecasting for electric vehicles. In terms of neural networks, Khan et al. [5] compare the characteristics of traditional linear regression, baggy tree regression, and neural networks in load forecasting through experimental analysis. The results intuitively show that the prediction error of artificial neural networks in load forecasting of electric vehicles is the lowest, especially in the peak period of one day. This further proves the effectiveness of neural networks in solving peak problem scheduling. Su et al. [10] propose a short-term charging load prediction method for charging stations based on LSTM and find that LSTM has high accuracy, can significantly shorten the training time, and reduce the demand for computing resources. Zhao et al. [15] propose a prediction method based on the NAR (Nonlinear AutoRegressive) model, and select Levenberg Model (LM) as the network training method. The results confirm the effectiveness of the NAR model in solving nonlinear problems with dynamic characteristics and time relationships. He et al. [3] propose an innovative prediction model called the spline interpolation-based gated cyclic unit, which demonstrates superior performance compared to the traditional GRU model in terms of deformation.

In a study of convolutional neural networks, Kiranyaz et al. [6] find that one-dimensional convolutional layers can efficiently handle time series problems and better extract spatiotemporal properties of the data. Therefore, CNN has been continuously applied in the load forecasting of electric vehicles. For example, Yan et al. [14] propose a charging load prediction method based on CNN-GAN (Convolutional Neural Network Generating Adversarial Neural Network) and semi-supervised regression, which achieves good prediction effect and further proves the feasibility of introducing CNN.

3 Structure

The overall framework for implementing load forecasting is depicted in Fig. 1. The process begins with data cleaning using a VAE-GAN approach, which imputes missing values accurately. Next, principal component analysis is applied to the data set to identify and select the most significant features. Finally, the high-quality data is employed for load forecasting training, resulting in the development of a predictive model.

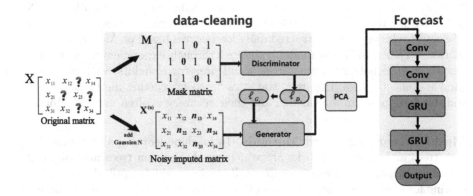

Fig. 1. Overall framework structure diagram.

3.1 Data Cleaning

In the field of data cleaning for load forecasting, researchers usually use statistical methods and deep learning methods. Statistical methods rely on assumptions about the probability distribution or model of the data, however, if these assumptions do not match reality, they can introduce biases that affect data accuracy. With the rise of deep learning, it is often applied for data cleaning.

Deep learning methods include Autoencoder (AE) and GAN. However, there are some limitations to both approaches. The AE has a relatively weak performance in generating data, especially compared to GAN. When GAN is used to optimize the deep generation model, the instability of the optimization makes it difficult for the generated model to fully reflect the diversity of the underlying observation data, which is called mode collapse.

For basic AE, the input-output relationship is deterministic. This means that, given data that satisfies a particular distribution, the decoder cannot generate the results we expect, and therefore lacks diversity and generalization ability. To overcome these problems, a Variational Autoencoder (VAE) is introduced. VAE can be used not only for data compression and reconstruction but also for generating new data samples. By learning the potential distribution of the data, VAE can sample the potential space of the learning to generate new samples similar to the original data.

In contrast, VAE has greater generative power, better latent spatial properties, and better data distribution modeling capabilities compared to AE, resulting in greater flexibility and expressiveness in many tasks.

To enhance data quality and prevent potential model crashes during the data cleaning process while leveraging the strengths of both VAE and GAN models, we adopt an innovative approach that amalgamates a hybrid algorithm of VAE and GAN, termed VAE-GAN. This novel method provides a robust and effective solution for data cleaning in the field of load forecasting.

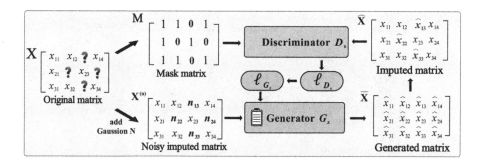

Fig. 2. Architecture Diagram of VAE-GAN.

VAE-GAN. As shown in Fig. 2, the framework of VAE-GAN consists of two parts: a generator and a discriminator.

The task of the generator is to generate data that is as similar as possible to the original data through the mechanism of a variational autoencoder. This is the main difference between VAE-GAN and GAN. The introduction of a VAE enables generators to generate data with a similar distribution to the original data more effectively. At the same time, the task of the discriminator is to distinguish between the generated data and the raw data, continuously improving the discriminant performance. There is an adversarial relationship between the generator and discriminator, where the generator is committed to deceiving the discriminator and generating more authentic data, while the discriminator seeks to improve the accuracy of identifying the generated data and the authentic data.

Finally, the VAE-GAN generator is used to generate default values for the data. This process integrates the iterative strategy of GAN, the idea of adversarial learning, and the advantages of AE, enabling VAE-GAN to perform excellently in generating data that matches the original data, especially in dealing with missing data and other scenarios. Next, we will give a specific example to introduce VAE-GAN from the two parts of the generator and discriminator.

Generator. Starting from the original data matrix X, regularize the data by introducing Gaussian noise N, and then input the noisy data $X^{(n)}$ into the generator. After training, generator $G_{(x)}$ can generate a data matrix $\overline{X}$ that is as similar as possible to the original data. This process utilizes prior knowledge of potential distributions (regularization by adding Gaussian noise) to ensure that the generated data not only conforms to the characteristics of the training data but also has a certain degree of generalization ability. The entire process aims to generate a data matrix that is more realistic and closer to the original data distribution.

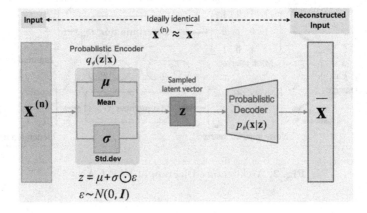

Fig. 3. Structure of Generator

As shown in Fig. 3, the input noisy data matrix $X^{(n)}$ is mapped to a low-dimensional space via an encoder to obtain a potential representation of z. During the encoding process, the model regularizes z based on the potential distribution $q_\phi(z|x)$. The prior regularization term L_{prior} in this step is to ensure that the data distribution in the latent space is close to a normal distribution. The purpose of the regular term is to enable the embedding of data mapped to a low-dimensional space to render a normal distribution. The latent representation R_v is decoded by the decoder to generate a data matrix $\overline{X}$, which aims to ensure that the generated missing values can successfully deceive the discriminator. The overall loss function of the generator G_x has the following form:

$$\ell_{G_x} = \ell_{\text{pro}} + \gamma \cdot \ell_{\text{VAE}} = \ell_{\text{pro}} + \gamma \cdot (\ell_{\text{rec}} + \ell_{\text{prior}}) \tag{1}$$

where γ is a hyperparameter that balances the reconstruction error with the importance of a prior regularization term. ℓ_{rec} is the reconstruction error of the predicted value. ℓ_{pro} is the probability that the updated discriminator D_x predicts the missing value generated by G_x, namely:

$$\ell_{\text{pro}} = -\mathbb{E}\left[(1 - M) \odot \log\left(D_x(G_x(X^{(n)}), M\right)\right] \tag{2}$$

Discriminator. First, a mask matrix is generated according to the original data X, where the elements corresponding to the default value are set to 0, and other elements are set to 1. The data matrix $\overline{X}$ generated by the generator is then filled to the default position of the original data matrix X, giving $\hat{X}$ as input to the discriminator. The loss function of the discriminator is as follows:

$$\ell_{D_x} = -\mathbb{E}\Big[M \odot \log\left(D_x\left(G_x\left(X^{(n)}\right), M\right)\right) \\ + (1 - M) \odot \log\left(1 - D_x\left(G_x\left(X^{(n)}\right), M\right)\right)\Big] \tag{3}$$

Thus, in small-lot iterations, the VAE-GAN model is divided into two distinct components, i.e. minimizing ℓ_{G_x} and maximizing ℓ_{D_x}. The specific formula is as follows:

$$\begin{aligned}\min_{G_x} \ell_{G_x} &= \min_{G_x}(\ell_{\text{pro}} + \gamma \cdot (\ell_{\text{rec}} + \ell_{\text{prior}})) \\ &= \min_{G_v} -\mathbb{E}\left[(1 - M) \odot \log(D_x(G_x(X^{(z)}), M))\right] \\ &+ \gamma \cdot \left(\mathbb{E}\left[M \odot (X^{(n)} - G_x(X^{(n)}))^2\right]\right) \\ &+ D_{\text{KL}}\left(q(z|X^{(n)})||\mathcal{N}(0,1)\right)\end{aligned} \tag{4}$$

and

$$\max_{D_x} \ell_{D_x} = \max_{D_x} -\mathbb{E}\Big[(M \odot \log(D_x(G_x(X^{(n)}), M)) \\ +(1 - M) \odot \log(1 - D_x(G_x(X^{(n)}), M))\Big] \tag{5}$$

where $q(z|X^{(n)})$ is the output of the G_x's encoder.

Therefore, VAE-GAN aims to both minimize generator losses and maximize discriminator losses to ensure that the generated data can both successfully fool the discriminator and remain as close to the real data distribution as possible. The overall process integrates encoding, decoding, regularization, and other steps to generate a data matrix with higher quality and closer to the original data.

3.2 Feature Extraction

As mentioned in Sect. 1, various factors affect the charging behavior of electric vehicles, such as user behavior, environmental, weather conditions, station location, and equipment status. The complex relationships between these factors make accurate predictions challenging. The original data also contains redundant information and noise. Feature extraction can select the most relevant features, reduce data dimensions, improve computational efficiency, and minimize overfitting. So here, we choose PCA for feature extraction. The principle is to project the data from high-dimensional space to low-dimensional space, then retain the principal components that contain more information.

3.3 Prediction Model

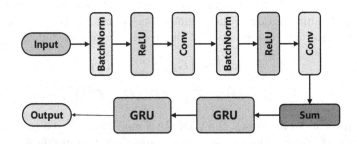

Fig. 4. The Structure of CGRU

The Convolutional Layers and GRU are used to predict the short-term load of electric vehicles. This is because the convolutional layer convolves the input data using a local receptive field, which allows the model to identify local patterns and features more intensively. For the charging load of electric vehicles, some local patterns may have an important impact on short-term prediction, and the introduction of a convolutional layer helps to capture this local information. Similarly, the use of convolutional layers can effectively reduce the number of parameters, thereby reducing the complexity of the model, reducing the risk of overfitting, and improving the generalization ability of the model. The reason for using a GRU is that it can better handle long-term dependencies. The short-term prediction of EV charging load can be influenced by the previous moment and the further moment, and the introduction of GRU helps the model to better capture this long-term dependence. In general, the combination of convolutional layers and GRU can flexibly capture information at different scales and levels. It can capture the temporal and spatial characteristics of the short-term load of electric vehicles more comprehensively and accurately, improve the forecasting performance, and adapt to complex data structures.

Prediction Model Structure. As shown in Fig. 4, the CNN-GRU model uses two convolutional layers. Through the local perception of input data, the convolutional layer can capture patterns in different times and spaces, which is crucial for understanding the spatio-temporal variation of EV charging load. Each layer of the convolutional layer contains multiple convolution nuclei, each of which is responsible for detecting different features in the input data. Secondly, a two-layer GRU is introduced, and the GRU's gating mechanism enables it to deal with the long-term dependence better, avoiding the gradient disappearance problem in the traditional RNN. By stacking two layers of GRU, this model can capture time series patterns in the data at a deeper level, thereby enhancing its sensitivity to complex changes.

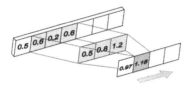

Fig. 5. The Structure of One-dimensional Convolution

The structure of the whole model covers multi-level feature extraction from space to time. Convolutional layers are used to capture patterns in space, while GRU focuses on learning dynamic changes in time series. This combination enables a more comprehensive understanding of the characteristics of EV charging loads, which makes the model perform better in short-term load forecasting tasks. Through the design of this model, we expect to provide a more accurate and reliable solution for the load prediction of electric vehicle charging piles. The following will introduce the specific structure of CNN and GRU.

Specific Structure of CNN. In Fig. 5, the convolution layer here is one-dimensional convolution, which can be used to obtain patterns and features in one-dimensional data, such as time series and signal lights. In one-dimensional convolution, the convolution kernel slides along each position of the one-dimensional data and performs convolution operations over each data point to extract features.

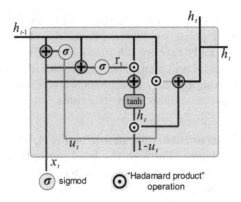

Fig. 6. The Structure of GRU

Specific Structure of GRU. In Fig. 6, GRU is a variant of RNN, and this structure can solve the problem of the disappearance of correlation gradient and the ineffective use of long-distance data features of traditional RNN [1]. Compared to standard RNN, GRU introduces the concept of updating gates and resetting gates, and the existence of these gates allows GRU to better capture long-term dependencies, but the input and output of GRU are the same as normal RNN. Therefore, the internal structure of the GRU consists of an update gate z_t, a reset gate r_t, a current input x_t, and a hidden state h_{t-1} passed down from previous nodes, which contains relevant information about the previous node. Combining x_t and h_{t-1}, the GRU gets the output of the currently hidden node and the hidden state h_{t-1} that is passed to the next node.

The Update Gate. The update gate in the GRU plays a crucial role in controlling the flow of information through the network during sequential data processing. Update gates help the GRU decide how much previous information should be passed on to the future and how much new information should be incorporated. Essentially, it regulates the balance between old and new information, allowing the network to selectively update its memory. Mathematically, the update gate is responsible for deciding which parts of the previously hidden state should be updated and which parts should be discarded. It is defined by the sigmoid activation function, and its value ranges from 0 to 1. If the update gate outputs a value close to 1, it indicates that the corresponding element in the hidden state should be updated, while a value close to 0 indicates that the information should be retained. At time t, calculate update gate u_t:

$$z_t = \sigma(W^{(z)}x_t + U^{(z)}h_{t-1}) \tag{6}$$

where x_t is the input vector at time t, and after a linear transformation, the data added with the linear change of h_{t-1} is put into the sigmoid activation function to obtain z_t.

The Reset Gate. The main function of the reset gate is to control whether the network ignores past information. When dealing with long sequences, RNNs may face problems such as disappearing gradients or exploding gradients. The introduction of reset gates helps models learn long-term dependencies more effectively. Its specific functions include forgetting past information, capturing long-term dependence" and mitigating gradient problems, etc. The specific calculation formula is as follows:

$$r_t = \sigma(W^{(r)}x_t + U^{(r)}h_{t-1}) \tag{7}$$

Current Memory Content. The new memory content will use the reset gate to store information related to the past. The calculated expression is

$$h'_t = \tanh(Wx_t + r_t \odot Uh_{t-i}) \tag{8}$$

The Final Memory of the Current Time Step. To calculate the current hidden load h_t, which performs two functions: on the one hand, it holds the information of the current cell, and on the other hand, it passes the information of the current cell to the next cell. This process requires the use of the update gate, which determines what information needs to be left in the current memory content h'_t and the previous time step h_{t-1}. This process can be expressed as:

$$h_t = z_t \odot h_{t-1} + (1 - z_t) \odot h'_t \tag{9}$$

Through this process, GRU is able to update the current memory content at each time step, while dynamically adjusting the retention and forgetting of memories through update gates and reset gates to better accommodate long-term dependencies in the sequence data.

4 Experiment

4.1 Experimental Setup

Dataset. We assess the effectiveness of our approach for gathering charging load data within smart grids. These meters are positioned within an accessible, communal charging facility, serving the needs of daytime commuters, afternoon or evening visitors, and late-night residents, as well as specialized user segments like taxi drivers, car-sharing services, or local logistics providers. For data collection, we choose the records of daily meter readings from January 1, 2021, to December 31, 2022, with a sampling interval of 15 min, yielding nearly 360,000 training samples. These datasets include detailed information such as meter numbers, meter power, contract capacity, operational capacity, composite electricity prices, data dates, and timestamps. In this dataset, we focus on the characteristics of charging loads, which are related to when the charging station serves. By analyzing and processing this data, we can better understand the patterns of charging load variations, thus providing more effective guidance and optimization suggestions for the operation of charging stations.

Baseline. We compare our method with three baselines.
(1) RNN: RNN is based on the principle of introducing recurrent connections at each time step, allowing the network to propagate information continuously and maintain its state. At each time step, the RNN receives the current input along with the previous time step's hidden state and generates output and a new hidden state. This recursive structure enables RNN to capture temporal dependencies in sequence data.
(2) LSTM: LSTM is a special type of RNN designed to address the vanishing gradient problem encountered by traditional RNNs when dealing with long sequences. LSTM harnesses gated mechanisms to manage the flow of information, overseeing input, output, and forgetting processes. This enables it to effectively tackle the issue of long-term dependencies in sequential data.
(3) GRU: GRU is another variant of RNN, similar to the LSTM model but with a simpler structure. The GRU model introduces update and reset gates to control the flow of information, addressing long-term dependency issues. Compared to the LSTM model, the GRU model has fewer parameters and is easier to train.

Hyperparameter Settings. We set $window\_size$ as 128, which specifies the size of the prediction window used for model training in time series data. $Input\_size$, which is set to 109, specifies the number of features that equals the number of features in the dataset. We set the learning rate to 0.0001 and use Adam as the optimizer to update the model parameters. The Adam optimizer combines the advantages of momentum and adaptive learning rate adjustment and is usually able to effectively optimize neural network models. The model will be trained for 20 epochs.

4.2 Comparison with Baselines

We evaluate the performance of three baseline models (RNN, GRU, and LSTM) and our proposed model (CNN-GAN).

Table 1. The Loss of Four Baselines.

Model	RNN	LSTM	GRU	CNN-GRU
Loss (MAE)	**0.0805**	**0.0115**	**0.0149**	**0.0083**

We use the Mean Absolute Error (MAE) [12] loss as the metric to measure the predictive performance of the models. Table 1 shows the final losses on the test set for these models after 20 epochs of training, while Fig. 7 shows the training loss curves of the whole training process. Based on the graph, it's clear that the CNN-GRU model demonstrates a lower MAE loss compared to the baseline model. This suggests that our model delivers superior predictive performance in time series generation tasks.

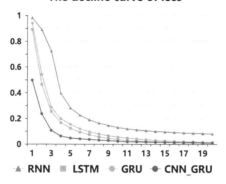

Fig. 7. The Decline Curve of Loss.

Our proposed CNN-GRU model exhibits significant advantages over traditional RNN, GRU, and LSTM models: the CNN-GRU model demonstrates lower MAE loss during training, proving its superior performance in load forecasting tasks. Compared to baseline models, the CNN-GRU model can better capture the underlying features of time series data, possesses stronger generalization capabilities, and is applicable to various types of time series prediction tasks.

4.3 Ablation Studies

To evaluate the feasibility of our proposed CNN-GRU model, we conduct ablation experiments comparing our model with its basic components. We remove the first two layers of one-dimensional convolution from the CNN-GRU model, retaining only two layers of GRU structures. We train both models using the same dataset and training settings and evaluate their performance on time series generation tasks.

Experiment Setup. In the ablation experiments, we train both models (CNN-GRU and the model containing only two layers of GRU) for the same number of epochs and record the corresponding Mean Absolute Error (MAE) loss during the training process.

Conclusion of the Ablation Experiment. Figure 8 shows the MAE loss curves of the two models during the training process. We observe that the CNN-GRU model exhibits lower MAE loss compared to the model without layers of convolution, here we call it OnlyGRU. Additionally, it demonstrates faster convergence speed and lower loss values in the later stages of training, indicating that our model possesses better performance and feasibility in the time series generation task of load forecasting.

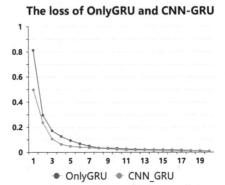

Fig. 8. The loss of OnlyGRU and CNN-GRU

4.4 Efficiency

We also compare the time efficiency of our CNN-GRU with other baseline models. Specifically, we record the time required for all models to complete one epoch of training.

Table 2. Model Run Time Record

Model	RNN	LSTM	GRU	CNN-GRU
Time(s)	404.8029	212.2857	267.5186	256.0589

As shown in Table 2, in terms of completing one epoch, our model trains faster compared to RNN and GRU. This indicates that under the same training conditions, our model may possess higher training efficiency and can learn the features of the data more quickly.

5 Conclusion

We propose a novel framework for short-term electric vehicle load forecasting to address issues of poor data quality and accuracy. By integrating data cleaning, feature extraction, and model prediction, our framework significantly enhances prediction accuracy, reduces computing time, and exhibits strong generalization ability. This approach effectively meets efficiency, stability, and cost-effectiveness requirements for load forecasting and management in electric vehicle charging stations.

Acknowledgement. This paper is supported by the science and technology project of State Grid Corporation of China: "Research on demand-side flexible resources portrait and aggregation technology based on multi-source data fusion" (Grand No.5700-202358310A-1-1-ZN)

References

1. Cho, K., et al.: Learning phrase representations using RNN encoder–decoder for statistical machine translation. arXiv:1406.1078 (2014)
2. Gao, Q., et al.:. Charging load forecasting of electric vehicle based on monte carlo and deep learning. In: 2019 IEEE Sustainable Power and Energy Conference (iSPEC) (2019)
3. He, Z., et al.: Short-term power load forecasting of multi-charging piles based on improved gate recurrent unit. IEEE Access **12**, 2490–2499 (2023)
4. Hu, X., Ferrera, E., Tomasi, R., Pastrone, C.: Short-term load forecasting with radial basis functions and singular spectrum analysis for residential electric vehicles recharging control. In: 2015 IEEE 15th International Conference on Environment and Electrical Engineering (EEEIC), pp. 1783–1788 (2015)
5. Khan, H., Khan, M.J., Qayyum, A.: Neural network-based load forecasting model for efficient charging of electric vehicles. In: 2022 7th Asia Conference on Power and Electrical Engineering (ACPEE), pp. 2068–2072 (2022)
6. Kiranyaz, S., Avci, O., Abdeljaber, O., Ince, T., Gabbouj, M., Inman, D.J.: 1d convolutional neural networks and applications: a survey. Mech. Syst. Signal Process. **151**, 107398 (2021)
7. Yi, L., Ting, X.T., Song, W., Yun, G., Hui, H., Ziwen, Q.: The load forecasting of charging stations based on support vector regression. In: 2023 5th International Conference on Power and Energy Technology (ICPET), pp. 991–995 (2023)
8. Madhukumar, M., Sebastian, A., Liang, X., Jamil, M., Shabbir, M.N.S.K.: Regression model-based short-term load forecasting for university campus load. IEEE Access **10**, 8891–8905 (2022)
9. Meng, Z., Xie, Y., Sun, J.: Short-term load forecasting using neural attention model based on EMD. Electr. Eng. **104**, 1857–1866 (2021)
10. Su, Z., et al.: Short-term load prediction of electric vehicle charging station based on long-short-term memory neural network. In: 2023 4th International Conference on Computer Engineering and Intelligent Control (ICCEIC), pp. 595–599 (2023)
11. Sun, X., Ouyang, Z., Yue, D.: Short-term load forecasting based on multivariate linear regression. In: 2017 IEEE Conference on Energy Internet and Energy System Integration (EI2), pp. 1–5 (2017)
12. Xie, Z., Wang, R., Wu, Z., Liu, T.: Short-term power load forecasting model based on fuzzy neural network using improved decision tree. In: 2019 IEEE Sustainable Power and Energy Conference (iSPEC), pp. 482–486 (2019)
13. Xu, L., Chen, Y., Wang, Y., Li, N., Zong, Q., Chen, K.: Research on electric vehicle ownership and load forecasting methods. In: 2020 5th International Conference on Mechanical, Control and Computer Engineering (ICMCCE), pp. 184–188 (2020)
14. Yan, W., Li, N., Shen, Y., Shi, L., Hu, B., Zhou, Z.: Charging load prediction of electric vehicles based on CNN-GAN and semi-supervised regression **42**(2), 83–89
15. Zhao, Y., et al.: Electric vehicle charging load prediction method based on nonlinear auto-regressive neural networks. In: 2023 4th International Conference on Computer Engineering and Intelligent Control (ICCEIC), pp. 600–605 (2023)
16. Zhu, J., Yang, Z., Guo, Y., Zhang, J., Yang, H.: Short-term load forecasting for electric vehicle charging stations based on deep learning approaches. Appl. Sci. **9**(9), 1723 (2019)

ERDSE

Audio-Guided Visual Knowledge Representation

Fei Yu, Zhiguo Wan[✉], and Yuehua Li[✉]

Research Center for Space Computing System, Zhejiang Lab,
Hangzhou 311121, China
{wanzhiguo,liyh}@zhejianglab.com

Abstract. Visual knowledge is primarily acquired through visual perception, but it is often exclusively represented in natural language, neglecting the collaborative nature of multisensory perception. To address this limitation, this paper proposes an audio-guided approach to visual knowledge representation. By integrating auditory cues into visual captioning, the model enhances environmental understanding through multisensory collaboration. Furthermore, the introduction of an audio-visual multimodal mutual information graph enriches the semantic content of visual captions. Additionally, while research on multimodal perception data is extensive, audio-visual datasets often lack fine-grained annotations. To address this issue, we construct a fine-grained multimodal dataset. Finally, experimental validation through multimodal-guided visual captioning and link prediction tasks demonstrates the effectiveness of this approach compared to existing methods.

Keywords: Visual Knowledge · Interpretable Expression · Multisensory Collaboration · Audio Guidance

1 Introduction

Multi-sensory integration, also known as multi-sensory collaboration or synesthesia, is crucial in achieving a holistic understanding of the world, emulating human perception via multiple senses like vision, hearing, and language. Current approaches to cross-media intelligence struggle with expressing multimodal perception effectively, often relying on separate processing and mapping of individual modalities, which overlooks the interactive synergy present in multi-sensory experiences. To overcome this limitation, Pan et al. [1,2] introduce visual knowledge-a paradigm that models and reasons about multiple sensory inputs tied to visual cues.

Multimodal perception plays a vital role in scene comprehension compared to unimodal perception. An instance illustrating this involves both visual and auditory senses: While visual data provides spatial and relational information about objects in a scene, auditory input complements it by adding functional and contextual details, like indicating rain through thunder sounds despite clear skies.

Existing multimodal representation learning methods predominantly focus on learning unified or coordinated spaces for different modalities. It is a challenging task due to the heterogeneity of multimodal data. Multimodal representations [3] can be categorized into two types: joint and coordinated representations. Joint representations combine unimodal signals into a single representation space, while coordinated representations process unimodal signals separately and enforce similarity constraints to bring them into a coordinated space. In the term of the applications in audiovisual text representation, the audiovisual text representation mainly utilizes the joint representations schema. For example, HMN [4], GLR [5], SGN [6], SwinBERT [7], and AVLFormer [8] are all multimodal representation learning models that utilize the joint representations schema. Although existing studies mainly focus on multimodal feature fusion and joint representation learning, there is less research on how to utilize multimodal information to enhance the visual knowledge expression ability.

Therefore, this paper proposes an audio-guided visual knowledge expression method, which enhances visual features through audio information to improve the expression ability of visual knowledge. Despite advances in multimodal fusion and joint representation, there is a gap in exploring how to harness multimodal cues to enrich visual knowledge expression. This paper introduces an audio-guided visual knowledge expression method that boosts visual feature representation using audio information.

Notably, while research on multimodal perception data is extensive, audio-visual datasets often lack fine-grained annotations. To address this, this study proposes a fine-grained multimodal representation that emphasizes visual elements alongside textual descriptions aligned with visual content, incorporating the influence of other sensory modalities. This approach constructs visual knowledge representations using multimodal triples, integrating common sense knowledge and developing a model for visual knowledge graph representation.

Contributions:

- Proposed a novel method for visual knowledge representation, addressing granularity limitations across multimodal data.
- Integrated diverse sensory inputs, including auditory cues, to construct a comprehensive representation of multimodal knowledge.
- Introduced multimodal triples alongside commonsense knowledge to create structured visual knowledge graphs capturing intricate relationships.
- Validated the proposed model through tasks like multimodal-guided visual captioning and link prediction on graphs, demonstrating its capacity to accurately interpret and generate nuanced information.

2 Related Work

2.1 Visual-Aware Multimodal Dataset

Currently available labelled visual-aware multimodal datasets can be categorized into three main types: audio-language, video-language, and multimodal-language, they aim to improve the machine's ability to understand and describe complex audio-visual scenes.

Audio-Language Datasets: AudioCaps [9] is the first large-scale audio caption dataset that captures real-world sounds in their natural environment. Clotho [10] enhances the diversity of audio content and subtitles by employing annotations from five different individuals for each audio clip. LAION-Audio-630K [11] is constructed by collecting audio and its corresponding text descriptions from multiple sources. WavCaps [12] uses ChatGPT to filter and rewrite the original descriptions to optimize the quality of captions.

Video-Language Datasets: MSVD [13] is one of the initial attempts, but it lacks audio modality to correspond to video content. MSR-VTT [14] provides supplementary audio data and expands the diversity and categories of videos. M-VAD [15] and MPII-MD [16] provide a large number of open-domain videos and high-quality text descriptions. YouCook [17] and YouCookII [18] focus on life-teaching videos, especially detailed descriptions of the cooking process. Charades [19] collects videos depicting daily family behaviours through crowdsourcing. Action Genome [20] annotates the relationship between objects within videos. VideoStory [21] is proposed specifically for social media videos.

Multimodal-Language Datasets: TVC [22] provides detailed descriptions of visual content and character dialogues within videos. Spoken Moments [23] describes video content through speech, simplifying text annotation. FAVDBench [8] emphasizes providing comprehensive text descriptions for videos that contain audio. Table 1 provides an overview of current major visual-aware multimodal datasets. Previous datasets often lack comprehensive annotations capturing intricate relationships between objects across multiple modalities. Building upon [20], our approach annotates object relationships to enhance the understanding of multimodal interplay. By annotating video and audio events as triples (subject, action, object), we capture deep semantic connections between visual, audio, and textual elements. This holistic annotation enables a comprehensive understanding of audio-visual content.

Table 1. The Current Major Visual-aware Multimodal Datasets. "Clips" refers to the number of video clips or audio clips included in the datasets. "Audio" indicates whether the datasets contain audio or not. "VC, AC, AVC" stand for video caption, audio caption, and audio-visual captions, respectively.

Dataset	Domain	Clips	Captions	Text source	Audio	VC	AC	AVC
AudioCaps [9] (2019)	Open	46K	46K	Manual	✓	–	✓	–
Clotho [10] (2020)	Open	5K	25K	Manual	✓	–	✓	–
MSVD [13] (2011)	Open	2K	70K	Manual	–	✓	–	–
MP-II Cooking [24] (2012)	Cooking	–	5.6K	Manual	✓	✓	–	–
YouCook [17] (2012)	Cooking	–	2.7K	Manual	✓	✓	–	–
YouCookII [18] (2018)	Cooking	15.4K	15.4K	Manual	✓	✓	–	–
M-VAD [15] (2015)	Movie	49K	56K	DVS	✓	✓	–	–
MPII-MD [16] (2015)	Movie	68K	68K	Script+DVS	✓	✓	–	–
MSR-VTT [14] (2016)	Open	10K	200K	Manual	✓	✓	–	–
Charades [19] (2016)	Human	10K	27.8K	Manual	✓	✓	–	–
Action Genome [20] (2020)	Human	10K	27.8K	Manual	✓	✓	–	–
ActivityNet Captions [25] (2017)	Open	100K	100K	Manual	✓	✓	–	–
VideoStory [21] (2018)	Social media	123K	123K	Manual	✓	✓	–	–
VATEX [26] (2019)	Open	41.3K	413K	Manual	✓	✓	–	–
HowTo100M [27] (2019)	Instructional	136K	136K	ASR transcription	✓	✓	–	–
TVC [22] (2020)	Social media	108K	262K	Manual	✓	✓	–	–
Spoken Moments [23] (2021)	Instructional	516K	516K	Manual	✓	✓	–	–
FAVDBench [8] (2023)	Open	11.4K	144K	Manual	✓	✓	✓	✓
Our dataset (Present)	Human	2K	15k	Manual	✓	✓	✓	✓

In the FAVDBench dataset [?], human activity videos constitute 12.8% of the total, with only 5 out of 11 categories representing indoor activities, accounting for just 5.3% of the total videos. Contrary to our published dataset, which focuses on human indoor activities, FAVDBench includes 3 indoor activities similar to our 5 categories. Our dataset encompasses 45 indoor activities, merging visual and audio partially labeled categories, where these 5 similar activities constitute only 11.1% of the total. As FAVDBench fails to cover all our categories, we released our dataset specifically for the visual-caption semantic enhancement task.

2.2 Multimodal Representation Learning

Multimodal representation learning aims to learn joint representations from data of different modalities, such as visual, audio, and text. It is a challenging task due to the heterogeneity of multimodal data. Multimodal representations [3] can be categorized into two types: joint and coordinated representations. Joint representations combine unimodal signals into a single representation space, while coordinated representations process unimodal signals separately and enforce similarity constraints to bring them into a coordinated space. To learn joint representations, three fusion strategies are commonly used: early fusion, late fusion, and hybrid fusion. Applications in Audiovisual Text Representation: In the context of audiovisual text representation, joint representations are widely used. For example, HMN [4] combines visual, motion, and object features with semantic information to generate subtitles. GLR [5] models visual information by considering multiple dimensions and generates associated video subtitles. SGN [6] uses partially decoded phrases to semantically group video frames and predicts words based on these groups. SwinBERT [7] passes video and caption representations to a multimodal transformer encoder to generate natural language sentences. AVLFormer [8] utilizes masked language modeling and autoregressive language modeling to help models learn visual and audio information for text generation. Dense Video Captioning: BMT [7] and MDVC [28] are two methods for dense video captioning, which involves localizing events in a video and generating textual descriptions for each event. BMT uses a dual-mode encoder to process audio and visual features and generates proposals and confidence scores. MDVC combines audio, visual, and speech information to generate subtitles for predicted proposals. In contrast, our proposed method addresses the granularity limitations in existing multimodal representation techniques by:

Introducing a novel approach for visual knowledge representation that integrates diverse sensory inputs, including auditory cues, to construct comprehensive multimodal knowledge representations. Proposing the use of multimodal triples alongside commonsense knowledge to create structured visual knowledge graphs that capture intricate relationships across modalities. Validating the effectiveness of our model through tasks such as multimodal-guided visual captioning and link prediction on graphs, demonstrating its capability to interpret and generate nuanced information beyond existing methods. This approach aims to advance the field by offering a more nuanced and integrated approach to multimodal representation learning, distinct from existing methods that primarily focus on fusion strategies and specific application tasks without necessarily addressing the comprehensive integration of sensory inputs and semantic relationships.

3 Preliminary

3.1 Problem Formulation

Let $D_\mathcal{M} = \{\mathcal{V}, \mathcal{A}, \mathcal{T}\}$ represent the dataset comprising visual, acoustic, and textual modalities, with the modalities denoted as $\mathcal{M} = \{v, a, text\}$. We denote the feature vector of each modality $m \in \mathcal{M}$ as $\mathbf{e}_m \in \mathbb{R}^{d_m}$, where d_m signifies the dimensionality of the features. The objective of the audio-guided visual knowledge representation (AVKR) is to establish a multimodal graph, wherein the graph consists of fundamental elements (triplets). The entities and relationships within these triplets are extracted from multimodal sources including visual, acoustic, and textual data. Given that visual knowledge acquisition is grounded in human common-sense knowledge, these fundamental elements undergo cognitive refinement through common-sense knowledge. As a result, primary elements primarily derived from visual perception, with auditory perception synergistically enhancing them, constitute the multimodal knowledge graph.

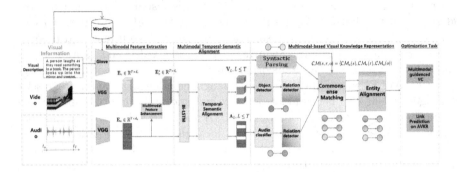

Fig. 1. The model flow of the proposed audio-guided visual knowledge representation framework (AVKR). AVKR includes three key components: (1) Multimodal feature extraction contains the multimodal feature initilization and Multimodal feature enhancement; (2) Multimodal Temporal-Semantic Alignment; (3) Multimodal-based visual knowledge representation.

4 Methodology

In this section, we outline AVKR's framework with three main components: Multimodal feature extraction enhances vision-audio features. Timing semantic alignment filters irrelevant information and modal noise. Knowledge representation creates a graphical portrayal of visual knowledge, inspired by human perception. The model is trained via multi-task joint optimization. The overall framework of AKVR is illustrated in Fig. 1.

4.1 Multimodal Feature Extraction

We propose a multimodal semantic extraction module to extract modality-specific features from multimodal data. By considering the temporal nature of visual and auditory perception data, we segment the data into discrete time intervals. This allows us to extract segment-based semantic features and facilitate subsequent analysis.

Currently, we possess audio-visual perception data, where text primarily serves for natural language understanding of visual perception. Therefore, such textual information is classified as visual data. When considering the temporal aspect of audio-visual perception, we primarily deal with the visual signals of videos and the auditory signals of audio. We represent audio-visual data as $\mathcal{D}_{va} = \{\mathcal{V}_T, \mathcal{A}_T\}$, where $\mathcal{V}_T$ and $\mathcal{A}_T$ denote videos and audio data with a duration of T seconds, respectively. Using a discrete segmentation method, we divide video data $\mathcal{V}_T$ into consecutive T video segments, each with a 1-second interval:

$\mathcal{V}_T = \{V_1, V_2, ..., V_T\}$. Each V_i consists of K frames obtained through uniform sampling, represented as $V_i = \{I_i\}_{i=1}^{K}$. To synchronize video frames I_i with corresponding audio segments, we extract audio clips between the timestamps of each video frame and the next frame I_{i+1}. This ensures precise synchronization and forms a set of audio-visual pairings $\mathcal{D}_{va} = \{(I_1, A_1), (I_2, A_2), ..., (I_{T_s}, A_{T_s})\}$, where $T_s = T \times K$.

Multimodal Feature Initialization. We employ pre-trained Convolutional Neural Networks (VGG-19 [29]) to extract visual initial features $\mathbf{E}_v = \{\mathbf{e}_{v,t}\}_{t=1}^{T_s} \in \mathbb{R}^{T_s \times k \times d_v}$ and audio features $\mathbf{E}_a = \{\mathbf{e}_{a,t}\}_{t=1}^{T_s} \in \mathbb{R}^{T_s \times d_a}$ from the set $\{(I_t, A_t)\}_{t=1}^{T_s}$, while textual features, represented as $\mathbf{E}_{text} = \{\mathbf{e}_{text,t}\}_{t=1}^{T_s} \in \mathbb{R}^{T_s \times d_{text}}$, are extracted using a pre-trained word embedding model (GloVe [30]). Here d_v represents the feature dimension for visual segments, d_a represents the feature dimension for audio segments, d_{text} represents the feature dimension for textual segments, and k represents the vectorized spatial dimension of each feature map.

Multimodal Feature Enhancement. Multimodal perception of physical environment knowledge refers to the knowledge about the physical environment that is acquired through the collaborative action of multiple senses. Visual knowledge is the core of multimodal perception of physical environment knowledge because it provides rich spatial information and object property information. The focus of this paper is to enhance the representation capability of visual knowledge by augmenting video information with audio. Specifically, we leverage audio features with complementary spectro-temporal properties to enhance the tempo rally aligned video features. We initialize the visual and auditory features as $\mathbf{E}_v \in \mathbb{R}^{T_s \times d_v}, \mathbf{E}_a \in \mathbb{R}^{T_s \times d_a}$, respectively, and draw inspiration from AGVA [31]to enhance visual information via audio information through an attention mechanism:

$$\mathbf{E}_v^\star = \mathcal{F}_{a \to v}(\mathbf{E}_a, \mathbf{E}_v) = \alpha_{en} \mathbf{E}_v, \qquad (1)$$

wherein α_{en} is the visual feature enhancement weight, which can be computed using an MLP with the following equation:

$$\alpha_{en} = \sigma(W_f \tanh(W_v g(\mathbf{E}_v) + W_a g(\mathbf{E}_a))). \qquad (2)$$

In our model, we introduce a non-linear transformation function $g(\cdot)$ for mapping the visual and auditory initialization features into a shared feature space. Specifically, we apply $g(\cdot)$ to the visual initialization feature $\mathbf{E}_v$ and the auditory initialization feature $\mathbf{E}_a$, obtaining the mapped feature representations $\hat{\mathbf{E}}_v = g(\mathbf{E}_v) \in \mathbb{R}^{T_s \times d}$ and $\hat{\mathbf{E}}_a = g(\mathbf{E}_a) \in \mathbb{R}^{T_s \times d}$ respectively. In this study, we utilize the softmax function as the specific activation function $\sigma(\cdot)$ in $g(\cdot)$. The function $g(\cdot)$ denotes the transformation of the input $\mathbf{E}_v$ and $\mathbf{E}_a$ through a dense layer utilizing a non-linear activation function. Similarly, obtain $\mathbf{E}_a^\star$ as above.

4.2 Multimodal Temporal-Semantic Alignment

The previous section focused on enhancing visual perception through auditory input, simulating multisensory synergy. However, in multimodal perception, temporal asynchrony arises due to differences in sensory modalities' processing speeds. For example, visual perception may precede auditory perception. To address this, we propose a temporal semantic alignment network for visual and auditory modalities. We begin with time-based initialization alignment, then introduce a multimodal semantic temporal alignment function to filter out segments with semantic and temporal asynchrony. This reduces irrelevant content and modality noise, enhancing understanding of the physical environment.

First, we perform temporal feature learning on the enhanced audio-visual features, resulting in $\hat{\mathbf{E}}_v^t = Bi\text{-}LSTM(\mathbf{E}_v^\star)$ and $\hat{\mathbf{E}}_a^t = Bi\text{-}LSTM(\mathbf{E}_a^\star)$.

$$sim_{av}^t = f(\frac{\mathcal{F}(\hat{\mathbf{E}}_v^t)\mathcal{F}(\hat{\mathbf{E}}_a^t)^\top}{\sqrt{d_k}}), \tag{3}$$

$$\mathbf{V}_L = \{\hat{\mathbf{E}}_v^t | sim_{av}^t > \tau\}, \tag{4}$$

$$\mathbf{A}_L = \{\hat{\mathbf{E}}_a^t | sim_{av}^t > \tau\}. \tag{5}$$

$$\mathcal{ST}(\hat{\mathbf{E}}_v^t, \hat{\mathbf{E}}_a^t) = \begin{cases} 1, & \text{if } sim_{av}^t > \tau; \\ 0, & \text{otherwise}, \end{cases} \tag{6}$$

wherein the function f is a composite function. To eliminate negatively correlated segments, we introduce the ReLU function to filter out negative temporal parts. Then, we perform row-wise "1 normalization" (dividing each element in a row by the sum of that row's elements, ensuring that the sum of each row's elements is equal to 1). This operation maps the values of the similarity matrix to a range between 0 and 1, facilitating subsequent computation and analysis. Through the temporal semantic alignment network, we can filter out irrelevant temporal segments and obtain the audio-visual temporal alignment with semantic meaning. Specifically, we can obtain the following results: $\mathbf{V}_L = \{\hat{\mathbf{E}}_v^t | sim_{av}^t > \tau\}$ and $\mathbf{A}_L = \{\hat{\mathbf{E}}_a^t | sim_{av}^t > \tau\}$. The sequence of audio-visual frames corresponding to the audio-visual sequence is represented as: $S_v = \{I_{v,n}\}_{n=1}^L$, $S_a = \{I_{a,n}\}_{n=1}^L$.

4.3 Multimodal-Based Visual Knowledge Representation

After ensuring the semantic and temporal consistency of audio-visual perception data, it's essential to logically express multimodal knowledge primarily focused on visual perception, aiming to comprehensively convey information about the perceptual environment and the semantic understanding derived from multisensory perception. We referenced the definition of visual knowledge proposed by Pan and Yang [1,2], and represented multimodal audio-visual text data as triplets, employing the subject-verb-object expression logic. By integrating this with a repository of common sense knowledge, the resulting fused multimodal knowledge constitutes visual knowledge.

Multimodal Triple Extraction. We represent the audio-visual-language multimodally using the logic of a triple structure, then introduce a common-sense knowledge base to combine the semantic representation of the audio-visual text multimodal with human common-sense knowledge. The resulting integrated multimodal knowledge is referred to as visual knowledge.

(1) Video Triple Extraction. Let $S_v = \{I_{v,n}\}_{n=1}^{L}$ express the frames in the visual segment, where n indicates the frame index and L denotes the total number of labeled frames. Through a pre-trained *spatial-temporal transformer* [32] model, a dynamic knowledge graph $G_v = \{G_n^v\}_{n=1}^{L}$ has been extracted from the visual segment. Here, $G_n^v = \{V_n, R_n\}$ represents the scene graph of the frame $I_{v,n}$, with V_n as the set of entities and R_n as the set of relations in the frame $I_{v,n}$. Entitie of V_n and relationships of R_n together form the set of triplets $\langle s_v^n, r_v^n, o_v^n \rangle$.

(2) Audio Triple Extraction. Utilizing a pre-trained *audio spectrogram transformer* model [33], we extract audio features from segment S_a and construct a knowledge graph $G_a = \{G_i^a\}_{i=1}^{M}$, where M denotes the total number of time frames in S_a. Each G_i^a represents the audio label o_a^i at time i. To derive subject-verb-object triples for the audio, we integrate specific activities that produce the audio with external common-sense knowledge from WordNet. Subsequently, we employ a depth-first search approach to obtain triples $\langle s_a^n, r_a^n, o_a^n \rangle$ based on the identified labels (objects) of the audio segments.

(3) Text Triple Extraction. We utilize *open information extraction* (OpenIE) technology for textual information extraction, to extract meaningful triples from the provided text S_l. Subsequently, we employed these extracted triples to build a text knowledge graph $G_l = \{G_i^l\}_{i=1}^{M}$, where G_i^l represents a set of triples $\langle s_l^i, r_l^i, o_l^i \rangle$ at time i.

Multimodal Triplet Representation with Commonsense Knowledge Rendering. In order to combine the semantic features of multimodal perception with common sense knowledge, we introduce a common sense knowledge base to reprocess the semantic knowledge of audiovisual text multimodal perception learned in the previous section, that is, to map the semantic features of audiovisual text to the common sense knowledge base. This operation not only realizes the role of common sense knowledge in the process of visual knowledge generation, but also realizes the semantic alignment and information enhancement of the semantic knowledge of multimodal perception.

Let $\mathcal{T}r_{avt} = \{\langle s_m, r_m, o_m \rangle\}_{m \in \mathcal{M}}$ be the set of multimodal triplets obtained from audiovisual text. Let $\mathcal{C}_{wn}$ be the set of concepts in WordNet and $\mathcal{R}_{wn}$ be the set of relations in WordNet. We define a mapping function $\mathcal{CM}$ maps each multimodal triplet to a WordNet triplet. The mapping function $\mathcal{CM}$ is defined as follows:

$$\mathcal{CM}(\langle s, r, o \rangle) = \langle \mathcal{CM}_s(s), \mathcal{CM}_r(r), \mathcal{CM}_o(o) \rangle, \langle s, r, o \rangle \in \mathcal{T}r_{avt}, \quad (7)$$

where $\mathcal{CM}_s : \mathcal{S}_{avt} \to \mathcal{CK}_s$ is a function that maps each subject in $\mathcal{T}r_{avt}$ to a concept in $\mathcal{CK}$, $\mathcal{CM}_r : \mathcal{R}_{avt} \to \mathcal{CK}_r$ is a function that maps each relation in $\mathcal{T}r_{avt}$ to a relation in $\mathcal{CK}_r$, and $\mathcal{CM}_o : \mathcal{O}_{avt} \to \mathcal{CK}_s$ is a function that maps each object in $\mathcal{T}r_{avt}$ to a concept in $\mathcal{CK}_s$. The triplet $\langle \mathcal{CM}_s(s), \mathcal{CM}_r(r), \mathcal{CM}_o(o) \rangle$ is then used as the representation of the multimodal triplet (s, r, o).

4.4 Visual Knowledge Representation

Based on the knowledge representation of audio-visual text modalities through common-sense knowledge rendering, the next step is to integrate the expressions of audio-visual text into the form of a multimodal knowledge graph, consisting primarily of multimodal triplets (s, r, o). In our approach, we first use the vocabulary and relationships from WordNet as input data, which is then integrated with the GloVe model. Specifically, we map the vocabulary and relationships from WordNet into the GloVe vector space and use a specific threshold to determine which relationships and words should be included in the model. Through this integration process, we train a pre-trained model specifically tailored to capture common-sense knowledge. This enriched GloVe model, infused with common-sense knowledge, is utilized to align entities within the multimodal triple representations. Ultimately, through this alignment process, we derive the final visual knowledge expression, culminating in a comprehensive multimodal knowledge graph.

Subsequently, cosine similarity is employed to measure the similarity between entities of distinct modalities at the same timestamp, computed by:

$$cosine(\mathbf{e}_i^t, \mathbf{e}_j^t) = \frac{\mathbf{e}_i^t \cdot \mathbf{e}_j^t}{\|\mathbf{e}_i^t\| \|\mathbf{e}_j^t\|}, \tag{8}$$

where, $\mathbf{e}_i^t, \mathbf{e}_j^t$ represent the subject vectors of different modalities $i, j \in \mathcal{M} = \{v, a, text\}$ at time t. A specific threshold is employed to determine the similarity of entity pairs. During the entity pair filtering process, if the similarity of certain pairs falls below the specified threshold, the commonsense network is utilized to query the relation between these entities, thereby assisting in refining the entity alignment for improved accuracy. Following entity alignment, the final visual knowledge graph $\mathcal{G}$ was constructed.

4.5 Optimization Objectives of AVKR

This paper presents a method for multimodal knowledge expression, focusing on specific physical environments where visual perception plays a crucial role, complemented by the collaborative perception of other senses (such as sound-guided visual knowledge expression). To train our knowledge expression model, we conduct training within the context of specific application tasks. Therefore, we primarily train the model on two tasks related to audio-visual text: *task 1-multimodal-guided visual captioning (MVC)*, and *task 2-link prediction on the constructed multimodal graph (LP)*. For these tasks, we design distinct loss functions to optimize performance.

For MVC: we follow the existing research methods in video captioning and design the loss function for training the multimodal-guided video captioning task as follows:

$$\mathcal{L}_{MVC} = \frac{1}{n}\sum_{i=1}^{n}(\mathbf{tr}_i - \mathbf{gt}_i)^2 + \lambda \sum j||w_j||^2, \quad (9)$$

wherein n represents the number of samples, which corresponds to the number of selected triples. $\mathbf{tr}_i$ denotes the vector of the ith selected triple, while $\mathbf{gt}_i$ represents the vector of the ground truth triple. λ is the regularization strength parameter, and w_j denotes the model parameters.

For LP: following the common practice, the training objective of the TransE model is to minimize the total energy of positive samples while maximizing the total energy of negative samples to ensure the maximum energy difference between positive and negative samples. To achieve this, a margin is introduced during training, ensuring that the energy of positive samples must be at least a fixed margin higher than the energy of negative samples.

$$\mathcal{L}_{LP} = \sum_{(s,r,o)\in\mathcal{G}}\sum_{(s',r,o')\in\mathcal{G}'}\max[\gamma + E(\mathbf{e}_s, \mathbf{e}_r, \mathbf{e}_o) - E(\mathbf{e}'_s, \mathbf{e}_r, \mathbf{e}'_o), 0], \quad (10)$$

where $E(\mathbf{e}_s, \mathbf{e}_r, \mathbf{e}_o) = d(\mathbf{e}_s + \mathbf{e}_r, \mathbf{e}_o) = ||\mathbf{e}_s + \mathbf{e}_r - \mathbf{e}_o||$, γ serves as a preset controlling parameter determining the energy differences between positive samples and negative samples.

5 Experiments

5.1 Experimental Settings

Datasets. In this paper, our experimental research focuses on the perceptual data acquired by visual sensors and audio sensors in home scenes. Specifically, we utilize the Action Genome dataset [20] as the perceptual data from visual sensors and annotate the sound-producing subjects and relationships in the videos based on the home scenes in the AG dataset. This allows us to create a multimodal dataset that combines audio, visual, and language information.

The **Visual-Auditory-Language Multimodal dataset (VALM**[1]**)** consists of 2K videos and 58,764 frame-level scene graph images. The dataset includes 36 entity categories, which are the same as the AG dataset. Among these categories, 33 of them are sound-producing entities, excluding the entities "doorway", "sandwich," and "shelf" which do not produce sound. The dataset also comprises 26 relationship categories, which are classified into attention relationships, spatial relationships, and contracting relationships. These relationships represent whether the subject is gazing at the object or not, the relative spatial position and the different modes of contact between entities, respectively.

---

[1] The dataset used in this work will be made available upon reasonable request to us.

Baselines. *Task 1-Multimodal Video Captioning (MVC)*: We choose HMN [4], GLR [5], SGN [6], BMT [7], SwinBERT [34], AVLFormer [8], and MDVC [28] as our comparison methods. Some of the methods involve the use of information from multiple modalities. *Task 2-Link Prediction (LP)*: To validate the usability of our proposed visual knowledge expression (multimodal graph), we compare it with existing methods that also utilize triplets as the main elements for dynamic knowledge graph representation, particularly focusing on the temporal aspect. *For temporal knowledge graph methods*, we consider the following baselines: TeSAT [35], CENET [36], TiRGN [37]; *For multimodal knowledge graph method*, we consider the following baseline: IMF [4]: IMF is an interactive model that utilizes bilinear fusion based on Tucker decomposition and contrastive learning to extract multimodal features.

Metrics. *For MVC*: We employ commonly used evaluation metrics in video captioning tasks, including BLEU@1 [38], METEOR [39], and ROUGE-L [40]. *For LP*: We adopt AUC, Micro-F1, and Macro-F1 to measure the model performance of link prediction on the visual knowledge graph representation. It is worth noting that Micro-F1 and Macro-F1 are designed for multi-class tasks in link prediction. Micro-F1 provides a comprehensive measure of the model's classification performance. Macro-F1 calculates the average F1 score over all categories, ensuring a balanced evaluation across different categories.

Implementation Details. In this paper, we conduct experiments based on our proposed multimodal visual-audio-language dataset. The visual feature part of the video is extracted by a pre-trained VGG-19 [29] model with 512-dimensional features. The audio feature part of the video is extracted by a pre-trained VGGish [41] model on the AudioSet dataset to extract 512-dimensional features, and in addition, the audio waveform is converted to a spectrogram, which is used as an input to the audio classification model AST. Adam is employed for parameter learning with a learning rate set to 0.0001, and the embedding dimension is set to 100. To ensure the validity of the dataset, we divided it into an 8:2 ratio for training and validating the model.

5.2 Overall Performance

The Experimental Results and Analysis for MVC. Table 2 presents the experimental results comparing our proposed method with other methods in the field of visual-caption semantic enhancement. We provide both the numerical improvement percentages and the corresponding reasons for these improvements.

Table 2. The Experimental Results for MVC with Different Methods.

Method	Visual	Audio	Text	BLEU@1	METEOR	ROUGE-L
HMN [4]	✓			1.16	2.71	6.57
GLR [5]	✓			4.23	5.27	11.78
SGN [6]	✓			1.65	3.43	9.84
BMT [7]	✓	✓		11.83	9.44	14.33
SwinBERT [34]	✓		✓	1.35	3.92	10.48
AVLFormer [8]	✓	✓	✓	11.11	8.52	13.84
Ours	✓	✓	✓	**18.02**	**12.79**	**26.47**

Our method demonstrates significant improvements in the task of visual-caption semantic enhancement when compared to other methods. Specifically, when compared to AVLFormer, our method achieves an impressive increase of 62.3% in BLEU@1, 50.2% in METEOR, and 47.7% in ROUGE-L scores. In addition to outperforming AVLFormer, our method also achieves substantial improvements compared to other methods in the table. For example, when compared to HMN, our method achieves an increase of 145.7% in BLEU@1, 371.8% in METEOR, and 303.6% in ROUGE-L scores. Similarly, when compared to GLR, SGN, BMT, and SwinBERT, our method achieves improvements of 326.1%, 994.6%, 52.7%, and 1234.8% respectively in the BLEU@1 metric.

In Fig. 2, we compare different video caption methods. In the results of BMT, each event in the video is described, where only the parts marked by green color are close to the truth, and it is clear that there is redundancy in the description. In the results of SwinBERT, a brief description of the character's activity is provided, but no further details are covered. In the results of MDVC method, some object objects are not described accurately enough. In the results of AVLFormer, it can be seen that it tends to complete descriptions. It is accurate in describing character features and sound aspects, but there are inaccuracies in describing character activities and object objects.

The Experimental Results and Analysis for LP. Our temporal knowledge graph model differs significantly from existing ones. Traditional temporal knowledge graph models typically rely on knowledge graph information before the $(t+1)$ timestamp to predict entities and relationships at the $(t+1)$ timestamp. However, our temporal knowledge graph model not only depends on knowledge graph information from previous time steps but also leverages known entity information at the $(t+1)$ timestamp. As a result, our model demonstrates significant advantages in temporal link prediction tasks, enabling more effective capture of dynamic changes in temporal knowledge graphs.

Based on the experimental results provided in Table 3, we can draw the following quantitative and qualitative analyses: Our method significantly outperforms other existing models across all evaluation metrics: the AUC improved from 0.7551 of the top-performing comparative model CENET to 0.9389, representing an increase of approximately 0.1838, which corresponds to roughly a 24% improvement. The Micro-F1 score increased from 0.8520 of CENET to 0.9521, raising by around 0.1001, equivalent to about a 11.7% enhancement. The Macro-F1 score rose from 0.6147 of CENET to 0.9337, an improvement of approximately 0.3190, translating to nearly a 52% boost.

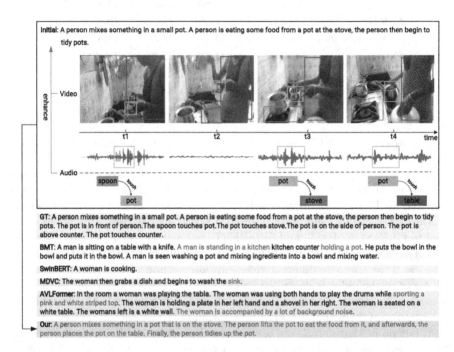

Fig. 2. In the example of qualitative results for video descriptions on our dataset video, the black box at the top of the figure represents our task to augment the initial caption based on visual and audio modal information. For each method, the parts of the video description that are close to the actual facts are marked in green. Additionally, the information from other modal enhancements in the video descriptions generated by our method is marked in red. (Color figure online)

Table 3. The Experimental Results for LP with Different Methods.

Model	Visual	Audio	Text	AUC	Micro	Macro
TeSAT	✓	✓	✓	0.6868	0.8708	0.5886
CENET	✓	✓	✓	0.7551	0.8520	0.6147
TiRGN	✓	✓	✓	0.7429	0.8144	0.5997
IMF	✓	✓	✓	0.4002	0.1831	0.1660
Ours	✓	✓	✓	**0.9389**	**0.9521**	**0.9337**

Overall, the table demonstrates the varying performance among models in handling multi-modal knowledge graph tasks. Our model particularly excels in the AUC and Macro-F1 metrics, which comprehensively assess class-balanced performance. This indicates that the new model performs better with small-sample categories, avoiding performance degradation caused by neglecting fine-grained information. IMF demonstrated subpar performance on multi-modal knowledge graphs, potentially due to its failure to fully consider the fine-grained information associated with each entity's multi-modal data, thereby introducing noise. Conversely, our model effectively mitigates this issue by meticulously extracting entity information from various modalities, such as recognizing a part of an image within a video frame as an entity, allowing for more precise capture of subtle features in multi-modal data and thus reducing noise interference. As a result, our model achieves superior performance.

In summary, our method exhibits outstanding performance in the experiments, especially showcasing significant improvements in measures of overall predictive capability and class-balanced performance. This notable enhancement primarily stems from the effective capture and utilization of fine-grained features in multi-modal data.

6 Conclusion

In conclusion, this paper introduces an innovative approach to multimodal data representation, with a particular emphasis on guiding visual knowledge representation using auditory cues. By integrating diverse sensory inputs and leveraging multimodal triples alongside commonsense knowledge, the proposed model constructs structured visual knowledge graphs that capture intricate relationships and contextual understanding, with a significant emphasis on incorporating auditory guidance. Through validation via multimodal tasks such as multimodal-guided visual captioning and link prediction, the effectiveness of the model in accurately interpreting and generating nuanced information across all modalities is demonstrated. This highlights its potential for advancing multimodal understanding in various application domains, particularly those where auditory cues play a crucial role.

Acknowledgement. We thank Jiang Liu and Hui Ji for their contributions to the organization of relevant experimental data and related work in this paper. We also acknowledge the support of Professor CHE Nan's and Professor LI Yuandi's research teams for their contributions to the construction of the VALM dataset.

References

1. Yang, Y., Zhuang, Y., Pan, Y.: Multiple knowledge representation for big data artificial intelligence: framework, applications, and case studies. Front. Inf. Technol. Electron. Eng. **22**(12), 1551–1558 (2021)
2. Pan, Y.: On visual knowledge. Front. Inf. Technol. Electron. Eng. **20**(8), 1021–1025 (2019)

3. Baltrušaitis, T., Ahuja, C., Morency, L.-P.: Multimodal machine learning: a survey and taxonomy. IEEE Trans. Pattern Anal. Mach. Intell. **41**(2), 423–443 (2018)
4. Ye, H., Li, G., Qi, Y., Wang, S., Huang, Q., Yang, M.H.: Hierarchical modular network for video captioning. In: Proceedings of the IEEE/CVF Conference on Computer Vision and Pattern Recognition, pp. 17939–17948 (2022)
5. Yan, L., et al.: Video captioning using global-local representation. IEEE Trans. Circuits Syst. Video Technol. **32**(10), 6642–6656 (2022)
6. Ryu, H., Kang, S., Kang, H., Yoo, C.D.: Semantic grouping network for video captioning. In: Proceedings of the AAAI Conference on Artificial Intelligence, vol. 35, pp. 2514–2522 (2021)
7. Iashin, V., Rahtu, E.: A better use of audio-visual cues: dense video captioning with bi-modal transformer. arXiv preprint arXiv:2005.08271 (2020)
8. Shen, X., et al.: Fine-grained audible video description. In: Proceedings of the IEEE/CVF Conference on Computer Vision and Pattern Recognition, pp. 10585–10596 (2023)
9. Kim, C.D., Kim, B., Lee, H., Kim, G.: Audiocaps: generating captions for audios in the wild. In: Proceedings of the 2019 Conference of the North American Chapter of the Association for Computational Linguistics: Human Language Technologies, Volume 1 (Long and Short Papers), pp. 119–132 (2019)
10. Drossos, K., Lipping, S., Virtanen, T.: Clotho: an audio captioning dataset. In: ICASSP 2020-2020 IEEE International Conference on Acoustics, Speech and Signal Processing (ICASSP), pp. 736–740. IEEE (2020)
11. Wu, Y., Chen, K., Zhang, T., Hui, Y., Berg-Kirkpatrick, T., Dubnov, S.: Large-scale contrastive language-audio pretraining with feature fusion and keyword-to-caption augmentation. In: ICASSP 2023-2023 IEEE International Conference on Acoustics, Speech and Signal Processing (ICASSP), pp. 1–5. IEEE (2023)
12. Mei, X., et al.: Wavcaps: a ChatGPT-assisted weakly-labelled audio captioning dataset for audio-language multimodal research. arXiv preprint arXiv:2303.17395 (2023)
13. Chen, D., Dolan, W.B.: Collecting highly parallel data for paraphrase evaluation. In: Proceedings of the 49th Annual Meeting of the Association for Computational Linguistics: Human Language Technologies, pp. 190–200 (2011)
14. Xu, J., Mei, T., Yao, T., Rui, Y.: MSR-VTT: a large video description dataset for bridging video and language. In: Proceedings of the IEEE Conference on Computer Vision and Pattern Recognition, pp. 5288–5296 (2016)
15. Torabi, A., Pal, C., Larochelle, H., Courville, A.: Using descriptive video services to create a large data source for video annotation research. arXiv preprint arXiv:1503.01070 (2015)
16. Rohrbach, A., Rohrbach, M., Tandon, N., Schiele, B.: A dataset for movie description. In: Proceedings of the IEEE Conference on Computer Vision and Pattern Recognition, pp. 3202–3212 (2015)
17. Das, P., Xu, C., Doell, R.F., Corso, J.J.: A thousand frames in just a few words: lingual description of videos through latent topics and sparse object stitching. In: Proceedings of the IEEE Conference on Computer Vision and Pattern Recognition, pp. 2634–2641 (2013)
18. Zhou, L., Xu, C., Corso, J.: Towards automatic learning of procedures from web instructional videos. In: Proceedings of the AAAI Conference on Artificial Intelligence, vol. 32 (2018)
19. Sigurdsson, G.A., Varol, G., Wang, X., Farhadi, A., Laptev, I., Gupta, A.: Hollywood in homes: crowdsourcing data collection for activity understanding. In: Com-

puter Vision–ECCV 2016: 14th European Conference, Amsterdam, The Netherlands, October 11–14, 2016, Proceedings, Part I 14, pp. 510–526. Springer (2016)
20. Ji, J., Krishna, R., Fei-Fei, L., Niebles, J.C.: Action genome: actions as compositions of spatio-temporal scene graphs. In: Proceedings of the IEEE/CVF Conference on Computer Vision and Pattern Recognition, pp. 10236–10247 (2020)
21. Gella, S., Lewis, M., Rohrbach, M.: A dataset for telling the stories of social media videos. In: Proceedings of the 2018 Conference on Empirical Methods in Natural Language Processing, pp. 968–974 (2018)
22. Lei, J., Yu, L., Berg, T.L., Bansal, M.: TVR: a large-scale dataset for video-subtitle moment retrieval. In: Computer Vision–ECCV 2020: 16th European Conference, Glasgow, UK, August 23–28, 2020, Proceedings, Part XXI 16, pp. 447–463. Springer (2020)
23. Monfort, M., et al.: Spoken moments: Learning joint audio-visual representations from video descriptions. In: Proceedings of the IEEE/CVF Conference on Computer Vision and Pattern Recognition, pp. 14871–14881 (2021)
24. Rohrbach, M., Amin, S., Andriluka, M., Schiele, B.: A database for fine grained activity detection of cooking activities. In: 2012 IEEE Conference on Computer Vision and Pattern Recognition, pp. 1194–1201. IEEE (2012)
25. Krishna, R., Hata, K., Ren, F., Fei-Fei, L., Niebles, R.C.: Dense-captioning events in videos. In: Proceedings of the IEEE International Conference on Computer Vision, pp. 706–715 (2017)
26. Wang, X., Wu, J., Chen, J., Li, L., Wang, Y.F., Wang, W.Y.: Vatex: a large-scale, high-quality multilingual dataset for video-and-language research. In: Proceedings of the IEEE/CVF International Conference on Computer Vision, pp. 4581–4591 (2019)
27. Miech, A., Zhukov, D., Alayrac, J.B., Tapaswi, M., Laptev, I., Sivic, J.: Howto100m: learning a text-video embedding by watching hundred million narrated video clips. In: Proceedings of the IEEE/CVF International Conference on Computer Vision, pp. 2630–2640 (2019)
28. Iashin, V., Rahtu, E.: Multi-modal dense video captioning. In: Proceedings of the IEEE/CVF Conference on Computer Vision and Pattern Recognition Workshops, pp. 958–959 (2020)
29. Simonyan, K., Zisserman, A.: Very deep convolutional networks for large-scale image recognition. arXiv preprint arXiv:1409.1556 (2014)
30. Pennington, J., Socher, R., Manning, C.D.: Glove: global vectors for word representation. In: Proceedings of the 2014 Conference on Empirical Methods in Natural Language Processing (EMNLP), pp. 1532–1543 (2014)
31. Tian, Y., Shi, J., Li, B., Duan, Z., Xu, C.: Audio-visual event localization in unconstrained videos. In: Proceedings of the European Conference on Computer Vision (ECCV), pp. 247–263 (2018)
32. Cong, Y., Liao, W., Ackermann, H., Rosenhahn, B., Yang, M.Y.: Spatial-temporal transformer for dynamic scene graph generation. In: Proceedings of the IEEE/CVF International Conference on Computer Vision, pp. 16372–16382 (2021)
33. Gong, Y., Chung, Y.A., Glass, J.: AST: audio spectrogram transformer. arXiv preprint arXiv:2104.01778 (2021)
34. Lin, K., et al.: Swinbert: end-to-end transformers with sparse attention for video captioning. In: Proceedings of the IEEE/CVF Conference on Computer Vision and Pattern Recognition, pp. 17949–17958 (2022)
35. Li, J., Su, X., Gao, G.: Teast: temporal knowledge graph embedding via archimedean spiral timeline. In: Proceedings of the 61st Annual Meeting of the

Association for Computational Linguistics (Volume 1: Long Papers), pp. 15460–15474 (2023)
36. Yi, X., Junjie, O., Hui, X., Luoyi, F.: Temporal knowledge graph reasoning with historical contrastive learning. In: Proceedings of the AAAI Conference on Artificial Intelligence, vol. 37, pp. 4765–4773 (2023)
37. Li, Y., Sun, S., Zhao, J.: Tirgn: time-guided recurrent graph network with local-global historical patterns for temporal knowledge graph reasoning. In: Proceedings of the Thirty-First International Joint Conference on Artificial Intelligence, IJCAI 2022, Vienna, Austria, 23–29 July 2022, pp. 2152–2158. ijcai. org (2022)
38. Papineni, K., Roukos, S., Ward, T., Zhu, W.J.: Bleu: a method for automatic evaluation of machine translation. In: Proceedings of the 40th Annual Meeting of the Association for Computational Linguistics, pp. 311–318 (2002)
39. Banerjee, S., Lavie, A.: Meteor: an automatic metric for MT evaluation with improved correlation with human judgments. In: Proceedings of the ACL Workshop on Intrinsic and Extrinsic Evaluation Measures for Machine Translation and/or Summarization, pp. 65–72 (2005)
40. Lin, C.Y., Och, F.J.: Automatic evaluation of machine translation quality using longest common subsequence and skip-bigram statistics. In: Proceedings of the 42nd Annual Meeting of the Association for Computational Linguistics (ACL-04), pp. 605–612 (2004)
41. Hershey, S., et al.: CNN architectures for large-scale audio classification. In: 2017 IEEE International Conference on Acoustics, Speech and Signal Processing (ICASSP), pp. 131–135. IEEE (2017)

Boundary Point Detection Combining Gravity and Outlier Detection Methods

Vijdan Khalique[1](✉), Hiroyuki Kitagawa[2], and Toshiyuki Amagasa[1]

[1] Center for Computational Sciences, University of Tsukuba, Tsukuba, Japan
khalique.vijdan@kde.cs.tsukuba.ac.jp, amagasa@cs.tsukuba.ac.jp
[2] International Institute for Integrative Sleep Medicine, University of Tsukuba, Tsukuba, Japan
kitagawa@cs.tsukuba.ac.jp

Abstract. Boundary point detection is the task of identifying points occurring at the boundary of a dense region in a dataset. It can reveal useful information about the system generating the data. Several data mining methods have been proposed to solve this problem. In our previous work, we proposed the Boundary Point Factor (BPF), which combined Gravity and an outlier detection method; Local Outlier Factor (LOF) to calculate the BPF score to detect the boundary points. The method is effective in a variety of real and synthetic datasets. However, one of the most crucial questions is whether other outlier detection methods can be used with Gravity for boundary point detection. In this work, we first investigate the favorable properties of LOF that make it suitable to be combined with Gravity for detecting boundary points. Next, the comparison of the commonly used outlier detection methods with the useful properties of LOF is performed on various datasets to demonstrate if these methods can be used for boundary point detection. Overall, it was found that, unlike LOF, a straightforward combination of other outlier detection methods with Gravity cannot be used, and sophisticated manipulation of the outlier scores generated by these methods may be needed to enable them to be used with Gravity for boundary point detection.

Keywords: Boundary point detection · Cluster boundary · Data mining

1 Introduction

In data mining, points occurring at the boundary of a cluster or dense region are often referred to as boundary points [15]. The detection of boundary points can be advantageous and has several applications such as data classification and mining [12], image processing [4], and disease diagnosis [11]. Overall, the boundary points represent the borderline or fringe behavior of the system generating the data. Similarly, the core points occurring near the cluster center or the outliers represent the common or abnormal behavior of the data generation process, respectively [3,6].

In the past, there have been many proposals for solving the problem of boundary point detection such as BORDER [15], BPDAD [11], BRIM [14], and BorderShift [4]. The problem with existing methods can be summarized as (a) the score calculated by some boundary point detection methods cannot discriminate boundary points from the outliers, (b) some methods require prior information or assumptions about the number of outliers in the dataset to tune the parameters, and (c) some methods do not perform well when the dataset contains one or more clusters of different shapes, sizes, or densities.

To address these problems, we proposed the Boundary Point Factor (BPF) method in our last works [8,9]. We introduced the concept of Gravity value which can be combined with the outlier score calculated by the outlier detection method Local Outlier Factor (LOF) to calculate the BPF score. The BPF score of boundary points tends to be larger than other points. We proposed two algorithms for detecting top-m boundary points from static and streaming data called *StaticBPF* and *StreamBPF*, respectively. The advantages of *BPF* can be given as (a) *BPF* can effectively discriminate boundary points from outliers and core points in datasets with clusters of different shapes, sizes, and densities, (b) it does not require prior assumptions about the data for tuning its parameter, and (c) it is easy to use with the LOF method as BPF and LOF can share the same parameter value.

Employing LOF in conjunction with Gravity for boundary point detection prompts an interesting and crucial question: *Can other outlier detection methods be employed with Gravity for identifying boundary points?* This question further probes the detailed properties of the LOF score that make it suitable for use with Gravity for boundary point detection. Therefore, in this paper, we first show the important properties of LOF scores. After that, the properties of the outlier scores calculated by other outlier detection methods are compared with the properties of LOF and we show if they can be used with Gravity for boundary point detection.

This study empirically demonstrates the characteristics of LOF scores across diverse datasets. Also, we test the effectiveness of boundary point detection by combining other outlier detection methods and Gravity. The key contributions of this work are given as follows:

1. Study of the favorable characteristics of the LOF score useful for combining it with Gravity for boundary point detection.
2. Empirical demonstration and evaluation of combining other outlier detection methods (DBOD, FastABOD, IF, and HBOS) with Gravity for boundary point detection.

2 Related Work

In the past many outlier detection methods have been proposed that exploit various data properties to calculate the outlier score to identify the outliers. The outlier detection methods can be classified according to the data properties they exploit such as distance, angle, density, isolation, and statistics [2,5]. Some of the representative methods belonging to these categories are DBOD [1], FastABOD [10], LOF [3], IF [13] and HBOS [7], respectively.

Limited attention has been devoted to boundary point detection. BORDER [15] exploits the observation that boundary points have less number of reverse neighbors than core points. However, BORDER [15] cannot discriminate outliers from the boundary points. BPDAD [11] has a similar problem as BORDER. BRIM [14] addresses this problem by considering the neighborhood in a given radius where the boundary points tend to have dense regions in a specific direction. However, BRIM cannot work on datasets with different clusters of shapes and densities. BorderShift [4], on the other hand, requires preliminary information about the data for tuning its parameters. BPF [8,9] overcomes these shortcomings by introducing Gravity and using it with LOF [3] to calculate the BPF score. BPF showed reasonable accuracy in boundary point detection on various real and synthetic datasets. However, it is important to study whether we can replace LOF with other outlier detection methods to combine with Gravity for boundary point detection. This research attempts to answer this interesting question.

3 Preliminaries

This section highlights the important definitions of Gravity, Local Outlier Factor, and Boundary Point Factor.

3.1 Local Outlier Factor (LOF)

LOF [3] is a density-based outlier detection method that compares the relative density of the target point with its k-nearest neighbors. Given a point p in the dataset D, let $N_k(p)$ represent the set of k-nearest neighbors of p. The LOF score of p ($LOF(p)$) can be calculated as

$$LOF(p) = \frac{\sum_{o \in N_k(p)} \frac{lrd_k(o)}{lrd_k(p)}}{|N_k(p)|}. \tag{1}$$

$lrd_k(p)$ in the above equation represents the local reachability distance which can be calculated as $lrd_k(p) = 1/(\frac{\sum_{o \in N_k(p)} reach\text{-}dist_k(p,o)}{|N_k(p)|})$, where $reach\text{-}dist_k(p,o) = max\{k\text{-}distance(o), d(p,o)\}$, $d(p,o)$ represents the distance between p and o and $k\text{-}distance(o)$ is the distance between o and its kth neighbor. The LOF scores of outliers are greater than 1, whereas core and boundary points have LOF scores close to 1.

3.2 Gravity

Intuitively, Gravity measures the distribution of the neighbors of the point. If the neighbors of a point is distributed in all directions, the Gravity value is expected to be low for such a point. Similarly, the Gravity value is expected to be high for a point whose neighbors are concentrated in a specific direction. The following equation defines the Gravity of a point p

$$G(p) = \frac{1}{|N_k(p)|} \left\| \sum_{o \in N_k(p)} \frac{\vec{po}}{\|\vec{po}\|} \right\|, \qquad (2)$$

where $\vec{po}$ is the vector from point p to o and $\|.\|$ is the vector norm.

The core points may have their neighbors distributed in all directions, and therefore they may have smaller Gravity values. On the other hand, the boundary points may have the nearest neighbors in a specific direction making their Gravity values larger than core points. The Gravity values of outliers depend on the distribution of points in the dataset.

3.3 Boundary Point Factor (BPF)

The Boundary Point Factor (BPF) combines the Gravity value and the LOF score as shown in the following equation to calculate the BPF score

$$BPF(p) = \frac{G(p)}{LOF(p)}. \qquad (3)$$

The boundary points tend to have larger BPF scores than the core points and outliers.

Table 1. Description of 2- and high dimensional datasets.

Dataset name	Data size (n)	#outliers	#boundary (m)
Gaussian	2100	114	234
Uniform	1600	68	252
Diamonds	3300	220	528
Rings	4200	172	567
Mix1	3800	262	838
Mix3	1800	127	484
10-d 1-Gaussian cluster	2200	157	597
10-d 3-Gaussian clusters	3300	372	939
20-d 3-Gaussian clusters	3300	401	997
50-d 3-Gaussian clusters	3300	400	1140

4 Results and Discussion

In this section, we demonstrate experimentally and present the discussion about the properties of LOF and combining outlier detection methods with Gravity for boundary point detection. Table 1 enlists the datasets used in the experiments. All of these datasets have been used in the evaluation of our previous works [8,9] except Gaussian, Uniform, and 10-d 1-Gaussian cluster datasets. Readers may refer to [9] for the details about obtaining the ground truth boundary points and outliers from these datasets, and the accuracy evaluation of *BPF*.

Firstly, the properties of LOF that make it usable with Gravity for boundary point detection are demonstrated experimentally. Secondly, based on the highlighted properties of LOF, other outlier detection methods are analyzed and subsequently, the experiment results of boundary point detection combining them with the Gravity are shown in Fig. 4 and Table 3.

4.1 Properties of LOF Score

In this section, important properties of LOF useful for boundary point detection with Gravity are demonstrated experimentally on the datasets shown in Table 1. In the pre-processing step, the data points in these datasets are arranged as core points, boundary points, and outliers according to the ground truth, and then each point is assigned a unique point ID. After that, the LOF scores, Gravity values, and BPF scores were calculated according to their respective definitions given in Sect. 3.

According to [3], the inliers (core and boundary points) have LOF scores close to 1, and the outliers have LOF scores much greater than 1. To show this, we first plot the distribution of LOF scores on 2- and high-dimensional synthetic datasets using histograms as shown in Figs. 1 and 2. The frequency axis is shown in the log scale to clearly show the bin heights of the core, boundary, and outlier points. It can be observed that typically the core and boundary points are assigned LOF scores close to 1, and the outliers are assigned LOF scores larger than 1.

Additionally, we plot the distribution of LOF, Gravity, and BPF scores of core, boundary, and outlier points in Gaussian and Uniform datasets as shown in Fig. 3. Although, similar Gravity, LOF, and BPF score distributions were observed for other 2- and high-dimensional datasets, we do not include the plots for other datasets due to space limitations.

Consider the LOF score distribution shown in Fig. 3. It can be seen that the LOF scores of core (green) points and boundary (red) points are close to 1, and their LOF scores are similar. However, the boundary points farther away from the dense region of the Gaussian or Uniform cluster may have relatively larger LOF scores than core points. Still, their LOF scores are smaller than the outliers. In contrast, the outliers have much larger LOF scores than boundary and core points. A similar LOF score distribution was observed in other datasets. Based on these observations, the useful properties of LOF scores can be summa-

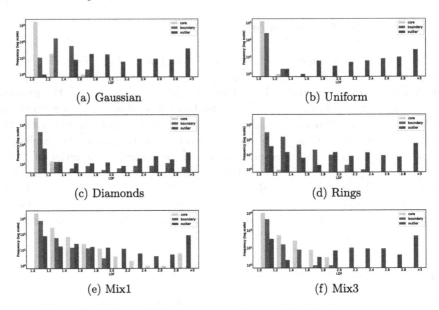

Fig. 1. Histograms showing the LOF scores in 2-d datasets.

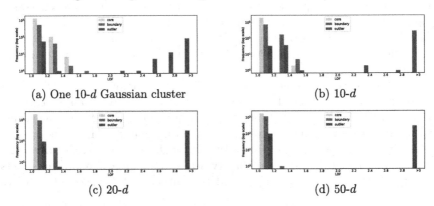

Fig. 2. Histograms showing the LOF scores in high-dimensional datasets.

rized as 1) the LOF scores of core and boundary points are similar and close to 1, and 2) the LOF score of outliers is relatively larger ($\gg 1$) than other points.

Similarly, the Gravity values of data points are also shown in these figures. Generally, the distribution of Gravity values on various datasets shows that the Gravity value of boundary points and outliers are larger than core points. Referring to Fig. 3, the Gravity value of the core points is generally small. It can be seen that some core points have Gravity values similar to boundary points because they are located close to the boundary region. On the other hand, the boundary points have Gravity values larger than or similar to core points and smaller than or similar to the outliers. However, isolated outliers in

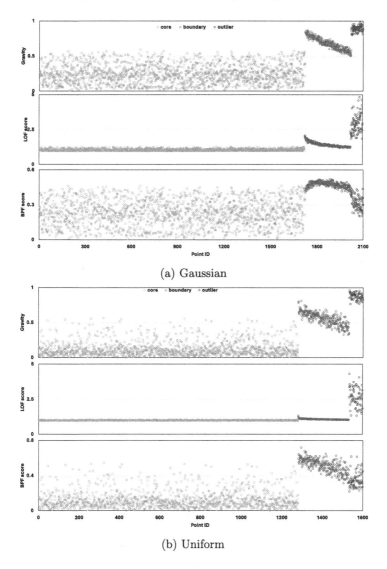

Fig. 3. Distribution of Gravity values, LOF scores, and BPF scores on 2-d datasets - Gaussian and Uniform obtained at $k = 100$.

the extreme corners have larger Gravity values as their nearest neighbors are in a specific direction. A similar trend was observed on other 2- and high-dimensional datasets.

Overall, the Gravity value is within $[0, 1]$ whereas the LOF score can take any value. Based on these properties of Gravity value and LOF score w.r.t. core points, outliers, and boundary points, calculating the BPF score as the ratio of Gravity value and LOF score ($BPF(p) = G(p)/LOF(p)$) results in smaller BPF scores of core points and outliers, and the BPF scores of boundary points are relatively large. The reasons can be summarized as follows.

1. Core points have LOF scores close to 1 and smaller Gravity values resulting in smaller BPF scores.
2. Outliers have relatively larger LOF scores ($\gg 1$), and Gravity values are similar to boundary points, therefore resulting in smaller BPF scores.
3. Boundary points have Gravity values larger than the core points and smaller than or similar to outliers, and LOF scores close to 1 (≈ 1) resulting in larger BPF scores than other points.

These properties and observations are expected to hold for a wide range of datasets. Based on the observations highlighted in this section, the next section demonstrates whether other outlier detection methods can be combined with Gravity for boundary point detection.

4.2 Combining Other Outlier Detection Methods with Gravity

As briefly mentioned in Sect. 2, several outlier detection methods calculate the outlier scores based on the properties of data. Therefore, the characteristics of the scores calculated by these methods may differ.

We analyze if other outlier detection methods have similar properties as LOF scores to check whether the outlier score calculated by them can be used with Gravity for boundary point detection. In our understanding, this is a reasonable and straightforward approach as outlier detection methods that similarly assign outlier scores as LOF can be effective for boundary point detection when combined with Gravity. For simplicity, we use 2- and 10-d 1-Gaussian datasets for analysis of outlier scores.

Firstly, the definitions of the outlier scores calculated by some of the commonly used outlier detection methods are presented. These outlier detection methods are (i) Distance-based outlier detection (*DBOD*), (ii) Fast angle-based outlier detection (*FastABOD*), (iii) Isolation forest (*IF*), and (iv) Histogram-based outlier score (*HBOS*). Following this, the definitions of the boundary scores calculated by combining the outlier scores with the Gravity values are given. Secondly, the quantitative and qualitative experiment results of boundary point detection are shown on 2- and high-dimensional datasets. In the qualitative results, the top-m boundary points detected by these methods are shown. The quantitative results are given as precision, the area under ROC curve (AUCROC), and the area under precision-recall curve (AUCPR). The parameter values of the outlier detection methods are set according to the recommendations given in their respective papers. For the Gravity, $k = 100$ is used until mentioned otherwise. Thirdly, the accuracy results are discussed by comparing outlier scores by other outlier detection methods with the LOF score. The following approach is applied to explain if an outlier detection method is useful with Gravity for boundary detection:

1. The outlier scores of all outlier detection methods are compared with LOF scores to check if they follow similar properties as LOF. The outlier scores by the other outlier detection methods (on the y-axis) and LOF scores (on the x-axis) are plotted as a scatter plot. Also, the core, boundary, and outlier points are assigned green, blue, and red colors, respectively, for comparison.
2. To demonstrate the difference between the outlier scores of outliers from the inliers (core and boundary points), the statistical z-score of outlier scores is calculated as $z = \frac{x-\mu}{\sigma}$, where x = raw outlier score, μ = mean, and σ = standard deviation. As LOF scores of outliers are much larger than the inliers, an outlier detection method exhibiting a similar property can be combined with Gravity for boundary point detection.

Table 2. Average z-score of outlier scores in 2-d and 10-d 1-Gaussian datasets.

Methods	Top x% outliers									
	2-d					10-d				
	1%	2%	3%	4%	5%	1%	2%	3%	4%	5%
LOF	6.59	5.64	4.89	4.22	3.68	5.47	5.15	4.88	4.61	4.09
kDist	6.07	5.29	4.65	4.06	3.58	5.4	5.06	4.79	4.52	4.02
AvgDist	6.28	5.45	4.77	4.15	3.64	5.42	5.08	4.80	4.53	4.03
FastABOD	3.98	2.14	1.46	1.10	0.89	3.24	2.98	2.81	2.67	2.77
IF	4.47	4.11	3.77	3.49	3.22	4.92	4.66	4.47	4.29	3.89
HBOS	4.98	4.42	3.95	3.58	3.29	5.28	4.89	4.64	4.38	3.9

Table 2 shows the average z-scores of the top-x% outliers having the highest LOF scores in the Gaussian datasets. It can be seen that the average z-score is large when considering the top 1% points with the highest LOF scores. This is because the difference in LOF scores of the top 1% outliers from the average LOF score is much larger. Similarly, as more top-x% points are considered, the overall average z-score drops as more points with smaller LOF scores are included in the top-x% list of outliers.

For comparison, we show the top-m boundary points detected by *BPF* in Fig. 4 with the boundary point detected by outlier detection methods combined with Gravity and the corresponding quantitative accuracy in Table 3.

Distance-Based Outlier Detection (*DBOD*). In *DBOD*, the outlier score can be calculated in two ways: i) *kDist* and ii) *AvgDist*. Given a point p, the boundary score can be calculated as $BoundaryScore(p) = \frac{G(p)}{DBOD(p)}$, where $DBOD(p)$ can be defined as the distance of p to its kth nearest neighbor ($kDist(p)$) or average distance of p with its k-nearest neighbors $AvgDist(p)$. The boundary point detection results of using *kDist* and *AvgDist* methods are shown in Table 3. Figure 4 visually shows the top-m boundary points detected by combining $kDist$ and Gravity at $k = 100$. We do not include the qualitative results of $AvgDist$ and $AvgDist$ vs LOF plots as they are almost the same as $kDist$.

Consider the accuracy of $kDist$ on Gaussian and Uniform datasets in Fig. 4. It can be observed that the boundary score can effectively detect boundaries in Uniform. On the other hand, a limited number of boundary points with low accuracy are detected in Gaussian at $m = 400$ and $k = 250$. In the more complex 2-d datasets like Diamonds, Rings, Mix1, and Mix3 in Fig. 4, the boundary score using $kDist$ and Gravity can detect the boundary points in the Diamonds dataset and the dense clusters of Mix1 and Mix3 dataset. However, it cannot detect many boundary points in the high-dimensional datasets and therefore has a low accuracy. This behavior can be attributed to the fundamental problem of DBOD's inability to detect outliers in datasets with clusters of different densities or sizes [3].

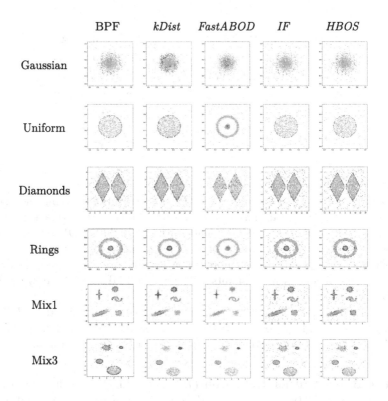

Fig. 4. Top-m boundary points detected by all methods.

Figure 5 shows the similarity between LOF score and scores by $kDist$. The points are plotted according to their $kDist$ outlier scores against the LOF scores.

Furthermore, LOF scores are marked on the x-axis, and outlier scores by $kDist$ are marked on the y-axis to show the distribution of outlier scores of individual methods. $kDist$ assign much higher outlier scores to the outliers than the inliers similar to LOF in both 2- and 10-d data. Figure 5 shows that $kDist$ and LOF assign outlier scores in a similar pattern to all points. Also, the z-scores of $kDist$ are similar to LOF as shown in Table 2. This shows that $kDist$ follows similar outlier scoring properties as LOF. Therefore, the detection of boundary points in the Diamonds dataset and the dense clusters of the other datasets was observed. Overall, $kDist$ can be used with Gravity for boundary point detection on datasets with one or more clusters of similar sizes with uniformly distributed points, whereas it may not perform well on datasets with clusters of different sizes and densities. The similar comments are applicable on $AvgDist$.

Table 3. Quantitative boundary point detection accuracy of all methods.

Methods		Gau.	Uni.	Dia.	Rin.	Mix1	Mix3	10d 1-Gau.	10d	20d	50d
BPF	Prec.	0.78	0.85	0.75	0.79	0.72	0.84	0.87	0.8	0.74	0.84
	ROC	0.98	0.99	0.96	0.97	0.91	0.96	0.97	0.94	0.83	0.94
	PR	0.83	0.92	0.75	0.71	0.73	0.82	0.93	0.8	0.83	0.78
kDist	Prec.	0.33	0.62	0.64	0.0	0.11	0.51	0.36	0.44	0.36	0.42
	ROC	0.65	0.92	0.91	0.32	0.39	0.75	0.61	0.67	0.55	0.65
	PR	0.24	0.52	0.59	0.09	0.17	0.5	0.33	0.44	0.34	0.45
AvgDist	Prec.	0.14	0.59	0.62	0	0.06	0.51	0.35	0.43	0.35	0.42
	ROC	0.56	0.89	0.9	0.33	0.36	0.75	0.6	0.65	0.54	0.63
	PR	0.14	0.53	0.52	0.09	0.16	0.51	0.32	0.39	0.32	0.43
FastABOD	Prec.	0	0.28	0.32	0.04	0.09	0.23	0	0.16	0.17	0.37
	ROC	0.1	0.56	0.74	0.34	0.32	0.54	0.1	0.46	0.44	0.47
	PR	0.06	0.23	0.32	0.09	0.16	0.26	0.16	0.25	0.25	0.30
IF	Prec.	0.58	0.68	0.19	0.12	0.27	0.26	0.79	0.48	0.46	0.47
	ROC	0.95	0.96	0.55	0.52	0.56	0.57	0.94	0.71	0.70	0.67
	PR	0.62	0.67	0.18	0.13	0.24	0.31	0.75	0.45	0.50	0.45
HBOS	Prec.	0.55	0.68	0.18	0.09	0.14	0.33	0.27	0.16	0.35	0.22
	ROC	0.95	0.95	0.54	0.48	0.41	0.63	0.51	0.28	0.43	0.32
	PR	0.57	0.64	0.18	0.12	0.18	0.37	0.28	0.19	0.32	0.26

Fast Angle-Based Outlier Detection (*FastABOD*). *FastABOD* is an angle-based outlier detection method that calculates the variance of distance weighted angle of a target point with all points in its k-nearest neighborhood [10]. The outliers have smaller outlier factors than other points. Given a point p and its k-nearest neighbor $N_k(p)$, *FastABOD* calculates the outlier score as

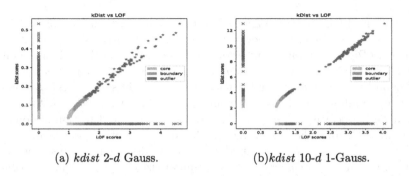

(a) *kdist* 2-*d* Gauss. (b) *kdist* 10-*d* 1-Gauss.

Fig. 5. Comparison of outlier scores by *kDist* with LOF.

$FastABOD(p) = VAR_{a,b \in N_k(p)} \left(\frac{1}{||\vec{pa}||\cdot||\vec{pb}||} \cdot \frac{\langle \vec{pa}, \vec{pb} \rangle}{||\vec{pa}||\cdot||\vec{pb}||} \right)$, where VAR represents statistical variance and $\langle . \rangle$ represents the dot product of two vectors. For simplicity of calculating boundary scores, the *FastABOD* scores are inversed (reciprocal) so that outliers have larger scores than other points. Hence, the boundary score of a point p can be calculated as $BoundaryScore(p) = G(p) \times FastABOD(p)$. The top-$m$ boundary points detected by this boundary score using $k = 100$ are shown in Fig. 4 and the accuracy is given in Table 3. The overall results suggest that a straightforward combination of *FastABOD* and Gravity cannot work for boundary point detection.

Figure 6 compares the outlier scores by *FastABOD* with LOF on 2- and 10-*d* 1-Gaussian datasets. It indicates a large fluctuation between the outlier scores of the outliers and other points. Comparing the z-score of top 1% outliers with the z-score of outliers below 1% also suggests that the fluctuation between the outlier scores of outliers is large. As shown in Fig. 6, the outlier scores of the core and boundary points are very close to 0, unlike the LOF scores which are close to 1. This results in large boundary scores of the core and boundary points. On the other hand, the boundary scores of the outliers are small because the outlier scores of the outliers are large. It can be further observed that there is a large fluctuation of outlier scores within the outliers as isolated outliers are assigned larger outlier scores. Table 3 and Fig. 4 show the resulting boundary point detection accuracy. Overall, the outlier scores by *FastABOD* are unstable, and a straightforward application of *FastABOD* and Gravity does not work for boundary point detection. Therefore, *FastABOD* scores may require more sophisticated manipulation which is not the scope of this research.

Isolation Forest (*IF*). Isolation Forest (*IF*) calculates the path length of a target point in the ensemble of t trees (*iTrees*) called *iForest* generated by randomly partitioning random samples of ψ points from the dataset where each node of *iTree* represents a random partition [13]. *IF* calculates the outlier score of a given point as $IF(p) = 2^{-\frac{E(h(p))}{c(n)}}$, where $h(p)$ is the path length of p in an *iTree*, $E(h(p))$ is the average path length, $c(n) = 2H(n-1) - (2(n-1)/n)$ and

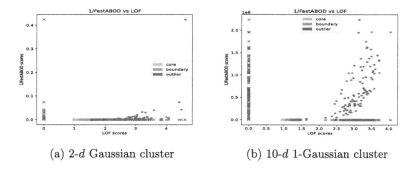

Fig. 6. Comparison of outlier score by $1/FastABOD$ and LOF.

$H(i)$ is the harmonic number which can be estimated by $ln(i) + 0.5772156649$ (Euler's constant). The *IF* method assigns a larger score to the outliers (usually > 0.5), and the range of outlier scores is [0,1]. The boundary score of a target point p can be calculated as $BoundaryScore(p) = \frac{G(p)}{IF(p)}$.

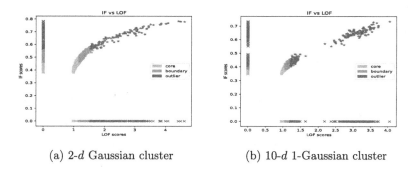

Fig. 7. Comparison of outlier score by *IF* and LOF.

Figure 4 shows the top-m boundary points detected by combining *IF* with Gravity, and the corresponding accuracy is given in Table 3. The number of trees $t = 100$, and sample size $\psi = 256$ for *IF* are set as recommended in [13], and $k = 100$ for Gravity. The results suggest that the calculated boundary score is less effective as outliers and boundary points cannot be discriminated.

Table 2 shows that the difference between the outlier scores of the outliers and the average outlier score is not as large compared with LOF. Furthermore, there is less similarity between the LOF scores and outlier scores by *IF* as shown in Fig. 7. Hence, *IF* does not follow the outlier scoring properties similar to LOF.

Histogram-Based Outlier Score (*HBOS*). Histogram-based outlier score (*HBOS*) is a statistical outlier detection method that constructs a histogram

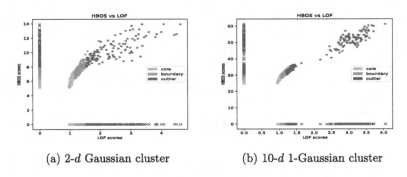

(a) 2-d Gaussian cluster (b) 10-d 1-Gaussian cluster

Fig. 8. Comparison of outlier score by *HBOS* and LOF.

for each dimension with a fixed number of bins. The outliers are expected to belong to bins with smaller heights. *HBOS* calculates the outlier score of all points in a dataset and assigns larger outlier scores to outliers than inliers. *HBOS* calculates the outlier score of given point p in a d-dimensional dataset as $HBOS(p) = \sum_{i=1}^{d} log(\frac{1}{hist_i(p)})$, where $hist_i(p)$ represent the height of histogram containing p. Hence, the boundary score of a given point p can be calculated as $BoundaryScore(p) = \frac{G(p)}{HBOS(p)}$.

In the experiments, the number of bins is set as 50. The top-m boundary points and the accuracy are in Fig. 4 and Table 3, respectively. The results show that the boundary score by using Gravity and *HBOS* cannot discriminate boundary points from the outliers. As a result, the top-m detected points contain outliers and boundary points resulting in low accuracy. The average z-scores in Tables 2 suggests that the difference of outlier scores of top outliers is not as large as LOF. Moreover, Fig. 8 shows that the outlier scores by *HBOS* are very different from LOF scores, especially in the case of outliers. Also, the range of outlier scores of core and boundary points fluctuates depending on the dataset, unlike LOF which assigns LOF scores to core and boundary points are close to 1.

5 Conclusion

This work addressed the question of whether the boundary point score calculated by combining Gravity with outlier detection methods (such as *DBOD*, *FastA-BOD*, *IF*, and *HBOS*) can effectively detect boundary points like *BPF*. *BPF* can effectively detect boundary points using LOF and Gravity which leads to exploring the possibility of replacing LOF with the other outlier detection methods. In the experiments, we first showed the properties of LOF useful for boundary point detection with Gravity. The core and boundary points had similar LOF scores close to 1 whereas outliers had much larger scores than 1. LOF when combined with Gravity gives a larger boundary point score to boundary points than others, and therefore made LOF suitable. Consequently, outlier detection methods exhibiting similar properties may be used with Gravity for boundary point detection. In the experiments, *kDist* and *AvgDist* were effective on the datasets with clusters of uniform densities and similar sizes. *FastABOD*, *IF*, and *HBOS* did not exhibit properties similar to LOF and cannot be straightforwardly combined with Gravity like LOF. They may require more sophisticated score manipulation to make them useful for boundary point detection with Gravity.

Acknowledgement. This work was partly supported by JSPS KAKENHI Grant Numbers JP23K28089, JP22K19802 and JP23K24949, JST CREST Grant Number JP-MJCR22M2, AMED Grant Number JP21zf0127005, and "Research and Development Project of the Enhanced infrastructures for Post-5G Information and Communication Systems" (JPNP20017), commissioned by the New Energy and Industrial Technology Development Organization (NEDO).

Disclosure of Interests. The authors have no competing interests to declare that are relevant to the content of this article.

References

1. Angiulli, F., Pizzuti, C.: Outlier mining in large high-dimensional data sets. IEEE Trans. Knowl. Data Eng. **17**(2), 203–215 (2005)
2. Boukerche, A., Zheng, L., Alfandi, O.: Outlier detection: methods, models, and classification. ACM Comput. Surv. (CSUR) **53**(3), 1–37 (2020)
3. Breunig, M.M., Kriegel, H.P., Ng, R.T., Sander, J.: LOF: identifying density-based local outliers. In: Proceedings of the 2000 ACM SIGMOD International Conference on Management of Data, pp. 93–104 (2000)
4. Cao, X., Qiu, B., Xu, G.: Bordershift: toward optimal meanshift vector for cluster boundary detection in high-dimensional data. Pattern Anal. Appl. **22**(3), 1015–1027 (2019)
5. Chandola, V., Banerjee, A., Kumar, V.: Anomaly detection: a survey. ACM Comput. Surv. (CSUR) **41**(3), 1–58 (2009)
6. Ester, M., Kriegel, H.P., Sander, J., Xu, X.: A density-based algorithm for discovering clusters in large spatial databases with noise. In: KDD, vol. 96, pp. 226–231 (1996)
7. Goldstein, M., Dengel, A.: Histogram-based outlier score (HBOS): a fast unsupervised anomaly detection algorithm. In: KI-2012: Poster and Demo Track, vol. 1, pp. 59–63 (2012)
8. Khalique, V., Kitagawa, H.: BPF: an effective cluster boundary points detection technique. In: International Conference on Database and Expert Systems Applications, pp. 404–416. Springer (2022)
9. Khalique, V., Kitagawa, H., Amagasa, T.: BPF: a novel cluster boundary points detection method for static and streaming data. Knowl. Inf. Syst. **65**(7), 2991–3022 (2023)
10. Kriegel, H.P., Schubert, M., Zimek, A.: Angle-based outlier detection in high-dimensional data. In: Proceedings of the 14th ACM SIGKDD International Conference on Knowledge Discovery and Data Mining, pp. 444–452 (2008)
11. Li, X., Wu, X., Lv, J., He, J., Gou, J., Li, M.: Automatic detection of boundary points based on local geometrical measures. Soft. Comput. **22**(11), 3663–3674 (2018)

12. Liang, Q., Zhang, J., Bah, M.J., Li, H., Chang, L., Kiran, R.U.: Effective and robust boundary-based outlier detection using generative adversarial networks. In: Strauss, C., Cuzzocrea, A., Kotsis, G., Tjoa, A.M., Khalil, I. (eds.) Database and Expert Systems Applications, pp. 174–187. Springer International Publishing, Cham (2022)
13. Liu, F.T., Ting, K.M., Zhou, Z.H.: Isolation forest. In: 2008 Eighth IEEE International Conference on Data Mining, pp. 413–422. IEEE (2008)
14. Qiu, B.Z., Yue, F., Shen, J.Y.: Brim: an efficient boundary points detecting algorithm. In: Pacific-Asia Conference on Knowledge Discovery and Data Mining, pp. 761–768. Springer (2007)
15. Xia, C., Hsu, W., Lee, M.L., Ooi, B.C.: Border: efficient computation of boundary points. IEEE Trans. Knowl. Data Eng. **18**(3), 289–303 (2006)

A Meta-learning Approach for Category-Aware Sequential Recommendation on POIs

Jia-Ling Koh[✉] [iD], Po-Jen Wen, and Wei Lai

National Taiwan Normal University, Taipei, Taiwan
jlkoh@csie.ntnu.edu.tw

Abstract. The goal of sequential recommendation is to gain valuable insights from previous interactions between users and items to predict the next item that the user maybe interest. In this research, an enhancement of the Meta Transitional Learning (*MetaTL*) framework is introduced, known as the Category-Aware Transitional Meta Learner (*CAT-ML*). The *CAT-ML* model combines a category-level transition meta-learner and an item-level transition meta-learner. By utilizing the category-level transition meta-learner, the proposed model effectively captures user behavior patterns by initially acquiring general features from behaviors at the category level. Subsequently, the feature representation obtained from category transitions is inputted into the item-level transition meta-learner, where an attention mechanism is employed to guide the extraction of behavior features from interactions at the item level. The experiments conducted on the Foursquare Check-in Dataset demonstrate that the *CAT-ML* model outperforms the *MetaTL* model, exhibiting improvements of 10.2% in top one item hit rate and 23.8% in category hit rate. Notably, the *CAT-ML* model demonstrates superior performance in scenarios involving cold-start users or new items in user history behavior, surpassing the *MetaTL* model by a significant margin.

Keywords: sequential recommendation · category-aware · meta-learning

1 Introduction

The Internet's rapid development has transformed people's lives, with various online platforms recording user actions for behavioral analysis. Predicting users' interests through sequential recommendation based on past interactions can save time and promote relevant items for enhancing user experience. The user's behavior sequence includes user and item IDs as input features for model training. Sequential recommendation models learn hidden representations from this sequence to predict the user's interest in candidate items for the next interaction. Previous sequential recommendation systems often required users to have a long sequence of interactions and used user ID for learning personoized embedding vectors. However, in real application, the sequential recommendation faced challenges in data sparsity and cold-start scenarios, thereby limiting its prediction effectiveness.

To overcome this challenge, incorporating item category features is suggested as a viable solution [2]. Item categories provide higher-level semantic information about interactions, conveying more comprehensive behavioral meanings. Moreover, category data tends to be denser than item data, enabling the system to better analyze user preferences and make accurate predictions regarding their next interactive item. In [10], a meta-learning framework tackled cold-start challenges in sequential recommendation using a transition model to learn item-to-item patterns forming features of user behavior. Findings showed that learning transition patterns between items was more transferable with limited data than learning sequential patterns.

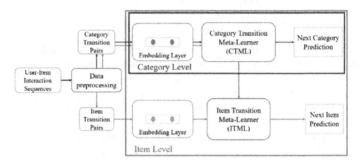

Fig. 1. The system architecture of the *CAT-ML* model.

This paper presents the *CAT-ML* model, a sequential recommendation system inspired by [2] and [10], which incorporates meta-learning techniques and considers item categories. The *CAT-ML* model consists of three main components: data preprocessing, category-level prediction, and item-level prediction (see Fig. 1). Initially, user-item interaction sequences are transformed into transition pairs of categories and items. The model learns category transition patterns and generates corresponding category-level transition vectors by a Category Transitional Meta-Learner (CTML). The CTML is trained with the goal to predict the next interaction category, relying on obtaining a representation vector of the user's historical category transition patterns. Moreover, the Item Transitional Meta-Learner (ITML) generates item-level transition vectors for continuous item pairs in an interaction sequence. The representation vector of a user obtained from CTML is utilized to influence the weights for aggregating the item transition vectors learned from ITML using an attention mechanism. For each user, the embedding vector of their most recent interacted item is added to the aggregated item transition vector, resulting in the representation vector for the next item. Subsequently, the model predicts the item that exhibits the highest similarity to this representation vector as the next recommended item.

The task of recommending the next point-of-interest (POI) is a type of sequential recommendation. Past research has focused on creating an encoder that incorporates both spatial and temporal information [15]. However, we believe that the transitions between consecutive points of interest (POIs) reflect user movement behaviors, which often carry valuable user intentions. In our study, we will evaluate the performance of the proposed *CAT-ML* model for next point-of-interest (POI) recommendation without utilizing private information such as location and time data of interactions.

The contributions of this study are as follows:

- We present the Category-Aware Transitional Meta Learner (*CAT-ML*), an extended version of the meta-learning framework of sequential recommendation to incorporate item category information into the training process.
- The experiments conducted on the Foursquare Check-in Dataset demonstrate the effectiveness of *CAT-ML* in cold-start POI sequential recommendation. The model shows significant improvements over the baseline model when recommending items or categories that users have not interacted with before.

2 Related Works

Recently, deep learning has increasingly been utilized in the development of sequential recommendation systems. One notable example is the Neural Collaborative Filtering (NCF), which employs a neural network to amalgamate hidden features of users and items, learned from matrix factorization. It then uses a multi-layer perceptron layer to forecast user ratings for items.

Different neural network architectures are used in sequential recommendation systems. Recurrent neural networks (RNNs), designed for sequential data, are employed to capture the chronological behavioral patterns of user interaction sequences. The GRU4Rec model [5] uses gated recurrent units (GRU) to encode these sequences, enabling prediction of the likelihood of the next interaction item.

Convolutional neural networks (CNNs) learn convolutional kernels to extract features. They are used in sequential recommendation models [8, 9, 14] to recognize feature patterns within the kernel's coverage range. The Caser model [8] is a significant example of a CNN-based sequential recommendation system.

Graph Neural Networks (GNNs) in sequential recommendation systems consider not only the sequence of interaction items but also indirect relationships within user-item interactions. GNN-based models construct a graph to represent relationships among users and items and use attention mechanisms within this graph to learn data representations. This approach enhances the learning of item and user representations by leveraging useful information from indirect interactions [7, 12].

Although various neural network models can be employed to extract user preferences and behavior characteristics, a model with a complex structure typically demands sufficient historical data of the user for personalization, leading to challenges in cold-start recommendation scenarios.

Meta-learning refers to the utilization of knowledge and skills gained from past learning experiences across diverse tasks to enable efficient adaptation to new tasks. A commonly used meta-learning algorithm known as Model-Agnostic Meta-Learning (MAML), which was introduced in [4], aims to optimize the initial parameters of a model architecture for application in different tasks. Through iterative training on multiple tasks and adjustment via gradient descent using the average loss function, the optimized parameters facilitate the creation of task-specific models even with limited training data.

A meta-learning-based recommender system, called MeLU [6], was first proposed by modifying the Model-Agnostic Meta-Learning (MAML) algorithm to estimate new user's preferences with a few consumed items. [11] introduced MetaCF, a meta-learning approach designed for recommendation systems, specifically targeting cold-start collaborative filtering challenges. MetaCF aims to learn a general model capable of handling new user scenarios with limited interactions. During the training phase, MetaCF employs a dynamic subgraph sampling technique centered around existing users to construct representative training tasks. This process effectively addresses the impact of limited interactions of new users. Additionally, they propose an updating strategy to optimize learning rates for fast adaptation. In [1], a meta-learning methodology is employed to support cross-market product recommendations. This approach utilizes parallel markets as auxiliary data to enhance the quality of recommended items for users in the target market. The overall process involves pre-training a market-agnostic model from a pair of markets, which is subsequently fine-tuned individually for each market to optimize the recommendation performance.

In contrast to previous approaches, the *MetaTL* model [10] introduces a novel solution for recommending items to cold-start users by utilizing the MAML algorithm to learn item transition patterns. Treating the cold-start sequential recommendation task as a few-shot learning challenge, the model simulates item pairs purchased by users with limited data by sampling transition pairs from frequently buying users. In the training stage, each user's item transitions are treated as individual tasks, allowing the model to learn initial parameters that can be applied across different users, facilitating the prediction of their next item transition using a meta-learning approach.

Our research emphasizes the significance of item categories in expressing higher-level behavioral meanings and observes that category-level patterns are denser than item-level patterns. To enhance the effective capture of behavioral representations in item transitions, we extend the *MetaTL* model by incorporating a category transition meta-learner. This integration enables the inclusion of category transition information alongside the item transition meta-learner, further enhancing the recommendation capabilities of the model on cold-start users.

3 Method

3.1 Problem Setup

The user-item interaction sequence is made up of several records. Each record contains the user ID, item ID, and category ID. The set $U = \{u_1, u_2, \ldots, u_P\}$ represents all users in the dataset. Meanwhile, $I = \{i_1, i_2, \ldots, i_N\}$ symbolizes all interaction items, and $C = \{c_1, c_2, \ldots, c_M\}$ denotes all item categories. It is assumed that each item $i_n \in I$ corresponds to a unique category $c_m \in C$.

A user-item interaction record is described as a triplet (u_n, i_j, c_m), where $u_n \in U$, $i_j \in I$, and $c_m \in C$. It signifies an interaction between user u_n and item i_j that falls under category c_m. For any user u, the user-item interaction sequence Seq_u represents the historical records of interactions between that user and various items. It is derived by chronologically arranging the interactions. In this paper, $Seq_{u,k}$ specifies the k-th user interaction record (u, i_k^u, c_k^u) in Seq_u. Meanwhile, $|Seq_u|$ is indicative of the interaction sequence's length. For this, $u \in U$, $i_k^u \in I$, $c_k^u \in C$, and $1 \leq k \leq |Seq_u|$.

Using the user-item interaction sequence Seq_u, a category/item transition pair is developed. This is achieved by choosing categories or items from consecutive interaction records, specifically between the i-th and $(i+1)$-th records, for a designated user u.

These transition pairs can be expressed individually as $Seq_{u,i}^C \to Seq_{u,i+1}^C$ and $Seq_{u,i}^I \to Seq_{u,i+1}^I$, where i spans from 1 to $|Seq_u| - 1$.

The goal of the next item prediction task is forecasting the item that is likely to feature in the subsequent interaction record of user u when provided with the user-item interaction sequence Seq_u.

3.2 Category-Aware Item-Level Transitional Meta Learner

This paper introduces a category-aware sequential recommendation model. The term "category-aware" signifies the model's capacity to utilize behavioral features derived from categories present in the user-item interaction sequence. As illustrated in Fig. 1, the model's system architecture encompasses three main components: 1) data preprocessing, 2) category-level prediction task, and 3) item-level prediction task. Subsequently, we delve into these three components in sequence and then introduce the two-stage training procedure for the model.

Data Preprocessing. To employ a meta-learning approach for training the model, we commence by sampling p users as learning tasks. For every selected user u_n (where $n = 1,..., p$), we extract $K + 1$ consecutive interaction records from their user-item interaction sequence Seq_{u_n}. This results in K interaction transition pairs for each task T_n. The data specific to each task T_n is further partitioned into a support set (S_n) and a query set (Q_n). Here, the support set S_n consists of the initial K-1 category/item transition pairs, while the query set Q_n comprises the last category/item transition pair.

Category-Level Prediction Task. Upon completion of the data preprocessing phase, we obtain category transition pairs. These pairs undergo an embedding layer to get embedding vectors for both categories within each pair. The derived category embedding vectors are subsequently channeled into the Category Transitional Meta-Learner (CTML). This module calculates a representation vector essential for predicting the subsequent item's category. The category from the item categories set C exhibiting maximal similarity to the category-level representation vector is chosen for prediction.

Item-Level Prediction Task. The item transition pairs, formulated during preprocessing, are input into the embedding layer of categories to get embedding vectors for both items within each pair. The item embedding vectors from the transition pairs, coupled with the category-level representation vector, sourced from CTML, are input into the Item Transitional Meta-Learner (ITML). Here, the module determines the item-level representation vector for predicting the next item. The prediction is determined by finding the item from the set of items I that exhibits the highest similarity to the item-level representation vector.

Two-Stage Model Training. Within this framework, the training of the Category Transitional Meta-Learner and the Item Transitional Meta-Learner is divided into two distinct phases. Initially, the Category Transitional Meta-Learner is trained using the category-level prediction task. In the subsequent phase, while instructing the Item Transitional Meta-Learner for item prediction, the parameter set of the Category Transitional Meta-Learner remains static. The pre-trained Category Transitional Meta-Learner is then employed to derive the category-level representation vector, which is then fed into the Item Transitional Meta-Learner.

3.3 The Proposed Model

Embedding Layer. The categories or items in transition pairs are initially passed through distinct embedding layers. Considering the i-th category transition pair, $Seq_{n,i}^C \rightarrow Seq_{n,i+1}^C$, both categories are converted into their respective category embedding vectors. These vectors have dimensionality d and are represented as e_i^C and e_{i+1}^C. They are obtained through the category embedding layer as $e_j^C = M_{emb}^C Seq_{n,j}^C$. In a similar vein, for the i-th item transition pair, $Seq_{n,i}^I \rightarrow Seq_{n,i+1}^I$, the items are mapped to their respective embedding vectors of dimensionality d, symbolized as e_i^I and e_{i+1}^I. This is achieved through the item embedding layer: $e_j^I = M_{emb}^I Seq_{n,j}^I$. In these formulas, M_{emb}^C and M_{emb}^I denote the trainable category and item embedding matrices, respectively. Their dimensions are defined as $M_{emb}^C \in R^{d*|C|}$ and $M_{emb}^I \in R^{d*|I|}$.

Category Transitional Meta-Learner (CTML).

Layers in CTML. The embedding vectors of both categories in each category transition pair are utilized as input. After concatenation of these vectors, the result is processed through L layers of a Multilayer Perceptron (MLP) to obtain a representation vector for the category transition. The computation formulas for each layer are illustrated in Eqs. 1 and 2.

$$t_i^{C^1} = \sigma(W_C^1(e_i^C || e_{i+1}^C) + b^{C^1}) \tag{1}$$

$$t_i^{C^l} = \sigma\left(W_C^l t_i^{C^{l-1}} + b^{C^l}\right), l = 2, \ldots, L \tag{2}$$

For a given user u_n, their category-level transition representation is deduced from the K prior consecutive transition pairs in their interaction sequence. The representation vectors of these individual pairs are averaged to produce the user's transition representation at the category level, symbolized as tr_n^C. This process is detailed in Eq. 3.

$$tr_n^C = \frac{1}{K} \sum_{i=1}^K t_i^{C^L} \tag{3}$$

The final category embedding vector in the user-item interaction sequence Seq_{u_n}, represented as e_K^C, is combined with the aggregated transition representation tr_n^C. This results in a representation vector for predicting the category of the subsequent item. The predicted category, c_{K+1}^n, is the category projected to be nearest to the vector $e_K^C + tr_n^C$.

Loss Function of CTML. To assess the effectiveness of the transition representation, tr_n^C, in capturing the transition pattern of $e_h^C \rightarrow e_t^C$, it is desirable for the distance between e_t^C and the estimated vector $e_h^C + tr_n^C$ to be small. The 2-norm distance function, defined in Eq. 4, calculates the value of $dist_C(e_h^C \rightarrow e_t^C)$.

$$dist_C\left(e_h^C \rightarrow e_t^C\right) = \left\| e_h^C + tr_n^C - e_t^C \right\|^2 \tag{4}$$

During the training process, for each category transition pair $Seq_{n,h}^C \to Seq_{n,t}^C$, a negative sampling pair $Seq_{n,h}^C \to Seq_{n,t'}^C$ is generated. Here, $Seq_{n,t'}^C$ is a randomly selected category that has not appeared in Seq_n. The loss function utilizes the margin loss, as shown in Eq. 5, to compute the loss values with a margin threshold value, r, for the predictions of each transition pair in the support set S_n.

$$L_{S_n}^C = \sum_{(e_h^C \to e_t^C) \in S_n} [r + \text{dist}_C\left(e_h^C \to e_t^C\right) - \text{dist}_C\left(e_h^C \to e_{t'}^C\right)]_+ \qquad (5)$$

During the item-level prediction task, the model parameters, θ^C, of the Category Transitional Meta-Learner (CTML) are kept frozen. The category transition pairs that correspond to the item transition pairs are input to obtain $e_K^C + tr_n^C$ as the category-level representation vector, denoted as h^C. This representation vector is then fed into the Item Transitional Meta-Learner (ITML) to assist in determining the next item prediction.

Meta-learning Process. The parameters of the CTML are represented by θ^C, and the function to be learned is denoted as f_{θ^C}. The primary objective when optimizing the model parameters through meta-learning is to minimize the loss value $L_{S_n}^C$, as defined in Eq. 5. To meet this goal, the model updates f_{θ^C} using gradient descent to reduce $L_{S_n}^C$. As a result, the model parameters are updated to $\theta_n^{C'}$, specially tailored for user u_n, as described in Eq. 6.

$$\theta_n^{C'} = \theta^C - \alpha \nabla L_{S_n}(f_{\theta^C}) \qquad (6)$$

Once the adjusted $\theta_n^{C'}$ is derived from S_n, the model $f_{\theta_n^{C'}}$ is used to predict tr_n^C in the query set. The model's performance on the query set Q_n is assessed by computing the loss value $L_{Q_n}^C(f_{\theta_n^{C'}})$, achieved by replacing S_n with Q_n in Eq. 5. The overarching goal of meta-learning in this scenario is to refine the parameters θ^C, so they serve as effective initialization parameters for each task. Accordingly, the model optimizes and modifies the parameter θ^C through stochastic gradient descent (SGD) by aggregating the loss values across multiple tasks, T_n, as shown in Eq. 7.

$$\theta^C = \theta^C - \beta \nabla_\theta \sum_{T_n \sim p(T)} L_{Q_n}^C\left(f_{\theta_n^{C'}}\right) \qquad (7)$$

Within this framework, $p(T)$ indicates the distribution of meta-training tasks. This distribution is formed by randomly sampling users to constitute the set of meta-training tasks, T_n.

Prediction Process of CTML. To predict the behavior of a user u_{test}, we begin the process by constructing a support set of u_{test}, S_{test}, derived from their historical item interactions. Subsequently, the model, f_{θ^C}, is fine-tuned using the transition pairs found in S_{test} as guided by Eq. 7. The fine-tuning results in an updated model, $f_{\theta_{test}^{C'}}$. With these adjusted model parameters, we can then derive the transition vector, tr_{test}^C. In the final step, this predicted transition vector, tr_{test}^C, is integrated with the embedding vector of the user's most recently interacted item category by vector addition, e_{last}^C, to obtain the category-level representation vector, h^C.

Item Transitional Meta-Learner (ITML).

Layers in ITML. For each item transition pair, the embedding layer is used to derive two item embedding vectors. These vectors are subsequently concatenated, and a Multilayer Perceptron (MLP) generates the representation vector for the item transition. Equations 8 and 9 detail the calculations for each layer:

$$t_i^{I^1} = \sigma(W_I^1(e_i^I || e_{i+1}^I) + b^{I^1}) \tag{8}$$

$$t_i^{I^l} = \sigma(W_I^l t_i^{I^{l-1}} + b^{I^l}) \tag{9}$$

For each user u_n, the representation vectors of the K consecutive item transition pairs undergo a weighted aggregation to produce the overall transition representation tr_n^I at the item level. Here, an attention mechanism determines the relative importance of each item transition pair. This mechanism incorporates the learned category-level representation vector, h^C, obtained from CTML with frozen model parameters. The relative importance of each transition pair is calculated using Eqs. 10 and 11, where $W^{att} \in \mathbb{R}^{|S_n|*(2d)}$ and $W^T \in \mathbb{R}^{1*|S_n|}$ represent the model learnable parameters.

$$a_i = W^T \tanh\left(W^{att} \begin{bmatrix} h^C \\ t_i^{I^L} \end{bmatrix}\right) \tag{10}$$

$$\alpha_i = \frac{\exp(a_i)}{\sum_{j=1}^{|S_n|} \exp(a_j)} \tag{11}$$

Subsequently, the representation vectors of each item transition pair are combined using weights α_i to compute the item-level transition representation for the user, as shown in Eq. 12.

$$tr_n^I = \sum_{i=1}^{|S_n|} \alpha_i t_i^{I^L} \tag{12}$$

Loss Function of ITML. Throughout the training process, the loss function adopts the margin loss with a margin threshold, r, to determine the loss for every transition pair prediction within the support set S_n. Equation 13 illustrates the loss function.

$$L_{S_n}^I = \sum_{(e_h^I \to e_t^I) \in S_n} [r + \text{dist}_I(e_h^I \to e_t^I) - \text{dist}_I(e_h^I \to e_{t'}^I)]_+ \tag{13}$$

For every transition pair $Seq_h^I \to Seq_t^I$, a negative transition sample $Seq_h^I \to Seq_{t'}^I$ for contrasting is generated. The sample $Seq_{t'}^I$ corresponds to an interaction item not previously seen in the user's interaction sequence and selected at random.

Meta-learning Process. The item transition meta-learner utilizes the MAML (Model-Agnostic Meta-Learning) algorithm, similar to the category transition meta-learner, optimizing the initial parameters θ^I for each user. When predicting for a user u_{test}, the model f_θ^I undergoes fine-tuning using transition pairs from u_{test}'s past interactions. This process refines the model, producing a model $f_{\theta_{test}^I}$ specifically tailored to u_{test}, which is then used to predict the item-level transition representation tr_{test}^I.

Prediction Process of ITML. Afterwards, the subsequent interaction representation at the item level, h^I, is derived by integrating the embedding vector e_{last}^I of the user's most recent interaction with the transition representation tr_{test}^I. The next predicted item, i_p, is chosen from all candidate items based on which has the smallest $dist_I \left(e_{last}^I \rightarrow e_p^I \right)$.

4 Performance Evaluation

4.1 Dataset and Evaluation Matric

The data used in this study comes from the Foursquare Global-scale Check-in dataset, collected between April 2012 and September 2013. This dataset includes check-in records from Foursquare users worldwide [13]. The data is organized into two tables: Check-ins and POIs. Each row in the Check-ins table represents a user's interaction record with a venue. This table includes columns for User ID, Venue ID, the UTC time of interaction, and the corresponding time zone. The POIs table provides detailed information about each venue, including Venue ID, longitude, latitude, venue category, and country code. In this dataset, Venue IDs correspond to interacted items, while venue categories denote item categories.

To assemble user-item interaction sequences, essential fields like User ID, Venue ID, and interaction time were extracted from the Check-ins table. These records were sorted by User ID and interaction time. Venue categories and country codes were sourced from the POIs table, using Venue ID for reference. In our experiments, we focused on the data associated with Country code = 'AE'. Statistical data on the number of users (#User), items (#Item), and item categories (#Category) within this dataset is presented in Table 1.

Table 1. Statistical information of the dataset.

#User	#Item	#Category
3181	10611	355

To evaluate the effectiveness of the recommendations, this study uses the hit rate for the top k recommended items ($I\_HR@k$) as the evaluation metric. Equations 14 and 15 define $I\_HR@k$. In this context, D_{test} denotes the test set, Seq_i refers to the history interaction sequence of the i-th user in the test set, $\hat{y}_i$ indicates the actual next item following Seq_i, and $Top_k(Seq_i)$ represents the set of the top k items that model predicts with the highest probabilities for the next item prediction.

$$I\_HR@k = \frac{1}{|D_{test}|} \sum_{i=1}^{|D_{test}|} I\_Hit_k(\hat{y}_i) \quad (14)$$

$$I\_Hit_k(\hat{y}_i) = \begin{cases} 1, & \text{if } \hat{y}_i \text{ in } Top_k(Seq_i) \\ 0, & \text{else} \end{cases} \quad (15)$$

Moreover, when the recommended item differs from the user's interaction item but belongs to the same category, the recommendation remains relevant. Consequently, we introduce an additional evaluation metric: category hit rate for the top k recommendations (C_HR@k). Equations 16 and 17 define C_HR@k, with category(t) symbolizing the category of item t.

$$C\_HR@k = \frac{1}{|D_{test}|} \sum_{i=1}^{|D_{test}|} C\_Hit_k(\hat{y}_i) \qquad (16)$$

$$C\_Hit_k(\hat{y}_i) = \begin{cases} 1, & \text{if } category(\hat{y}_i) \text{ in } \{category(t) | t \text{ in } Top_n(Seq_i)\} \\ 0, & else \end{cases} \qquad (17)$$

4.2 Result of Performance Evaluation

The performance evaluation is twofold. Firstly, is contrasts the recommendation performance of the *CAT-ML* model against *MetaTL* across various scenarios: general recommendations, cold-start user recommendations, and suggestions for previously uninteracted items or categories. Secondly, it delves into the models' implementations of category-level and item-level meta-learners.

Experiment 1-1. Performance comparison of *CAT-ML* and *MetaTL* in general cases.

The recommendation performance of both models, *CAT-ML* and *MetaTL*, is shown in Table 2. Across all metrics, whether evaluating the item hit rate or the category hit rate, the *CAT-ML* model consistently surpasses *MetaTL*.

Notably, *CAT-ML* shows a significant 10.2% improvement over *MetaTL* in the item hit rate (*I-HR@1*) and a 23.8% increase in the category hit rate (*C_HR@1*). This demonstrates the effectiveness of integrating category information in the *CAT-ML* model, resulting in enhanced recommendation outcomes.

Table 2. Hit rates of *CAT-ML* and *MetaTL* in general test cases.

	I_HR@1	I_HR@5	I_HR@10	C_HR@1	C_HR@5	C_HR@10
MetaTL	0.225	0.474	0.581	0.281	0.592	0.739
CAT-ML	**0.248**	**0.498**	**0.602**	**0.348**	**0.651**	**0.764**
improvement	+10.2%	+5.1%	+3.6%	+23.8%	+9.9%	+3.3%

Experiment 1-2. Performance comparison of *CAT-ML* and *MetaTL* for cold-start users.

This experiment assesses the *CAT-ML* and *MetaTL* models on a distinct test set of cold-start users. From the primary test set, users with an interaction sequence length of three are included, resulting in a set where users have only two past interactions. The goal is to predict the third interaction.

Table 3 lists the hit rates for both models. While performance drops for both models due to scarce transition data of cold-start users, *CAT-ML* remains superior in recommendations for cold-start users compared to *MetaTL*. Comparing the performance drop

Table 3. Hit rates of *CAT-ML* and *MetaTL* for cold-start users.

	I_HR@1	I_HR@5	I_HR@10	C_HR@1	C_HR@5	C_HR@10
MetaTL	0.181	0.380	0.467	0.218	0.482	0.664
CAT-ML	**0.197**	**0.413**	**0.541**	**0.312**	**0.582**	**0.733**

between the two models for cold-start users versus general cases, the *MetaTL* model consistently displays a 20% decline in item hit rates. On the other hand, while the *CAT-ML* model shows a comparable drop in *I_HR@1* to *MetaTL*, its reductions in *I_HR@5* and *I_HR@10* are limited to 17% and 15%. Notably, even with limited user interaction data, *CAT-ML* retains item hit rates over 50% for the top 10 items.

Diving into the category hit rates, *CAT-ML*'s decline is just 10% for *C_HR@1* and a minimal 4% for *C_HR@10* when contrasted with *MetaTL*, highlighting its superior performance. These results emphasize that in scenarios with limited interaction information for cold-start users, the *CAT-ML* model leverages the category transition meta-learner to provide valuable insights to the item transition meta-learner, resulting in effective prediction of the next recommended item, even for cold-start users.

Experiment 1-3. Performance comparison of *CAT-ML* and *MetaTL* on previously un-interacted items or categories.

In some recommendation scenarios, systems aim to suggest items or categories with which users have not previously interacted. This experiment focuses on (1) items never previously interacted with and (2) items from previously un-interacted categories.

In this experiment, we assessed and compared the recommendation performance of both the CAT-ML and MetaTL models using the un-interacted item test set. We selected users from the original test set with the specific task of recommending items with which they had not previously interacted. For generating the top-n recommendations, denoted as $Top_n(Seq_i)$, we chose candidate items that were absent from the user's prior interaction sequence, Seq_i.

Table 4. Hit rates of *CAT-ML* and *MetaTL* on the un-interacted item test set.

	I_HR@1	I_HR@5	I_HR@10	C_HR@1	C_HR@5	C_HR@10
MetaTL	0.208	0.391	0.544	0.257	0.573	0.725
CAT-ML	**0.227**	**0.473**	**0.589**	**0.340**	**0.628**	**0.766**

The results from the comparison between the *CAT-ML* model and the *MetaTL* model on the un-interacted item test set are detailed in Table 4. In this test scenario, where users lack similar item transitions from their previous interactions, both models have lower hit rates than those observed in general scenarios, as shown in Experiment 1-1. Nevertheless, *CAT-ML* outperformed *MetaTL* by achieving superior hit rates.

Upon analyzing the dropping in prediction effectiveness for general cases, it's evident that *MetaTL*'s performance decreased significantly, showing an 18% reduction at

I_HR@5. In contrast, the *CAT-ML* model experienced a relatively modest reduction of 5%. Additionally, for *CAT-ML*, the *I_HR@10* metric on the un-interacted item test set saw only a 2% decrease, maintaining a commendable rate of 59%. In terms of category recommendation hit rates, *CAT-ML* marked a minor decline of 2% at *C_HR@1*, and interestingly, even outperformed its results from general test cases at *C_HR@10*.

The results highlight that when recommending unfamiliar items to users, the *CAT-ML* model effectively utilizes the category transition meta-learner to inform the item transition meta-learner. This strategy allows the model to bridge the gap caused by the lack of item transition patterns, subsequently improving the accuracy of the upcoming item recommendation.

To evaluate the ability of the *CAT-ML* model and the *MetaTL* model in recommending previously un-interacted item categories, we utilized a subset of the test data. This subset exclusively comprises users whose next item predictions pertain to unfamiliar categories. We name this subset as the test set for new category recommendation. For the predictions concerning $Top_n(Seq_i)$, items from previously encountered categories within Seq_i were excluded.

Table 5. Hit rates of the two models on the test set for new category recommendation.

	I_HR@1	*I_HR@5*	*I_HR@10*	*C_HR@1*	*C_HR@5*	*C_HR@10*
MetaTL	0.127	0.272	0.333	0.152	0.341	0.489
CAT-ML	**0.165**	**0.351**	**0.441**	**0.252**	**0.433**	**0.524**

Table 5 presents the performance of both the *CAT-ML* and *MetaTL* models on the test set for new category recommendation. This dataset posed a greater challenge than the one used for recommending un-interacted items, mainly because users hadn't previously engaged with items from analogous category transitions. As illustrated in Table 5, it reveals a significant drop in *MetaTL*'s performance, with declines of nearly 40% for both the item and category hit rates. In contrast, the *CAT-ML* model experienced a milder reduction of around 30% across both these metrics. Impressively, even in the absence of historical data on similar category transitions, the *CAT-ML* model attained a category hit rate of 0.252 at *C_HR@1*, underscoring the robustness of its category transition meta-learner.

These results show that the *CAT-ML* model can learn from observed category transitions and use them to predict new category transitions, thereby enabling it to recommend the next item that aligns with the desired item's category.

Experiment 2. In this experiment, the performance of recommendation systems is evaluated based on the integration of category-level and item-level meta-learners, choosing between sequence models or transition models.

Historically, sequential recommendation research has leaned heavily on RNN-based models, particularly to capture behavioral features within interaction sequences. In this experiment, we substitute the category-level and item-level meta-learners with sequence models, specifically using bi-GRU models, to extract sequence representations from user interaction sequences.

To differentiate between the models, we've introduced specific designations. "CSML" and "CTML" specify whether the category-level meta-learner employs a sequence model or a transition model via meta-learning. Likewise, "ISML" and "ITML" determine if the item-level meta-learner is constructed using a sequence model or a transition model through the meta-learning framework. Table 6 showcases the various system architectures that merge both category-level and item-level meta-learners, also highlighting their respective performances.

Table 6. Hit rates of the combined meta-learners by sequence/transition models.

	I_HR@1	I_HR@5	I_HR@10	C_HR@1	C_HR@5	C_HR@10
(1) CSML+ISML	0.193	0.432	0.534	0.259	0.523	0.651
(2) CTML+ISML	0.201	0.450	0.550	0.274	0.575	0.703
(3) CSML+ITML	0.242	0.481	0.590	0.298	0.614	0.736
(4) CTML+ITML (*CAT-ML*)	**0.248**	**0.498**	**0.602**	**0.348**	**0.651**	**0.764**

When comparing the combination models (1) CSML+ISML and (2) CTML+ISML, we observe enhanced effectiveness in category recommendations by switching the category-level model from learning sequence patterns to transition patterns (CTML+ISML). This enhancement also marginally improves item recommendation hit rates. Upon comparing model (1) CSML+ISML with model (3) CSML+ITML, it's evident that the combination model (3) attains significant improvements in item recommendation hit rates over combination model (1).

This indicates that learning item patterns through transition learning is more effective in capturing user behavior characteristics than learning through category learning. A plausible explanation is that the transition learning model on items more directly provides effective user behavior features. The combination model (4) proposed in this study, namely *CAT-ML*, adopts transition learning models for both tasks, enabling them to learn better user behavior characteristics and achieve the highest recommendation effectiveness.

Summarization of Experiments

The *CAT-ML* model achieves a hit rate of 54.1% for the top 10 recommended items for cold-start users, marking a notable improvement of 15.8% over *MetaTL*. Moreover, when suggesting items or categories that users have not previously interacted with, the *CAT-ML* model exhibits a more substantial enhancement compared to *MetaTL*. Specifically, the *CAT-ML* model demonstrates a remarkable 65% improvement in the category hit rate (C_HR@1) when recommending un-interacted categories, outperforming *MetaTL*. This improvement is credited to the integration of the category-level and item-level meta-learners in the *CAT-ML* model. Both learners utilize transition learning models, which, as evidence suggests, are more adept at ensuring recommendation effectiveness than traditional sequence models.

5 Conclusion

In this paper, we introduce *CAT-ML*, a meta-learning-based sequential recommendation model that incorporates item categories. *CAT-ML* enhances the *MetaTL*, a transition-based meta-learner, by encoding category transition pairs to encapsulate general user behaviors, as well as encoding item transition pairs to capture individual preferences. Through two distinct transition meta-learners, one for category-level transitions and another for item-level transitions, the model derives a comprehensive representation of user behavior from transition pairs. An attention mechanism guides the combination of item-level transition representation based on the predicted category-level representation. Owing to its meta-learning framework, *CAT-ML*'s offers personalized recommendations even with sparse data. Results of the experiments on the Foursquare Check-in Dataset demonstrate *CAT-ML*'s superiority over *MetaTL*, making a hit rate improvement of 10.2% for top-ranked POI item recommendations and 23.8% for item category recommendations. *CAT-ML* is particularly effective for addressing cold-start users and suggesting new items for users, making significant progress over *MetaTL* in these scenarios.

User interaction sequences might feature identical category transition pairs at varied times of the day or on different days of the week, each reflecting distinct intentions. Furthermore, the time intervals between transitions can convey varied implications. Given this, there lies an opportunity to utilize temporal information as supplementary features for model learning. Such an approach can enrich POI recommendations, aligning them more closely with a user's real-time interests and requirements, thus facilitating context-aware recommendations.

References

1. Bonab, H., Aliannejadi, M., Vardasbi, A., Kanoulas, E., Allan, J.: Cross-market product recommendation. In: Proceedings of the 30th ACM International Conference on Information and Knowledge Management (CIKM'21), pp. 110–119 (2021)
2. Cai, R., Wu, J., San, A., Wang, C., Wang, H.: Category-aware collaborative sequential recommendation. In: Proceedings of the 44th International ACM SIGIR Conference on Research and Development in Information Retrieval (SIGIR'21), pp. 388–397 (2021)
3. He, X., Liao, L., Zhang, H., Nie, L., Hu, X., Chua, T.S.: Neural collaborative filtering. In: Proceedings of the 26th International Conference on World Wide Web (WWW'17), pp. 173–182 (2017)
4. Finn, C., Abbeel, P., Levine, S.: Model-agnostic meta-learning for fast adaptation of deep networks. In: Proceedings of the 34th International Conference on Machine Learning, pp. 1126–1135 (2017)
5. Hidasi, B., Karatzoglou, A., Baltrunas, L., Tikk, D.: Session-based recommendations with recurrent neural networks. In: Proceedings of the 4th International Conference on Learning Representations, pp. 1–10 (2016)
6. Lee, H., Im, J., Jang, S., Cho, H., Chung, S.: MeLU: meta-learned user preference estimator for cold-start recommendation. In: Proceedings of the 25th ACM SIGKDD International Conference on Knowledge Discovery & Data Mining (KDD'2019), pp. 1073–1082 (2019)
7. Qiu, R., Li, J., Huang, Z., Yin, H.: Rethinking the item order in session-based recommendation with graph neural networks. In: Proceedings of the 28th ACM International Conference on Information and Knowledge Management (CIKM'19), pp. 579–588 (2019)

8. Tang, J., Wang, K.: Personalized top-n sequential recommendation via convolutional sequence embedding. In: Proceedings of the 11th ACM International Conference on Web Search and Data Mining (WSDM'18), pp. 565–573 (2018)
9. Tuan, T.X., Phuong, T.M.: 3D convolutional networks for session-based recommendation with content features. In: Proceedings of the 11th ACM Conference on Recommender Systems (RecSys'17), pp. 138–146 (2017)
10. Wang, J., Ding, K., Caverlee, J.: Sequential recommendation for cold-start users with meta transitional learning. In: Proceedings of the 44th International ACM SIGIR Conference on Research and Development in Information Retrieval (SIGIR'21), pp.1783–1787 (2021)
11. Wei, T., et al.: Fast adaptation for cold-start collaborative filtering with meta-learning. In: Proceedings of the IEEE International Conference on Data Mining (ICDM'20) (2020)
12. Wu, S., Tang, Y., Zhu, Y., Wang, L., Xie, X., Tan, T.: Session-based recommendation with graph neural networks. In: Proceedings of the AAAI Conference on Artificial Intelligence, vol. 33, no. 1, pp. 346–353 (2019)
13. Yang, D., Zhang, D., Qu, B.: Participatory cultural mapping based on collective behavior data in location-based social networks. ACM Trans. Intell. Syst. Technol. **7**(3), 1–23 (2016)
14. Yuan, F., Karatzoglou, A., Arapakis, I., Jose, J.M., He, X.: A simple convolutional generative network for next item recommendation. In: Proceedings of the 12th ACM International Conference on Web Search and Data Mining (WSDM'19), pp. 582–590 (2019)
15. Zhang, L., et al.: An interactive multi-task learning framework for next POI recommendation with uncertain check-ins. In: Proceedings of the 29th International Conference on International Joint Conferences on Artificial Intelligence (IJCAI'20), pp. 3551–3557 (2020)

Automatic Post-editing of Speech Recognition System Output Using Large Language Models

Sheng Li[1], Jiyi Li[2(✉)], and Yang Cao[3]

[1] National Institute of Information and Communications Technology, Kyoto, Japan
sheng.li@nict.go.jp
[2] University of Yamanashi, Kofu, Japan
jyli@yamanashi.ac.jp
[3] Tokyo Institute of Technology, Tokyo, Japan
cao@c.titech.ac.jp

Abstract. This paper explores the integration of automatic speech recognition (ASR) with large language models (LLMs), aiming to validate the effectiveness of this combination, particularly for automatic post-editing (PE) tasks. Initially, we investigate the use of LLMs for ASR PE error correction, performing second-pass rescoring on the output transcriptions generated by the ASR system, using both N-best decoding hypotheses and lattices. Subsequently, we examine the combination of ASR outputs from various systems using LLMs, addressing a classic system combination task. Experimental results demonstrate that LLMs can offer substantial assistance in automatic PE.

Keywords: automatic post-editing (PE) · automatic speech recognition (ASR) · large language models (LLMs)

1 Introduction

An automatic post-editing (PE) task involves using computational tools and algorithms to automatically revise and improve the quality of written content, such as documents, articles, or translations. This process typically includes correcting grammar, syntax, style, and coherence errors and adjusting language to meet specific requirements or preferences. Automatic PE tasks aim to streamline the editing process, saving time and effort while enhancing the overall readability and effectiveness of the text.

PE tasks also edit automatic speech recognition (ASR) outputs. When speech is transcribed into text using speech recognition technology, errors can occur due to variations in accents, background noise, or misinterpretations of words. Automatic PE algorithms can be applied to these transcriptions to correct inaccuracies, improve grammar and syntax, and ensure the text accurately reflects the spoken content. This process helps enhance the quality and readability of the transcribed text, making it more useful for applications such as captioning, transcription services, or voice-controlled systems.

In recent years, large pretrained neural network-based language models (LMs) have been increasingly employed in automatic speech recognition (ASR)-related post-editing (PE) tasks. Zhang et al. [40] introduced a transformer-based spelling corrector to reduce substitution errors in Mandarin speech recognition. Later, Zhang et al. [39] improved BERT's [8] effectiveness in detecting spelling errors by using a soft-masking technique that links the error detector and corrector. Futami et al. [11] generated soft labels for ASR training through knowledge distillation from BERT. Other studies, such as those by Salazar et al. [31], and Shin et al. [32], have explored enhancing ASR rescoring using BERT. Additionally, BERT has been successfully applied in multimodal studies, including vision-language pretraining [17,18,20,41] and voice-language pretraining [1,14,36].

A speech recognition system that integrates large language models (LLMs) was suggested more recently by [4]. The present study expands upon the rescoring technique expounded in [4], and assesses its efficacy for Japanese ASR problems in multiple configurations, encompassing N-best, lattice, and system combination.

2 Related Work

2.1 A Brief Introduction of Speech Recognition

The lexicon, language model, and acoustic model must all be optimized in conventional ASR systems, such as those that use deep neural networks in conjunction with hidden Markov models (DNN-HMM) [7] or hybrid Gaussian mixture models and hidden Markov models (GMM-HMM) [28]. However, end-to-end (E2E) models integrate all of these components into a single neural network, simplifying the process of developing ASR systems. These models tackle the sequence labeling problem, converting variable-length speech frame inputs into label outputs such as phones, characters, syllables, or words. They have demonstrated promising results in automatic speech recognition (ASR) tasks. Notable end-to-end (E2E) models that have been recently studied include connectionist temporal classification (CTC) [12,21], attention-based models [3,5], lattice-free maximum mutual information (LFMMI) [13] models, and CTC+attention jointly trained (CTC/Attention) [37,38] models.

Additionally, the transformer model has been used with promising results in end-to-end (E2E) speech recognition systems [10,34,42,43]. Self-supervised learning (SSL) models have gained popularity recently, including Wav2Vec 2.0 [2]. These models are first trained on unlabeled speech data using self-supervised learning, and then they are refined on labeled data by applying the connectionist temporal classification (CTC) objective for the ASR problem.

Recently, speech recognition models trained with vast amounts of speech data have become popular in industry and academia. They are called the speech foundation model. No individual can train from scratch with such large costs of data and machines. In the meantime, these foundation models simplify the development of downstream tasks because we only need to fine-tune these pretrained models for downstream tasks. Here are several speech foundation models listed as follows:

1. OpenAI's Whisper [23] adopts the standard Transformer encoder-decoder architecture [35] and scales the training data to 680k h of proprietary labeled audio. It has strong ASR, speech translation (ST), and language identification (LID) performance.
2. CMU's OWSM model reproduced Whisper-style training using public data [24]. The latest OWSM v3.1 models [25] follow the encoder-decoder architecture. The recent model, OWSM-CTC [26], also switched to a Wav2vec-style encoder-based structure.
3. Meta's Massively Multilingual Speech (MMS) ASR model [27] supports over 1,000 languages by utilizing religious texts and self-supervised learning for speech recognition tasks, using less data than existing technologies like Whisper. It is a Wav2vec-style encoder-based ASR.

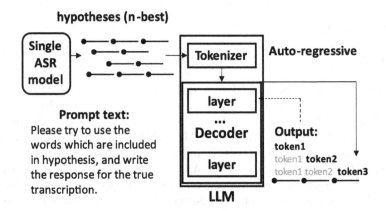

Fig. 1. The flowchart of the method rescoring N-best hypotheses.

2.2 LMs for ASR Task

Incorporating a language model (LM) into automatic speech recognition (ASR) systems enhances speech recognition performance. Typically, ASR combined with an LM employs two strategies: first-pass decoding and second-pass rescoring.

The problem in ASR is framed as a noisy channel model, where X is the speech signal and W is the matching text, using Bayes' rule, $P(W|X) = P(X|W)P(W)$. The two distributions of $P(X|W)$ and $P(W)$ are referred to as the "acoustic model" and "language model," respectively. The LM is trained

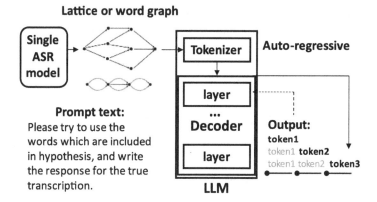

Fig. 2. The flowchart of the method rescoring lattice hypotheses.

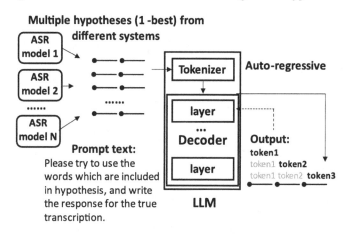

Fig. 3. The flowchart of the method combines hypotheses from different systems

independently on text data and is only utilized for decoding [16]. Weighted finite-state transducer (WFST) based decoding integrates n-gram LMs into the decoding network for efficient first-pass decoding [22]. However, employing larger n-gram LMs could lead to an excessively large decoding graph. To get around this, researchers either improved the compilation method [9] or used second-pass rescoring in offline and on-the-fly settings [19,30]. Scores could be interpreted as pseudo-likelihoods in the hybrid DNN-HMM model era by deducting a suitable prior, allowing the same framework for decoding and rescoring to be used. The Bayesian formulation was thus still relevant.

Models in End-to-End (E2E) ASR systems approximate $P(W|X)$. Using methods like shallow fusion or cold fusion, language models can still be merged during first-pass decoding [6,33]. Using beam search on the ASR model, hypotheses are created for second-pass rescoring. An externally trained LM is then used to rerank the hypotheses. A two-pass E2E ASR model with a shared encoder between a full-context LAS decoder and a streaming RNN-T model has been suggested [29]. Further studies have looked into how transformers and BERT may be used to improve ASR rescoring [11,31,32,39,40]. Large language models (LLMs) have been proposed to be integrated into voice recognition systems more recently by [4].

3 Post-editing Using LLM for ASR System

This paper employs a large language model (LLM) for post-editing (PE) error correction, specifically using second-pass rescoring on the output transcriptions generated by the ASR system (N-best decoding hypotheses), as illustrated in Fig. 1.

To optimize the procedure without changing a pre-trained model's whole set of parameters, we present LoRA [15]. To approach the entire parameter updates, LoRA does this by inserting a neural module with a few extra trainable parameters. This enables effective N-best to transcription mapping learning without changing the previously learned LLM parameters. Our approach allows the model to learn from new data while maintaining the old LLM knowledge by embedding trainable low-rank decomposition matrices into the current layers of LLMs. To be more precise, LoRA uses matrix multiplication to reparameterize each model layer by injecting low-rank decomposition matrices, as Fig. 1 illustrates. As a result, task-specific tuning does not affect the representations produced by the LLM, and the adaptor module acquires the capacity to forecast the genuine transcription from the N-best hypotheses. With the help of this effective training strategy, we can make use of a large-scale language model that should be able to understand the task description and identify correlations within the N-best list.

We also introduce two modifications to the N-best hypotheses-based method.

First, we replace N-best hypotheses with lattice, a more compact ASR output format (as shown in Fig. 2). In speech recognition, a lattice is a graph representation of multiple potential word sequences generated during decoding. It captures various hypotheses of the spoken utterance, allowing for more accurate recognition by considering alternative paths and word sequences. Lattices are useful for handling ambiguity and improving recognition accuracy, especially in noisy or complex speech environments.

Second, we use the 1-best hypotheses output from different ASR systems instead of the N-best hypotheses from only one system (as shown in Fig. 3).

4 Experiments

4.1 Experimental Settings

The following experimental parameters are used to fine-tune a Japanese language model in a low-resource context.

1. LLM used: The japanese-Llama-2-7b model is utilized in this experiment[1]. This model is most likely a modified version of Meta AI's Large Language Model (LLaMA), specifically tuned for analyzing Japanese language input.
2. Whisper-Large-v3, the ASR model, produces 10-best results (CER%=12.91 for 1-best). Kaldi (nbest-to-lattice toolkit) is used to transform these 10-best outputs into lattice format. We also utilized the Meta MMS model, the CMU OWMC model, and the previous iterations of Whisper-Large (v1 and v2) in the system combination experiment.
3. 8-bit training is used to complete the training on an NVIDIA Tesla V100 GPU. The learning rate of 1e−4, batch size of 64, LoRA rank 4, and 15 epochs are the hyperparameters that need to be adjusted.
4. Dataset: SPREDS-U1-ja[2] is the Japanese dataset that we utilize from SPREDS-U1. Nine hundred Japanese sentences are used in the fine-tuning procedure. one hundred distinct Japanese sentences for assessment.
5. Evaluations: The Character Error Rate (CER%) is calculated using NIST-SCTK.

4.2 Experimental Results

Table 1 shows the results of the experiments using N-best hypotheses and lattices. Experiments show both data structures can effectively improve the performance of the original system output, reflected in the CER%. However, currently, the N-best hypotheses-based method has the best result. A possible reason is that the lattice-based method causes information loss in data format conversion.

We do a paired sample t-test on the best model's (Epoch 11) data for the N-best PE experiment. The statistical significance of the improvement in CER% is indicated by the paired sample t-test.

Table 2 shows the performance changes before and after using the LLM-based system combination. It has the current best result of PE correction. The leap forward in performance is due to the diversity of different system hypotheses.

Table 1. Evaluation of LLM-based PE Correction Model (CER%)

before correction	after (N-best)	after (lattice)
12.91	**7.77**	9.91

---

[1] huggingface.co/elyza/ELYZA-japanese-Llama-2-7b.
[2] ast-astrec.nict.go.jp/en/release/SPREDS-U1.

Table 2. PE error correction performance by combining different systems evaluated with Character Error Rate (CER%)

before using LLM error correction					after
MMS	OWSM v3.1	Whisper v1	Whisper v2	Whisper v3	Combined
32.41	10.18	9.29	9.20	12.91	**7.07**

5 Conclusion

This paper proves that using LLM-based PE can effectively improve the performance of a speech recognition system with different settings, such as N-best, lattice input, and system combination. We plan to carry out additional trials in the future to show the method's efficacy in a wider variety of scenarios.

Acknowledgement. The work is partially supported by JSPS KAKENHI No. JP23K11227, and JP23K28092.

References

1. Baevski, A., Mohamed, A.: Effectiveness of self-supervised pre-training for ASR. In: Proceedings of the IEEE-ICASSP, pp. 7694–7698 (2020)
2. Baevski, A., Zhou, Y., Mohamed, A., Auli, M.: wav2vec 2.0: a framework for self-supervised learning of speech representations. In: Proceedings of the NeurIPS, pp. 12449–12460 (2020)
3. Chan, W., Jaitly, N., Le, Q., Vinyals, O.: Listen, attend and spell: a neural network for large vocabulary conversational speech recognition. In: Proceedings of the IEEE-ICASSP (2016)
4. Chen, C., Hu, Y., Yang, C.H.H., Siniscalchi, S.M., Chen, P.Y., Chng, E.S.: HyPoradise: an open baseline for generative speech recognition with large language models. arXiv abs/2309.15701 (2023)
5. Chorowski, J., Bahdanau, D., Serdyuk, D., Cho, K., Bengio, Y.: Attention-based models for speech recognition. In: Proceedings of the NeurIPS (2015)
6. Chorowski, J., Jaitly, N.: Towards better decoding and language model integration in sequence to sequence models. In: Proceedings of the INTERSPEECH (2016)
7. Dahl, G., Yu, D., Deng, L., Acero, A.: Context dependent pre-trained deep neural networks for large vocabulary speech recognition. IEEE Trans. ASLP **20**(1), 30–42 (2012)
8. Devlin, J., Chang, M.W., Lee, K., Toutanova, K.: BERT: pre-training of deep bidirectional transformers for language understanding. In: Proceedings of the NAACL, pp. 4171–4186 (2019)
9. Dixon, P.R., Hori, C., Kashioka, H.: A specialized WFST approach for class models and dynamic vocabulary. In: Proceedings of the INTERSPEECH, pp. 1075–1078 (2012)
10. Dong, L., Xu, S., Xu, B.: Speech-transformer: a no-recurrence sequence-to-sequence model for speech recognition. In: Proceedings of the IEEE-ICASSP (2018)

11. Futami, H., Inaguma, H., Ueno, S., Mimura, M., Sakai, S., Kawahara, T.: Distilling the knowledge of BERT for sequence-to-sequence ASR. CoRR abs/2008.03822 (2020)
12. Graves, A., Jaitly, N.: Towards end-to-end speech recognition with recurrent neural networks. In: Proceedings of the ICML (2014)
13. Hadian, H., Sameti, H., Povey, D., Khudanpur, S.: End-to-end speech recognition using lattice-free MMI. In: Proceedings of the INTERSPEECH (2018)
14. Hsu, W.N., Tsai, Y.H.H., Bolte, B., Salakhutdinov, R., Mohamed, A.: HuBERT: how much can a bad teacher benefit ASR pre-training? In: Proceedings of the IEEE-ICASSP, pp. 6533–6537 (2021)
15. Hu, J.E., et al.: LoRA: low-rank adaptation of large language models. arXiv abs/2106.09685 (2021)
16. Jelinek, F.: Continuous speech recognition by statistical methods. Proc. IEEE **64**, 532–556 (1976)
17. Li, L.H., Yatskar, M., Yin, D., Hsieh, C.J., Chang, K.W.: VisualBERT: a simple and performant baseline for vision and language. arXiv preprint arXiv:1908.03557 (2019)
18. Li, X., et al.: Oscar: object-semantics aligned pre-training for vision-language tasks. In: Proceedings of the ECCV, pp. 121–137 (2020)
19. Ljolje, A., et al.: Efficient general lattice generation and rescoring. In: EUROSPEECH (1999)
20. Lu, J., Batra, D., Parikh, D., Lee, S.: ViLBERT: pretraining task-agnostic visiolinguistic representations for vision-and-language tasks. arXiv preprint arXiv:1908.02265 (2019)
21. Miao, Y., Gowayyed, M., Metze, F.: EESEN: end-to-end speech recognition using deep RNN models and WFST-based decoding. In: Proceedings of the IEEE-ASRU, pp. 167–174 (2015)
22. Mohri, M., et al.: Speech recognition with weighted finite-state transducers (2008)
23. OpenAI: Whisper: Robust speech recognition via large-scale weak supervision (2023)
24. Peng, Y., et al.: Reproducing whisper-style training using an open-source toolkit and publicly available data. In: Proceedings of the IEEE-ASRU, pp. 1–8 (2023)
25. Peng, Y., et al.: OWSM v3.1: better and faster open whisper-style speech models based on e-branchformer. arXiv abs/2401.16658 (2024)
26. Peng, Y., Sudo, Y., Shakeel, M., Watanabe, S.: OWSM-CTC: an open encoder-only speech foundation model for speech recognition, translation, and language identification (2024)
27. Pratap, V., et al.: Scaling speech technology to 1,000+ languages. CoRR abs/2305.13516 (2023)
28. Rabiner, L.: A tutorial on hidden Markov models and selected applications in speech recognition. Proc. IEEE **77**(2), 257–286 (1988)
29. Sainath, T.N., et al.: Two-pass end-to-end speech recognition. arXiv abs/1908.10992 (2019)
30. Sak, H., et al.: On-the-fly lattice rescoring for real-time automatic speech recognition. In: Proceedings of the INTERSPEECH (2010)
31. Salazar, J., Liang, D., Nguyen, T.Q., Kirchhoff, K.: Masked language model scoring. arXiv:1910.14659 (2019)
32. Shin, J., Lee, Y., Jung, K.: Effective sentence scoring method using BERT for speech recognition. In: Proceedings of the ACML, pp. 1081–1093 (2019)
33. Sriram, A., et al.: Cold fusion: training Seq2Seq models together with language models. In: Proceedings of the INTERSPEECH (2017)

34. Zhou, S., Dong, L., Xu, S., Xu, B.: Syllable-based sequence-to-sequence speech recognition with the transformer in mandarin Chinese. In: Proceedings of the INTERSPEECH (2018)
35. Vaswani, A., et al.: Attention is all you need. In: Proceedings of the NeurIPS 30 (2017)
36. Wang, C., Wu, Y., Liu, S., Zhou, M., Yang, Z.: Curriculum pre-training for end-to-end speech translation. arXiv preprint arXiv:2004.10093 (2020)
37. Watanabe, S., et al.: ESPnet: end-to-end speech processing toolkit. In: Proceedings of the INTERSPEECH (2018)
38. Watanabe, S., Hori, T., Kim, S., Hershey, J.R., Hayashi, T.: Hybrid CTC/attention architecture for end-to-end speech recognition. IEEE J. Sel. Top. Signal Process. **11**(8), 1240–1253 (2017)
39. Zhang, S., Huang, H., Liu, J., Li, H.: Spelling error correction with soft-masked BERT. arXiv preprint arXiv:2005.07421 (2020)
40. Zhang, S., Lei, M., Yan, Z.: Investigation of transformer based spelling correction model for CTC-based end-to-end mandarin speech recognition. In: Proceedings of the Interspeech, pp. 2180–2184 (2019)
41. Zhou, L., Palangi, H., Zhang, L., Hu, H., Corso, J., Gao, J.: Unified vision-language pre-training for image captioning and VQA. In: Proceedings of the AAAI, pp. 13041–13049 (2020)
42. Zhou, S., Dong, L., Xu, S., Xu, B.: A comparison of modeling units in sequence-to-sequence speech recognition with the transformer on mandarin Chinese. CoRR abs/1805.06239 (2018)
43. Zhou, S., Xu, S., Xu, B.: Multilingual end-to-end speech recognition with a single transformer on low-resource languages. CoRR abs/1806.05059 (2018)

Comparative Analysis with Multiple Large-Scale Language Models for Automatic Generation of Funny Dialogues

Amon Shimozaki[✉], Yousuke Tsuge, Tatsuya Kitamura, Tomohiro Umetani, and Akiyo Nadamoto

Konan University, Okamoto 8-9-1 Higashinada-ku, Kobe, Japan
{m2324007,m2424002}@s.konan-u.ac.jp,
{t-kitamu,umetani,nadamoto}@konan-u.ac.jp

Abstract. In recent years, the widespread use of large-scale language models, such as ChatGPT, has facilitated the generation of various documents. Moreover, numerous studies have been conducted on automatic dialogue generation using large-scale dialogue models, with accuracy improving daily. Most automatic dialogue generation targets chats, Q&A, manuals, and so on. However, automatically generating dialogues incorporating humor, such as those in Manzai scenarios, remains challenging. In this study, we explore the potential for generating dialogues that include humor by employing several existing large-scale language models. Specifically, we focus on Manzai, a form of Japanese comedic content, as a case study for humorous dialogue. For this purpose, we utilized models such as Llama2, Llama2-Chat, and ChatGPT to generate a Manzai scenario automatically. Additionally, we fine-tuned Llama2 and Llama2-Chat with various datasets to automatically generate humorous dialogues and compare the outcomes.

Keywords: LLM · Dialogue · Humor · Generative AI

1 Introduction

In recent years, which is based on large-scale language models exemplified by ChatGPT[1], has rapidly gained popularity. People apply generative AI not only in research fields but also in various business fields. Generative AI has facilitated the generation of a wide variety of documents. Furthermore, extensive research has been conducted on automatic dialogue generation using generative AI, and its accuracy is improving daily. Most automatic dialogue generation targets chats, Q&A sessions, and manuals. However, it is widely acknowledged as challenging to automatically generate dialogues that include humor, such as

---

[1] https://chat.openai.com/.

Manzai dialogue, as noted by Carlos [5]. However, humor and jokes are often integral to people's daily conversations. Dialogues infused with humor not only enrich lives but also contribute positively to health. This study investigates the feasibility of generating dialogues with humor using several existing large-scale language models.

On the other hand, we have studied the generative automatically Manzai scenario [6,12]. The system is rule-based, but some parts use AI techniques. Manzai is a traditional Japanese stand-up comedy that involves two performers. One plays the role of the "funny man" called "Boke", who says something silly or incorrect, and the other is the "straight man" called "Tsukkomi", who corrects or "covers" the funny man's mistakes in a humorous way. The feature of the automatic Manzai dialogue generation is that it automatically generates a Manzai dialogue according to the user's preferences and provides Manzai performances using robots and chat applications. Generally, a Manzai scenario has a theme, and a humorous dialogue is generated for that theme.

Therefore, we believe that Manzai dialogue is appropriate as a humorous dialogue, and we attempt to generate Manzai dialogue as a dialogue generation with humor using multiple large-scale language models. Figure 1 presents an example of a Kobe-themed Manzai dialogue generated by Chat-GPT4 as of February 2024. At a cursory glance, the system seems to produce a Manzai dialogue. However, it becomes evident that the dialogue, while effectively themed around Kobe, lacks the humorous or comedic elements. Furthermore, it does not exhibit the characteristic interplay between the "Boke" and "Tsukkomi" that is quintessential to Manzai. Therefore, we attempt to generate Manzai dialogue utilizing existing models and identify the inherent challenges.

In this paper, we aim to compare the automatic generation of Manzai dialogue using multiple large-scale language models to identify the current challenges in generating Manzai dialogues, including the integration of humorous lines. Specifically, we attempt to automatically generate a script for a Manzai dialogue using ChatGPT, Llama2, and Llama2-Chat. Furthermore, Llama2 and Llama2-Chat perform fine-tuning using various data, automatically generate Manzai dialogues, and perform a comparative analysis of the results. Finally, we conduct user experiment using generated funny dialogue.

2 Related Work

There are many studies of dialogue generation using LLM. Odede et al. [10] develop JayBot, an LLM-based chatbot system that aims to improve the user experience for university personnel in the UK. Mukherjee et al. [9] develop a chatbot called ImpactBot aimed at promoting critical thinking among researchers. Zhao et al. [13] develop FluGCF, a model for fluent dialogue generation. Lee et al. [1] propose the system called DAPIE. It is user-friendly design for children. Montagna et al. [8] propose an LLM-based chatbot system capable of addressing privacy and security concerns vital for supporting chronic disease management. Ashby et al. [2] propose generating quests considering player personalization

Prompt	The theme of the Manzai is about Kobe. The Manzai features a Boke and a Tsukkomi. The dialogue is in Kansai dialect. Use "Nande ya nen" (The English word for no way!) when inserting a retort.
Boke	Hey, Grandma, let's talk about Kobe today, shall we?
Tsukkomi	Oh, Kobe, huh! Well, when you think of Kobe, what comes to mind?
Boke	Of course, it's Kobe beef! It's absolutely delicious!
Tsukkomi	That's right! That tenderness and juiciness are irresistible! Speaking of Kobe beef, it's expensive, though.
Boke	Seriously! To eat such delicious meat, you have to check your wallet first.
Tsukkomi	But Kobe isn't just known for beef, right?
Boke	That's right! Kobe has many famous tourist spots too, you know.
Tsukkomi	Yeah! Kobe Port, Meriken Park, the famous night view of Luminarie, there are so many attractions!
Boke	Absolutely! The night view of Kobe Port is romantic, Luminarie is beautiful, and Meriken Park is perfect for a leisurely stroll.
Tsukkomi	That's right, and Kobe also has famous sports teams!
Boke	Oh, right! It's Vissel Kobe, isn't it? With Iniesta on the team, it's gaining attention.
Tsukkomi	That's it! Many grandpas and grandmas must be cheering for Vissel Kobe, too.
Boke	Exactly! Kobe has plenty of delicious food and fun places to explore. It's one of the joys of retirement, isn't it?
Tsukkomi	That's right! Kobe is such an attractive city!
Boke	Absolutely! Even in old age, you can have a great time in Kobe.
Tsukkomi	That's it! It's Kobe, Nande ya nen! (Audience laughter)

Fig. 1. Example of generative Manzai dialogues using ChatGPT

within role-playing games (RPGs), along with generating dialogues between players and in-game NPCs before the occurrence of the quests. Li et al. [7] propose a Mixed Knowledge Enhanced Empathetic Dialogue Generation Model (MKEMP) to generate expressions of empathy in dialogue systems. Qiu et al. [11] propose the Syntax Recognition Dialogue Generation (SynDG) model, which incorporates syntactic information for generating responses with correct grammar.

These researches target ordinary dialogue, but our research aims to generate funny dialogue.

There are some research studies about humor. Go et al. [4] propose a method for automatically generating humorous dialogues between a user and an avatar in real time. The system extracts humorous words related to the user's input topic using concept distance, rhyme, and familiarity, and then generates a rule-based dialogue based on the extracted humorous words. They generate rule-based humorous dialogue, while our study differs from generating funny dialogue using LLM.

There are some researches generating humor using LLM. Carlos at al. [5] generate and compare humor in various LLMs through prompt engineering. They say that existing LLM models can not generate humor using prompt engineering. Zhong et al. [14] propose a system that generates humorous sentences from humorous images using Oogiri datasets as fine-tuning. The research does not generate humorous dialogue. In contrast, our research aims to generate humorous dialogue, marking a distinct focus within the field.

3 Comparison Models

3.1 Using Existing Models

In this study, we conduct a comparative experiment to ascertain whether current models can generate optimal Manzai dialogues. We evaluate the performance of various large-scale language models in the automatic generation of Manzai dialogues. The models selected for this comparison are Llama2, Llama2-Chat, and ChatGPT, with a focus on Japanese dialogue to honor Manzai's cultural specificity.

- Llama2

    We utilize Swallow[2], which, as of February 2024, is the highest-performing model in generating Japanese among the Llama2 variants available for research, commercial use, and fine-tuning. Due to GPU memory constraints during fine-tuning, we employ Swallow-13B-hf[3], a publicly accessible model within the Swallow series. Swallow-13B-hf is pre-trained on 173 million web pages, encompassing 312.1 billion words.

- Llama2-Chat

    We use Llama2-Chat, capable of fine-tuning and highly accurate in Japanese dialogue. In this study, we use Youri-7b-chat[4], which has high accuracy in generating Japanese among the models for dialogue. Youri-7b-chat's training data includes Databricks Dolly data (English), Japanese Databricks Dolly data (Japanese), Anthropic HH RLHF data (Japanese translation), FLAN Instruction Tuning data (Japanese translation), and the Izumi lab LLM Japanese dataset (Japanese).

- Chat-GPT4

    GPT-4 Turbo[5] represents the pinnacle of generation capabilities within the ChatGPT model series. We utilize gpt-4-0125-preview[6], the most recent model iteration available. Although this model is designed for multilingual use, it does not support fine-tuning.

---

[2] https://tokyotech-llm.github.io/swallow-llama.
[3] https://huggingface.co/tokyotech-llm/Swallow-13b-hf.
[4] https://huggingface.co/rinna/youri-7b-chat.
[5] https://openai.com/gpt-4.
[6] https://openai.com/product.

Both Llama2 and Llama2-Chat undergo training via Lora learning to minimize memory usage and focus on trainable layers. The training parameters are set to a batch size of 8, a learning rate of 0.0002, using the AdamW optimizer, over 10 epochs.

3.2 Comparison Model Types

Llama2 and Llama2-Chat models are evaluated against fine-tuned versions using the generated Manzai data. We conduct four distinct types of comparative experiments:

1. Without fine-tuning.
2. Fine-tuning with only Manzai data.
3. Fine-tuning both the Manzai data and the chatting data.
4. Fine-tuning the Manzai data with the speakers' roles.

In our comparative analysis, Model 1 serves as the baseline, juxtaposed against three models: Llama2, Llama2-Chat, and ChatGPT. For subsequent comparisons, Models 2-4, we narrow the focus to two models, Llama2 and Llama2-Chat. We use Nagoya University Corpus [3] as chatting data. The corpus transcribes 129 conversations and approximately 100 h of chat between native Japanese speakers. Table 1 shows the comparison models; we use 9 models in this paper. Table 1 shows the number of each model. The number is that Before the hyphen is the number of the fine-tuning type, and after the hyphen is the number of the existing model.

Table 1. The comparison models

Type	Llama2	Llama2-Chat	ChatGPT-4
No Fine-tuning	1-1	1-2	1-3
Manzai only	2-1	2-2	–
Manzai and Chat	3-1	3-2	–
Manzai and Role	4-1	4-2	–

4 Generating Manzai Data Set

In comparison to models 2-4, fine-tuning with Manzai data is necessary to generate Manzai performances in a large-scale language model. Fine-tuning is a data-intensive process, and there is a notable scarcity of textual Manzai dialogue data, as most existing Manzai content is in video format. Consequently, there is a pressing need to transcribe these dialogues into text. Additionally, this transcription process must include not only the speech text but also information about the speakers, which presents a significant challenge. In this study, we propose an automatic transcription system for Manzai dialogues, which is automatic annotation. The flow of the proposed system is shown in Fig. 2.

We utilized a corpus of video clips to amass a comprehensive collection of Manzai dialogues. The corpus comprises 120 videos, totaling approximately 700 min of content.

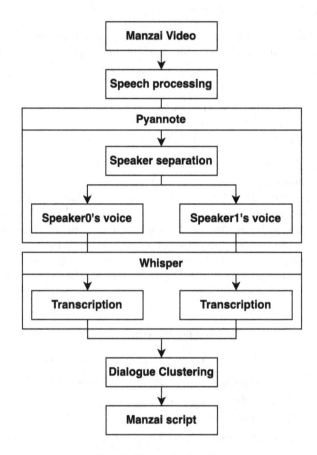

Fig. 2. Transcription flow

Speaker Separation

Manzai is performed by two comedians, Boke and Tsukkomi.

To fine-tune the Manzai dialog, it is necessary to separate who is speaking the line.

In this study, we use Pyannote[7] to separate the speakers. However, although it is possible to separate the speaker's role, it is difficult to annotate it because the role of the "Boke" and the role of the "Tsukkomi" may differ depending on

---

[7] https://huggingface.co/pyannote/speaker-diarization.

the Manzai dialogue. Therefore, in this study, the speakers are set to 0 and 1 so that each Manzai dialogue is completed within each Manzai dialogue.

Transcription

In this study, we employ the Whisper model[8] for the automatic transcription of Manzai dialogue from video to text data. Whisper offers various model sizes, and our comparative analysis of the small, medium, and large-v3 models revealed that the small model produced numerous typographical errors, necessitating extensive correction efforts. Conversely, the large-v3 model, despite requiring more time to process the dataset, significantly reduced the overall workload and expedited the dataset creation process. Therefore, we opt for the large-v3 model for this research. The transcription process begins with the extraction of speech data from each speaker within the Manzai videos. This speech data is then transcribed, and the Manzai dialogues are reconstructed based on the output timestamps.

Post-processing

The current methodology for transcribing and distinguishing speakers in Manzai dialogue encounters challenges due to the degradation of dialects and the quality of video data. These factors can impede the accuracy of speaker separation and transcription fidelity, necessitating further refinement of the system to accommodate dialectical variations and enhance video data processing. Furthermore, Manzai has a speech technique called "kabuse". Boke and Tsukkomi speak very fast, and the back-and-forth banter, where Tsukkomi's quick retort or "cover" follows the boke's setup, is a key element of the humor in Manzai. At this time, Tsukkomi speak; while Boke still speakers this speech technique is called "Kabuse". In this way, in Manzai, "Kabuse" makes the speaker's speech fast and overlapping in many places. Therefore, automatic transcription software cannot completely transcribe them; we must correct them manually. Finally, the output scripts manually correct the speaker and speech information. The script consists of each utterance's start time, end time, speaker, and lines, as shown in Table 2.

Table 2. Example of the results of transcription

Start Time	End Time	Speaker	Dialogue
00:10:0	00:11:3	0	Hi!
00:11:3	00:12:8	1	Nice to meet you.
...			
04:34:1	04:35:4	1	Get a job!
04:35:2	04:36:7	0	Enough already! Thank you very much.

---

[8] https://huggingface.co/openai/whisper-large-v3.

Sys prompt	The following are instructions explaining the task. Write a response that appropriately meets the requirements.
Prompt	###instructions: You are a funny comedian. Generate a Boke and Tukkomi dialogue following the conversation content below. ###Conversation content: Theme Person A: Boke Person B: Tsukkomi

Fig. 3. Prompt

Dialogue Clustering

In general, speakers take turns speaking in a conversation. However, in Manzai, the Boke and the Tsukkomi do not necessarily speak alternately. For example, there are many cases in which the Boke speaks continuously, and then the Tsukkomi speaks, forming a single conversation. The opposite also happens. Therefore, we must separate dialogue semantically. In this study, therefore, we manually cluster Manzai dialogues.

5 Feature Analysis of Generated Results

5.1 Condition of Generations

Models
We generated 27 types of dialogue (show Table 1).

Prompts. The accuracy of sentence generation can be significantly enhanced by carefully crafting the input prompts for large-scale language models. We consider when it would be possible to generate Manzai dialogues by considering prompts from existing large-scale language models. Therefore, we generated multiple prompts for existing large-scale language models. Figure 3 shows the prompt that was created.

We use the same prompts for the dialog generated by four models.

5.2 Feature Analysis of Each Model

Tables 3, 4 and 5 shows the results of generates.

Model1: Without Fine-Tuning
Each Tables 3, 4 and 5(1) shows the results generated by the models without fine-tuning. The results show that all models can generate dialogues about the theme, however they are general dialogues without humor or jokes.

ChatGPT generates dialogues close to normal everyday dialogues about the theme, but Llama2 and Llama2-Chat are unnatural dialogues about the theme.

Furthermore, a comparison of Llama2 and Llama2-Chat shows that Llama2-Chat generates a more conversational tone. This indicates that using a model without fine-tuning to generate Manzai dialogue is not useful. On the other hand, ChatGPT is the most useful model for generating chat dialogues without humor.

Model2: Fine-Tuning with only Manzai Data

Each Tables 3, 4 and 5(2) shows the results of generation model2. The results show that both models are able to generate "Tsukkomi" such as "Where exactly? (Dokoya-nen in Japanese)" and "No way! (Nandeya-nen in Japanese)". These Japanese words are typical lines of Manzai dialogue. On the other hand, neither model was able to generate the "boke" (a technique called "hazushi" that uses words different from the words that normally come) as seen in Manzai. Furthermore, although the prompt input was set to "A" as a "Boke" and "B" as a "Tskkomi", the actual generated results often showed that "A" was speaking as a "Tskkomi". The reason is that the roles of the speakers in the Manzai dialogue data were not annotated, and the data could not be training. In addition, the output of Llama2 and Llama2-Chat tended to be shorter than the output from experiment1 without fine-tuning, and the output content was monotonous. We consider that the reason is that many Manzai dialogues contain short phrases such as "Un" (a kind of "aizuchi" in Japanese).

Model3: Fine-Tuning Both the Manzai Data and the Chatting Data

We generated the dialogue that included both Manzai data and the chatting data in the fine-tuning because the results of model 2, in which only the Manzai data was used for fine-tuning, showed that each utterance in the dialogue was shorter than in model 1. However, the results indicated that the length of a single utterance during dialogue was not significantly different from the results of model 2.

Furthermore, Model 3 does not produce any tsukkomi compared to Model 2. We consider that this is due to the large amount of additional chatting data compared to the Manzai data. Therefore, it is necessary to consider the ratio of chatting data to be added to the Manzai data.

Model4: Fine-Tuning the Manzai Data with the Speakers' Roles

We decide to define the speaker's role in model 3 due to the ambiguity regarding the speaker's role in model 2. The results demonstrate that Llama2 and Llama2-Chat are capable of generating humor, jokes, and Tsukkomi. This suggests that role annotation can produce dialogues that are more humorous than those generated by the Manzai model. Consequently, annotations prove to be beneficial for learning Manzai data. However, the jokes and Tsukkomi produced through annotations did not elicit laughter to the same extent as traditional Manzai data.

We believe that the reason is that there are different techniques used in Manzai jokes and Tsukkomi, and these techniques have not been well learned.

Table 3. Results of theme of

No.	Type	Llama2	Llama2-Chat	ChatGPT-4
Theme: Kobe				
(1)	Pre-traing	A: Kobe City, located in the southeastern part of Hyogo Prefecture, is a port town. B: That's right. A: Kobe City has prospered as a trading port since ancient times, and is still known today as an international port city. B: That's right.	A: There are many tourist attractions in Kobe, you know. B: Oh, that sounds great. A: Kobe is a city comparable to Osaka and Kyoto in Japan. B: Hahaha!	A: Hey, let's go on a trip to Kobe sometime. B: Huh? Kobe? Are we going to meet cows? A: Don't be silly! I'm not talking about Kobe beef, I mean going to Kobe. B: Well, if we go to Kobe, I definitely want to eat Kobe beef at least!
(2)	Manzai only	A: Where are we? B: We're in a big town in the west of Japan. A: Where exactly? (Dokoya-nen in Japanese)	A: Hey, hey, did you know? The hotel across from Kobe Harbor used to be a passenger ship. B: Huh, really? I had no idea. A: You didn't know? That hotel used to be a white passenger ship and was built before World War II. B: Wow, really? That's amazing.	–
(3)	Manzai and Chat	A: What do you think of Kobe City? B: Kobe City? A: Yeah! B: Hmm, I haven't really been there much.	B: Where are you from? A: I'm from Kobe. B: What's Kobe like?	–
(4)	Manzai and Role	A: Um, well, first of all, I thought we could talk about Kobe Memorial Hall, which is our home ground. B: Yeah, yeah, yeah. Why didn't you mention that first? That's quite sudden, really. Well, never mind, let's talk about Kobe Memorial Hall then.	A: I live in Kobe. B: What are the people in Kobe like? A: People from Kanagawa may have a joking expression, but people from Kobe seem to be joking seriously. B: What does that mean?	–

Table 4. Results of theme of Karaage

No.	Type	Llama2	Llama2-Chat	ChatGPT-4
	Theme: Karaage			
(1)	Pre-traing	A: Shall we have fried chicken for dinner tonight? B: Sounds good! Fried chicken is really delicious, isn't it? A: Fried chicken is great, right? Personally, I prefer it not too crispy, but how about you, B? B: I actually prefer it crispy.	B: Do you like fried chicken? A: Yes, I love it. B: What kind of fried chicken do you like? A: I don't really have any particular preferences.	A: Well then, shall we talk about fried chicken, B? B: Sounds good, fried chicken talk is always okay with me. A: I don't really think fried chicken is the best dish in the world, after all. B: Ah, in that case, we mustn't forget the fleeting nature of fried chicken, right?
(2)	Manzai only	B: Why do you always want to eat fried chicken? A: Because it's delicious! B: A, that's just a lump of meat. A: B, don't say things like that.	B: Um, about this fried chicken, is it homemade? A: No, it's from a convenience store bento. B: Oh, I see. A: What's with that expression?	–
(3)	Manzai and Chat	A: Fried chicken is delicious, isn't it? B: Yeah, it is! A: But fried chicken has a lot of oil, so I can't eat too much of it. B: Oh, really?	A: Hey, you know, fried chicken doesn't really taste that different, right? B: It's not like that at all. A: Huh, what does it taste like then? B: Well, it's coated with batter, so it tastes like the batter.	–
(4)	Manzai and Role	A: What's your favorite food? B: I like chicken. A: Which part of chicken breast do you like, B? B: I like chicken wings with bones, but I don't like the wing part. A: Do you eat chicken tails and wings? B: No way.	A: What are you making? B: I'm making fried chicken using chicken breast. A: Your breast is too small. B: Your mind is too small.	–

Table 5. Results of theme of baseball team (Hanshin Tigers)

No.	Type	Llama2	Llama2-Chat	ChatGPT-4
		Theme: Baseball team (Hanshin Tigers)		
(1)	Pre-traing	A: The Hanshin Tigers won yesterday too, right? B: Huh? The Hanshin Tigers lost yesterday. A: Why is that? B: Because yesterday, the Hanshin Tigers didn't hit a single home run.	B: Today is the day the Hanshin Tigers play! A: Yeah, that's right. B: They'll win! A: Yeah, that's right.	A: Hey, you know, the Hanshin Tigers, they're obviously themed around tigers, right? B: Well, yeah, so what? A: If that's the case, then the team mascot must obviously be a tiger, right? B: Well, yeah, but what's your point?
(2)	Manzai only	A: How did the Hanshin Tigers game go today? B: I didn't watch the Hanshin game. A: You didn't watch it? B: Do you know the result of today's game?	A: What if the Hanshin Tigers win the championship? B: It's too early to say. A: No way! (Nandeya-nen in Japanese) hesitation? B: Do you really think they can win this year? A: (ř ř) Yes!	–
(3)	Manzai and Chat	A: What's going on with the Hanshin Tigers this year? B: There's no doubt they'll win the championship this year! A: Why are you saying that? B: No, really, there's no doubt the Hanshin Tigers will win the championship this year!	B: Are you a Hanshin Tigers fan, A? A: Yes, I am. B: As a Yomiuri Giants fan, I can't forgive that. A: Do you know what day it is today?	–
(4)	Manzai and Role	A: The Hanshin Tigers are strong this year, aren't they? B: Did you watch today's game? A: Nah, I didn't. I'm a Hanshin Tigers fan, you know. B: No way! (Nandeya-nen in Japanese)	B: Why did Hanshin lose? A: It's your fault! B: Why's that? B: It's Ramos's fault. A: Who's Ramos? B: You don't even know?	–

6 Experiment

6.1 Condition of the Experiment

We conducted a user experiment to determine whether the generated dialogue was humorous. The data is the 27 Manzai dialogues in the experiments in Tables 3, 4 and 5. The following number of data is the same as Table 1. We use crowdsourcing to conduct the experiments. The subjects evaluated each dialogue on a 5-point scale, with ratings ranging from 5 (good) to 1 (bad). We asked the following three questions from the perspective of establishing dialogue and whether it was interesting.

Q1. Are the generated dialogues correct?
Q2. Are the dialogues funny?
Q3. Are the dialogues Manzai-like dialogue?

6.2 Results and Discussion

Results and Discussion Q1

Figure 4 shows the results of Q1. From the Table 4 70% of the subjects judged that Llama2-Chat (1-2) and Llame2-Chat (2-2), a fine-tuning with only Manzai dialog, were dialogues. According to our research, only about 30% of the subjects perceived a successful dialog between Llame2 and Llame2 chat. We believe this is because creating humorous dialog can be challenging. Humorous dialogues are not always correct as dialogues because they often give unexpected responses.

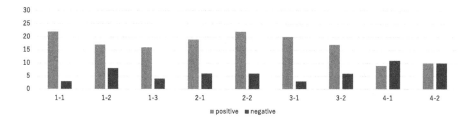

Fig. 4. Results of Q1

Results and Discussion Q2

In all models, subjects found fewer than 50% of the dialogues funny, suggesting that the models do not consistently produce humorous dialogues. However, Llama2-Chat (4-2) and the Manzai role model were considered to generate the funniest dialogues. The results of 2-1 and 2-2 indicate that fine-tuning the Manzai dialogue facilitates the generation of funny dialogues. On the other hand, using chatting data to fine-tune Llama2 and Llama2-Chat from 3-1 and 3-2 showed that it was difficult to generate funny dialogues as in the pre-training models 1-1 and 1-2. We consider that it is since the chatting data is too large compared to the Manzai data. Furthermore, subjects judged that ChatGPT (1-3) produced the most funny dialogue. However, the results show that ChatGPT cannot generate Boke or Tsukkomi. This indicates that the subjects found ChatGPT engaging due to its ability to generate more in-depth content relevant to the topic, which captivated their interest (Fig. 5).

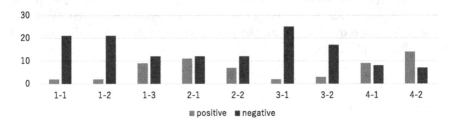

Fig. 5. Results of Q2

Results and Discussion Q3

Over 50% of the subjects rated dialogues from models 2-1, 4-1, and 4-2 as resembling Manzai. Notably, more than 60% of the subjects identified dialogues from model 4-2 as Manzai-like. These findings suggest that the optimal model is the one fine-tuned using Llama2-chat with Manzai data and roles. Additionally, it was observed that the models without fine-tuning and those fine-tuned with only chat data yielded the least favorable outcomes (Fig. 6).

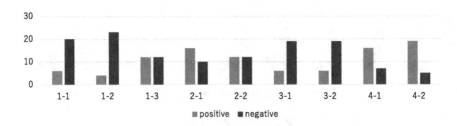

Fig. 6. Results of Q3

Total Discussion

The experimental results reveal that the Llama2-Chat model, incorporating Manzai dialogue data and roles, was deemed the most engaging by participants. However, fewer than 50% of the participants found the dialogues produced by this model humorous. This suggests that none of the models in this study consistently generated funny dialogues. The findings from RQ1 and RQ2, particularly model 4-2, show no correlation between the ability to generate funny dialogues and the production of correct dialogues. These outcomes highlight the need for future research and development of models capable of creating humor. Additionally, the study suggests that accuracy in dialogue generation need not be a priority when the goal is to produce humorous dialogue.

7 Conclusion

In this study, we aim to automatically generate funny dialogue using Manzai dialogue as the first stage of funny dialogue generation using large-scale language models.

First, we propose generating text-based Manzai dialogue from actual Manzai videos. Next, we analyzed three existing large-scale language models with 4 types of fine-tuning: one without fine-tuning, one with Manzai dialogue data only, one with Manzai dialogue data and chatting data, and one with Manzai dialogue data and speaker's role. In addition, we conducted user experiments using dialogue data generated by these models. The results show that Llama2-Chat, which annotates the speaker's role with Manzai data, currently generates the most interesting dialogue. The results indicate that humorous models are needed. Moreover, when we study generating funny dialogues, we need not be concerned about generating correct dialogues.

In the near future, we will study the following points.

- Fine-tuning more data of Manzai.
- Improving prompts during training and generation.
- Considering the level of detail of label type.

Acknowledgements. This work was partially supported by Research Institute of Konan University.

References

1. CHI '23: Proceedings of the 2023 CHI Conference on Human Factors in Computing Systems (2023)
2. Ashby, T., Webb, B.K., Knapp, G., Searle, J., Fulda, N.: Personalized quest and dialogue generation in role-playing games: a knowledge graph- and language model-based approach. In: Proceedings of the 2023 CHI Conference on Human Factors in Computing Systems (2023)
3. Fujimura, I., Shoju, C., Mieko, O.: Lexical and grammatical features of spoken and written Japanese in contrast: exploring a lexical profiling approach to comparing spoken and written corpora. In: Proceedings of the VIIth GSCP International Conference. Speech and Corpora, pp. 393–398 (2012)
4. Go, K., Onishi, T., Ogushi, A., Miyata, A.: Conversational agents replying with a Manzai-style joke. In: Proceedings of the 33rd Australian Conference on Human-Computer Interaction, pp. 221–230 (2022)
5. Gómez-Rodríguez, C., Williams, P.: A confederacy of models: a comprehensive evaluation of LLMs on creative writing (2023)

6. Haraguchi, K., Yane, K., Sato, A., Aramaki, E., Miyashiro, I., Nadamoto, A.: Chat-type Manzai application: mobile daily comedy presentations based on automatically generated Manzai scenarios. In: Proceedings of the 18th International Conference on Advances in Mobile Computing & Multimedia, pp. 143–152 (2021)
7. Li, X., Wang, G., Wang, Y., Zhou, Q.: Mixed knowledge-enhance empathetic dialogue generation. In: Proceedings of the 2023 International Conference on Electronics, Computers and Communication Technology, pp. 77–81 (2024)
8. Montagna, S., Ferretti, S., Klopfenstein, L.C., Florio, A., Pengo, M.F.: Data decentralisation of LLM-based chatbot systems in chronic disease self-management. In: Proceedings of the 2023 ACM Conference on Information Technology for Social Good, pp. 205–212 (2023)
9. Mukherjee, A., Santana, V.F.D., Baria, A.: ImpactBot: chatbot leveraging language models to automate feedback and promote critical thinking around impact statements. In: Extended Abstracts of the 2023 CHI Conference on Human Factors in Computing Systems (2023)
10. Odede, J., Frommholz, I.: JayBot – aiding university students and admission with an LLM-based chatbot. In: Proceedings of the 2024 Conference on Human Information Interaction and Retrieval, pp. 391–395 (2024)
11. Qiu, J., Zhang, H., Yang, Y.: SynDG: syntax-aware dialogue generation. In: Proceedings of the 2023 9th International Conference on Computing and Artificial Intelligence, pp. 504–510 (2023)
12. Shimozaki, A., Yamamoto, J., Nadamoto, A.: Automatic generation of funny-dialog based on cuisine recipes. In: Advances in Networked-Based Information Systems, pp. 216–226 (2023)
13. Zhao, Y., Cheng, B., Huang, Y., Wan, Z.: FluGCF: a fluent dialogue generation model with coherent concept entity flow. IEEE/ACM Trans. Audio Speech Lang. Process. **32**, 853–867 (2024)
14. Zhong, S., et al.: Let's think outside the box: exploring leap-of-thought in large language models with creative humor generation (2023)

Effectiveness of the Programmed Visual Contents Comparison Method for Two Phase Collaborative Learning in Computer Programming Education: A Case Study

Thanh Ha Nguyen[1](✉)[iD], Yi Sun[2], Takeshi Nishida[1], Xiaonan Wang[1], Kazuhiro Ohtsuki[1], and Hidenari Kiyomitsu[1][iD]

[1] Kobe University, Kobe, Japan
{hant,217c303c}@stu.kobe-u.ac.jp, tnishida@people.kobe-u.ac.jp,
{ohtsuki,kiyomitu}@kobe-u.ac.jp
[2] Kobe Institute of Computing, Kobe, Japan
sun@kic.ac.jp

Abstract. This paper introduces the Programmed Visual Content Comparison (PVCC) for collaborative programming learning. The PVCC is based on comparing two displayed visual objects produced by programming. This feature enables candidates to take the tests in a lighthearted and fun manner because they only judge "which is more difficult." The test based on the PVCC method works well to assess a candidate's programming abilities related to Panoramic Understanding of Programming (PUP). We conducted a collaborative learning experiment with students in a programming class using the PVCC method. This experiment method contains two phases of activity. First, each candidate answers a problem with the reason for the judge. In the second (collaborative phase), candidates discuss their answers and reasons in a group and then make the group answer with its explanation. The result from the experiment suggests a potential that collaborative learning facilitated by the PVCC method is beneficial for students in programming education by improving understanding and problem-solving skills in programming.

Keywords: computer science education · programming training · collaborative learning · software engineering

1 Introduction

In recent years, software development has seen an increase in situations where parts of the program are black-boxed, such as sample code copying, development tools, and existing libraries [1]. It has been noted that there is a growing number of software developers who struggle with coding [3]. In such circumstances, the ability to understand the context of a program, in other words, having a Panoramic Understanding of Programming (PUP), is more important than memorizing programming language syntax. Here, context refers to algorithmic understanding and the ability to comprehend program composition and design from a broader perspective.

To measure the panoramic understanding of programming, we proposed the Programmed Visual Contents Comparison (PVCC) method and developed a self-contained understanding assessment system that can be taken individually online [4, 5]. With this method, by comparing two or more displayed pictures produced by programming samples (a Problem), a student must decide which one of the programs producing those pictures is more difficult to build by programming or if the difficulty is similar for all of them. In a previous study [6], we examined the applicability of this assessment system to measure the understanding of all skill levels of employees in an IT company, specifically those engaged in software development. The results showed that the PVCC method can assess the panoramic understanding of programmers and non-programming IT company employees. Furthermore, we demonstrated that the nature of the test, which allows participants to give simple answers regarding the difficulty of visual content, makes it enjoyable and suitable for cooperative learning, as it feels more like a quiz than a burdensome test [6].

We also designed and prototyped a question-authoring system for learning programming skills based on the PVCC method [7]. This system enables educators and content creators to design questions targeting different programming concepts, allowing learners to engage in focused and targeted assessments. The question-authoring system enhances the learning experience by providing learners with tailored questions that challenge their panoramic understanding of programming.

We examine the practical applicability of the Programmed Visual Contents Comparison (PVCC) method to collaborative learning in computer programming education. In cooperative learning, learners first create answers to test questions individually. Next, the learners look at everyone's answers, discuss the reasons for each answer, and narrow the answer to one through discussion. We experimented with a small group of students at a graduate school [8]. The findings from this experiment provide compelling evidence that collaborative learning using the PVCC method is highly advantageous for students in programming education.

This paper introduces a case study of using collaborative learning in programming education with the PVCC method on a larger group of students. This experiment was conducted in four class sessions. This collaborative learning method is one of the strong candidates for transitioning the educational format from individual exercises to problem-solving.

2 PVCC Method and Collaborative Learning

2.1 The Basis of PVCC

The Programmed Visual Content Comparison method involves viewing two visual contents (static images, animations, interactive images that change with mouse operations) created by relatively similar programs, as shown in Fig. 1, and imagining the programming involved in creating them. It asks the viewer to decide among three options: which program was more difficult, less difficult, or of equal difficulty to create. Imagining the entire programming structure of a piece of visual content requires considerable knowledge and can be challenging for beginners to intermediate learners. However, when comparing differences, it is possible to focus only on the differing aspects without understanding the whole picture.

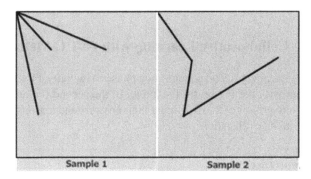

Fig. 1. A PVCC Question with/without *previous position storing* concept

"Previous position storing" is the core programming concept of the question shown in Fig. 1, and the correct answer is the sample on the right. The reason is that the right one stores the previous mouse position in a variable. When the mouse clicks, a line is drawn between the previous and current positions, while the left one draws a line between a fixed point and the mouse position.

The answering method is straightforward, giving a unique question: "If you were to make any of the previously displayed samples using a programming language, which one do you think is the most difficult?" and four answer options: "Sample 1" (the left one), "Sample 2" (the right one), "Both same" and "Not sure."

2.2 PVCC Test Authoring System for Collaborative Learning

A PVCC test comprises a carefully selected sequence of questions, each featuring a pair of programming examples to generate visual content. The source code for each example is either written from scratch, derived from coding sample books, or sourced online. Significantly, the conveyance of a programming concept hinges not just on the explicit content of a program but also on its structural comparison across various samples. Thus, attaching metadata to each programming sample is critical to furnish additional context.

Furthermore, this metadata encapsulation extends to the characteristics of each question derived from the sample pairings, offering more profound insights. Learners interact with only the visual outcomes and are unaware of the underlying source code. The objective of each question is to evaluate a learner's comprehension of specific programming concepts, ensuring that the visual disparities within a question are discernible to those with the requisite expertise.

To facilitate the organization of these tests, we created a prototype of an authoring system for the PVCC test [7]. This system serves as a comprehensive tool for arranging and administrating test content. It enables programming instructors to catalog each question component as a sample, complete with detailed metadata and descriptions. Users can pair samples to formulate questions and compile these into customized test sets for various assessments or research initiatives, offering a tailored approach to evaluating programming competencies. In addition, we allow the system to make question components with a single visual content. In some cases of collaborative learning, comparing a pair of visual contents is not necessary [6].

3 Two Phase Collaborative Learning with PVCC Method

This collaborative learning method contains two phases of activity. First, each candidate answers a problem with the reason for the judge. In the second (collaborative phase), candidates break into groups of 2 to 3, discuss their answers and reasons, and then make a group answer with its explanation.

3.1 Flow to Group Discussion

The initial phase, where participants individually answer a problem and provide their reasoning for the judge, plays a pivotal role in the collaborative learning process. This method ensures that each member brings a distinct perspective and a well-considered rationale to the discussion. When these individual insights and justifications are brought into the collaborative phase, they significantly enrich the group discussion. The diversity of viewpoints encourages a more profound exploration of the subject matter, promotes critical thinking, and facilitates a more comprehensive understanding. Moreover, this method fosters an environment where participants learn to value different perspectives, enhancing their ability to reach a consensus through constructive debate. In essence, the individual preparation in phase one acts as a springboard, propelling the group towards more meaningful and productive exchanges in the second phase.

3.2 Experiment

The experiment, conducted in a programming class at a university, was meticulously planned. Four class sessions were dedicated to it, each designed to accommodate more than 30 students. Before each experiment day, the students were divided into groups of 2–3 people. In some instances, students were absent from the experimental session, resulting in groups needing more members to conduct the discussion. In such cases, we merged two groups to have sufficient participants.

After independently answering the PVCC questions from the test system, the prearranged members of each group would gather and discuss to come up with the group's answers to the same initial questions. There were about 14–15 groups in each experimental session. Figure 2 shows the number of participants and groups on each experiment day.

4 A Case Study and Discussion

This chapter discusses the experiment's results. On days one and two, two questions were introduced to participants, while on days three and four, only one question was introduced.

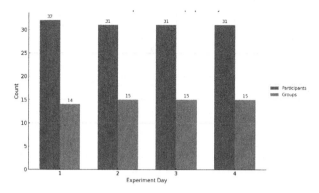

Fig. 2. The number of Participants and Groups of experiment per day

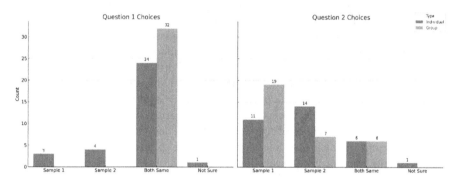

Fig. 3. Answer of Individuals and Groups in day 1 of Experiment

4.1 Discussion

Figure 3 illustrates the answers of individuals and groups on day one. The y-axis represents the number of persons. The results show the distribution of individual and group answers for the two questions presented in the experiment. For Question 1, most individuals and groups chose "Both same" as their answer, which matches the correct answer. This result indicates a high level of consensus in the perception of this question among members and their groups.

Meanwhile, Question 2 shows more diversity in individual answers, with choices distributed between "Sample 1," "Both same," and "Sample 2." However, the correct answer for Question 2 is "Sample 1," indicating that not all individuals accurately identified the correct answer, or their group did not always reach the correct consensus.

Figure 4 shows several observations regarding the responses to Questions 1 and 2 on day two. Both individual and group responses show a clear preference for specific options over others. For instance, "Sample 1" and "Both the same" appear more popular across the questions, indicating a trend in how the questions were interpreted or the samples' content.

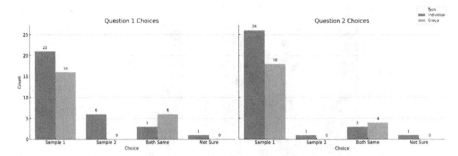

Fig. 4. Answer of Individuals and Groups in day 2 of Experiment

In both questions, there is a slight difference in the distribution of responses between individual and group decisions. Group discussions or consensus mechanisms might shift the preference slightly compared to individual choices.

The presence of "Not sure" responses across the questions, more so in individual answers for Question 2, might indicate the level of confidence or clarity participants had regarding the questions. It suggests that Question 2 might have been perceived as more challenging or ambiguous. The reduction in "Not sure" responses from individual to group might reflect the effect of group discussion in reducing uncertainty, as participants might have been persuaded or reached a consensus through the exchange of ideas.

The transition from individual to group responses shows an interesting dynamic. Group discussions might lead to convergence toward specific answers, reducing the diversity seen in individual responses. It could be a sign of social influence or a natural tendency towards consensus-building in group settings.

Figure 5 illustrates the results of day three and day four. Based on the results, the following evaluations are confirmed.

On day three, "Sample 1" was indeed the correct answer most frequently (as suggested by the dominance of this option in both individual and group responses), this indicates a high level of accuracy in both individual decision-making and group consensus. The effectiveness of group discussions in reaching the correct conclusion can be inferred from the prevalence of the correct answer among group decisions. The reduction of "Not sure" responses and the concentration of answers on the correct option suggest that participants were confident and accurate in their responses; it implies that the questions or samples provided were clear enough to reach decisive and correct conclusions.

The distribution of responses on day four, being more spread out across the options, including "Not sure," suggests a more challenging question or less clear samples. If the correct answer still aligns with "Sample 1" (the most popular choice), it suggests that, despite increased difficulty, a majority were able to identify the correct answer, albeit with less certainty. The presence of more varied responses, including "Not sure," in group decisions may indicate challenges in reaching a consensus or correctly identifying the correct answer. This variability could reflect a healthy debate within groups or uncertainty about the correctness of the decision, highlighting the complexity of the question or ambiguity in the provided information.

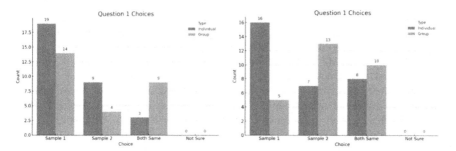

Fig. 5. Answer of Individuals and Groups in day 3 (left) and day 4 (right)

4.2 Findings

The analysis of responses over four days highlights the complexity of group and individual decision-making processes, especially in the context of varying question clarity and difficulty. On days when questions were straightforward, both individual and group responses showed a high degree of accuracy, underscoring the effectiveness of collaboration in discerning the correct answers. On the first two days of the experiment, some individuals chose "Not sure" during their individual response phase. However, no cases of this option were chosen after the group discussion. Therefore, group discussions clarified doubts, and the members could gain new knowledge from collaborative learning, compared to answering independently at an individual level. Furthermore, in the subsequent days of the experiment, there were no instances of choosing "Not sure," even in the individual response phase, indicating that the members made progress through collaborative learning. The members confidently selected the options they believed were correct instead of meaningless.

5 Conclusion

In this paper, we discuss the results of a case study of using collaborative learning in programming education with the PVCC method. The collaborative learning with the PVCC method is a selection for transitioning the educational format from individual exercises to problem-solving. The approach offered by the PVCC method transforms the testing experience into an enjoyable and engaging activity, comparable to quiz-solving, for programming learners. It bypasses the need for extensive programming syntax knowledge by concentrating on assessing problem difficulty levels.

Acknowledgments. This work was partly supported by JSPS KAKENHI Grant Number 19K03000 and 19K03030.

References

1. Burnett, M.M., Myers, B.: Future of end-user software engineering: beyond the silos. In: Proceedings of the Future of Software Engineering Conference (FOSE 2014), pp. 201–211 (2014)

2. Brandt, J., Guo, P.J., Lewenstein, J., Klemmer, S.R.: Opportunistic programming: how rapid ideation and prototyping occur in practice. In: Proceedings of the 4th International Workshop on End-User Software Engineering (WEUSE '08), pp.1–5 (2008)
3. LaToza, T.D., Myers, B.: Hard-to-answer questions about code. In: Proceeding of PLATEAU '10, pp.1–6 (2010). Article No. 8
4. Calderon, D.M., Man, K., Kiyomitsu, H., Ohtsuki, K., Miyamoto, Y., Sun, Y.: An evaluation method for panoramic understanding of programming by comparison with visual examples. In: 2015 IEEE Frontiers in Education Conference (FIE), El Paso, TX, USA, pp. 1–8 (2015). https://doi.org/10.1109/FIE.2015.7344104
5. Calderon, D.M., et al.: Measurement range increment in a method for evaluating Panoramic Understanding of Programming. In: 2016 IEEE Frontiers in Education Conference (FIE), Erie, PA, USA, pp. 1–8 (2016). https://doi.org/10.1109/FIE.2016.7757489
6. Kiyomitsu, H., et al.: An approach for evaluating IT employees' programming ability using the programed visual contents comparison method. In: Proceedings of the 2018 IEEE International Conference on Teaching, Assessment, and Learning for Engineering (TALE2018), Wollongong, New South Wales, Australia, pp. 423–430 (2018). https://doi.org/10.1109/TALE.2018.8615401
7. Kiyomitsu, H., Ha Nguyen, T., Ohtsuki, K., Carderon, D.M., Shao, S., Shigyo, K.: Question authoring for learning programming skills based on the programmed visual content comparison method. In: 2019 IEEE International Conference on Engineering, Technology and Education (TALE), Yogyakarta, Indonesia, pp. 1–6 (2019). https://doi.org/10.1109/TALE48000.2019.9225958
8. Ha Nguyen, T., Kiyomitsu, H., Sun, Y., Nishida, T., Ohtsuki, K.: Collaborative learning in programming education with the programmed visual contents comparison method. In: 2024 12th International Conference on Information and Education Technology (ICIET), Yamaguchi, Japan, pp. 467–471 (2024). https://doi.org/10.1109/ICIET60671.2024.10542766

Generating Achievement Relationship Graph Between Actions for Alternative Solution Recommendation

Tsukasa Hirano[1], Yoshiyuki Shoji[1,2(✉)] [iD], Takehiro Yamamoto[3] [iD], and Kouzou Ohara[2]

[1] Aoyama Gakuin University, Sagamihara, Kanagawa 252-5258, Japan
hirano@sw.it.aoyama.ac.jp
[2] Shizuoka University, Hamamatsu, Shizuoka 432-8011, Japan
shojiy@inf.shizuoka.ac.jp, ohara@it.aoyama.ac.jp
[3] University of Hyogo, Kobe, Hyogo 651-2197, Japan
t.yamamoto@sis.u-hyogo.ac.jp

Abstract. This paper proposes a method for generating a graph of the purpose and alternative actions for a given specific action name. For instance, the purpose of "concentrating on the lecture" can be achieved by taking actions such as "preventing sleepiness," "taking sugar," or "changing the lecture." In this case, the actions "concentrating on the lecture" and "preventing sleepiness" have a means-end (*i.e.*, achieved-by) relationship, and "preventing sleepiness" and "taking sugar" have substitutable relationships. Our method presents these achievable and substitutable actions as a graph to give people various choices and help them make decisions. The proposed method extracts descriptions of actions from product review data using a large-scale language model and assigns action names. The relationship between action names is calculated based on graph computation and language patterns in web documents. The results of subject experiments show that the proposed method can find many pairs of action names in an achievement relationship.

Keywords: Purpose and Means · Entailment · Ontology Construction

1 Introduction

Individuals engage in various actions daily to achieve specific purposes. For instance, some people often reduce carbohydrate intake to keep them slim. In such cases, most people fall into two types of tunnel vision when choosing actions to fulfill their purposes.

The first type is the tunnel vision of the means. Some might overly focus on one method, such as reducing carbohydrate intake to keep them slim, which can sometimes be detrimental to health. Ideally, other actions such as jogging or weight training should also be considered, but they might not come to mind. Similarly, another type of tunnel vision can occur regarding the purpose itself.

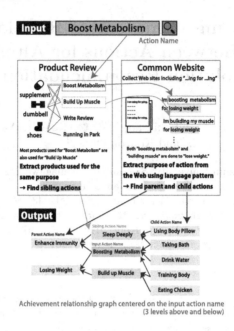

Fig. 1. System input/output and achievement relationship graph between actions

For instance, an excessive focus on keeping them slim could harm their health. If the reason for wanting to stay slim is to be more attractive, there are other means to achieve this, such as changing their attire or makeup, or increasing social interactions. If the purpose is to continue wearing favorite clothes, options like altering the clothing or purchasing the same designs in larger sizes exist. However, for individuals experiencing tunnel vision, conceiving these alternative actions can be challenging.

Traditional recommendation systems commonly adopt collaborative filtering [1]. Most collaborative filtering tends to recommend items similar to the selected product. When the user aims to "keep them slim," and buy one diet food, such recommendations will focus on food-related solutions overly.

In reality, alternative means exist to achieve "keep them slim," such as products related to "doing running" or "engaging in weight training," like exercise equipment or running shoes. Furthermore, if the underlying reason for wanting to keep slim is to be more attractive, then products that enhance one's appearance, such as "cosmetics" or "hair wax," could also effectively address this purpose. These alternatives highlight the limitations of traditional recommendation systems in recognizing and suggesting a broader range of solutions that align with the user's ultimate purposes.

In this research, we propose a method to collect action names related to a given action name and generate a sub-graph based on achievement relationships. Figure 1 shows an example of a graph representing achievement relationships. When the action name "Boost Metabolism" is given, the system outputs sibling actions such as "Sleep Deeply" and "Build up Muscle." Additionally, "Enhance Immunity" and "Losing Weight", which can be achieved by the input action, are outputted as parent actions. In this context, parent actions, which can be achieved through their child actions, are connected by a relationship termed "achieved by." This achievement relationship is commonly called the accomplishment of purposes, where the parent node is relatively considered the "purpose" and the child node is the "means."

Several studies have already been conducted to extract related behaviors as siblings or parents and children [2,4]. Our study aims to construct such studies as local graphs, focusing on product graphs and web documents.

We used product review data and web documents to implement such an algorithm (see the middle part of Fig. 1). We then determined the parent-child relationship for the extracted action names using the linguistic patterns of their appearance in the Web documents. We conducted subject evaluations on the graphs thus created to determine the validity and effectiveness of the algorithm.

2 Generating Achievement Relationship Graph Between Actions

This section describes a method for generating a graph representing the achievement relationship when an action name is input, consisting of parent, child, and sibling nodes of the action. To realize such an algorithm, we extract and abstract the names of actions from product review data using a language model. Using the obtained action names, we extract actions that form sibling relationships from product data and actions that form parent-child relationships from web search data. A graph representing the achievement relationship is constructed by linking these extracted actions with parent-child and sibling relationships.

2.1 Extracting Actions from Product Review Text

First, we extract actions from our product reviews. Many products are purchased to accomplish some purpose. For example, dumbbells are purchased to strengthen muscles, and cups are purchased to drink water. Here, reviews on the same product contain actions with similar purposes. Therefore, we create pairs of product and action names from the review data.

Our algorithm extracts only those sentences from the many reviews of a product that relate to actions taken with that product. To do so, the algorithm employs a fine-tuned LLM. Our approach was to create a crowdsourced dataset and fine-tune BERT. To do so, first, all sentences in the dataset are separated by punctuation marks (*i.e.*, ".", "!", "?"). Then, randomly selected sentences were presented to the crowdworkers and labeled if they contained actions.

The labeled text was fed to BERT, and a classifier was trained to classify whether a sentence contained a purpose as 0 or 1. The classifier thus learned was used to extract sentences containing actions from all sentences in the dataset.

2.2 Abstracting Action Names and Eliminating Spelling Inconsistencies

Next, the action sentences written in natural sentences are converted into action names consisting of a few short words, and the names of actions with different notations are grouped. Because the sentences containing actions extracted by the BERT classifier are natural sentences, there can be an infinite number of notations, even when they represent the same action. Therefore, we use abstract summarization techniques to turn action sentences into action sentences.

We utilized LLM and crowdsourcing in this process as well. Initially, we gave the crowdsourcers a task to read and summarize an action sentence into an action name. A large number of pairs of action sentences and action names were created. Next, we fine-tuned T5 (Text-to-Text Transfer Transformer), an LLM good at generating text from input text. T5 can be fine-tuned by giving pairs of input and output texts. As a technique during fine-tuning, T5 accepts fine tuning with prefixes and prompts, so the training input was "Summary: The action name of the *[Action Sentence]* is" and the output was "*[Action Name]*".

Thus trained, T5 can input any sentence and summarize it into a few-word action name. However, these action names may still be different notations for the same action. For example, "having a good sleep" and "sleeping well" are different expressions with the same meaning.

Therefore, we grouped semantically close actions under the same action name. To do so, we used a similarity machine called Sentence BERT. Sentence-BERT can, given arbitrary sentences, vectorize them and calculate their similarity. In this experiment, action names above a certain threshold were considered identical.

2.3 Product-Action Name Graph Processing for Sibling Action Discovery

The previous section extracted action sentences from the review sentences and summarized them as action names. This section generates a graph of the product and action names.

Actions that can be performed by using a certain product are similar. For example, if an alarm clock is used for "getting up early" and "waking up," both commonly achieve a common solution to a purpose related to getting up in the morning. To make this more generalized, products that perform similar actions are similar. Using this relationship, sibling actions can be extracted as actions to achieve the same objective.

On the other hand, it is not always possible to substitute between different actions that could once have been accomplished with the same commodity. Therefore, we performed a graphical process to extract only the names of actions characteristically described in common within the reviews of many products.

The data is first cleansed to create a two-part graph consisting of product and action names. As the cleansing, we removed product names that were too frequent and rare. Some action names could be related to any product. For example, the actions "write a review" and "open a package" could be included in a product review on a shopping site. We removed action names associated with more than one percent of the goods in the dataset. The products, such as assorted household goods, could be used for any purpose. We removed commodities associated with more than 15% of the actions. We also manually created blacklists to remove commodities and action names containing specific words (i.e., commodities containing "set" and commodities containing "deliver").

Next, the created graph is weighted, and RandomWalk with Restart is applied to extract the sibling actions. The graph's adjacency consisting of product and action names can be represented by the $|I|+|A|$-dimensional matrix G, where I is the set of all products, and A is a set of all actions. The rows of this matrix represent links from actions to products, and the columns represent links from products to actions, within 0 or 1.

After weighting this matrix by frequency, it is transformed into a transition establishment matrix. Some action can be achieved with many products. The relationship between these products and actions is relatively thin. Weighting is performed to consider the strength of the ties between products and actions. Here, the weighted adjacency matrix $G\prime$ can be expressed as

$$G'_{i<|A|j>|I|} = \frac{G_{ij}}{\sum_{k>|I|}^{|A|} G_{kj}} \quad (1)$$

by penalizing the square of the number of outgoing links from an action to an item. Converting this matrix to a transition matrix (i.e., normalizing the sum of the weights per column into 1), it can be expressed as

$$G''_{ij} = \frac{G'_{ij}}{\sum_{k=1}^{|I|+|A|} G'_{kj}}. \quad (2)$$

Apply RandomWalk with Restart (RWR) to this transition matrix graph. Assuming that an arbitrary action name is given as input from the user, consider the case of collecting the sibling action names. Given an input action name, we designate a one-hot vector of dimension $|A|$ as q, where the dimension corresponding to the input node is set to 1, and all other dimensions are set to 0. Here, c represents the random jump rate.

In the nth iteration of the RWR calculation, the probability of a surfer staying at a node starting from the input action name can be expressed as $p_n = (1-c)G''p_{n-1} + cq$, where p_{n-1} is the vector of the surfer's staying probabilities at each node from the previous iteration. The final vector of the surfer's staying probabilities for each node, p_∞, can be determined through the power method's convergence calculation. This graph calculation enabled ranking the action names in the data set by the value of the corresponding dimension of P_∞, given an arbitrary input action name.

2.4 Parent-Child Action Detection Using Web Documents and Linguistic Patterns

The parent and child action names are collected from web documents using these extracted sibling action names. In many cases, the relationship between multiple action names is noted in a general document. For example, the sentence "I did *[Child Action]* to do *[Parent Action]*" is common. We created several linguistic patterns to represent these relationships and collected sentences that follow the patterns from web documents.

In this case, it is practically impossible to download all web documents in advance and extract sentences matching the pattern. Therefore, in this experiment, a limited number of documents were collected using a Web search engine. Since the input action names or sibling action names have already been extracted, we assigned them to the language patterns. Then, a wildcard query is used to search the Web. For example, if the input document were "get up in the morning," our algorithm would execute the bidirectional search queries "to get up in the morning, I do *" and "I do * to get up in the morning." Apply this search brute force to all candidate action names, with all language patterns (Note that the actual search was conducted in Japanese, so the language pattern differs from the notation in the paper).

The search results, including action sentences, are summarized into the action name using the T5 summarizer used in Subsect. 2.1 to resolve spelling inconsistencies. We ranked the obtained action names by the frequency of commonly obtained actions and considered the top ones as parent and child actions.

3 Evaluation Experiment

We evaluated the method's usefulness by conducting an experiment in which participants labeled actions on an actual Rakuten Ichiba dataset. In the experiment, we evaluated sibling and parent-child actions in response to action name queries.

3.1 Experimental Settings

First, we used cleansed Rakuten Ichiba product review data as the product review data set. We constructed a dataset consisting of 607,096 products belonging to a specific category and the reviews for them.

Next, we cleansed the data and constructed a product name-action name graph. We obtained 27,881 product nodes and 14,414 action nodes. We obtained a ranking of sibling actions by applying the RWR algorithm to this graph. We set the random jump rate to 0.8 and adopted the top five ranking cases as sibling behaviors.

To discover parent and child behaviors, we used Google WebSearch API to search by the top 15 sibling behavior names and input behavior names to obtain search snippets for the top 30 results.

Table 1. Evaluation result of generated sibling action names

	ActionName-ness	Relevance	Siblingness
Proposed method	**4.71**	**2.11**	**1.81**
Weighting only	**4.71**	2.01	1.74
Node merging only	4.67	2.03	1.74
Co-occurrence	4.51	1.95	1.71

Table 2. Evaluation result of generated parent action names

	ActionName-ness	Relevance	Substitutability	Purpose-likeness	Means-likeness
Proposed method	**4.94**	**3.24**	**2.92**	**2.46**	**3.24**
Weighting only	4.58	2.63	2.18	2.18	2.61
Node merging only	4.73	**2.85**	2.11	**2.11**	2.89
Co-occurrence	4.46	1.95	1.59	1.59	1.86

Table 3. Evaluation result of generated child action names

	ActionName-ness	Relevance	Substitutability	Purpose-likeness	Means-likeness
Proposed method	4.53	**1.87**	**1.77**	**1.38**	**1.97**
Weighting only	4.44	1.80	1.58	1.43	1.93
Node merging only	4.49	**2.01**	1.69	**1.46**	**2.04**
Co-occurrence	**4.56**	1.84	1.54	1.37	2.00

To ascertain the effectiveness of each procedure of the proposed method, four comparative methods were developed:

- **Weighting+node merging** (proposed method),
- **Weighting only** (Method without the Eliminating Spelling Inconsistencies process in Subsect. 2.2),
- **Node merging only** (Method without Edge Weighting in Subsect. 2.3),
- **Co-occurrence** (baseline. No graph processing is performed. It simply lists the names of actions attached to the same product in order of frequency).

Sibling and parent-child actions were generated for each method for the 15 pre-prepared queries. For each query, each output was randomly mixed and presented to two participants. Participants labeled each action with five criteria: Action Name-ness, Relevance to the query, Substitutability, Purpose-likeness, and Means-likeness on a four-point scale.

3.2 Experimental Result

Table 1 represents the evaluation results of action names in sibling relationships, and Tables 2 and 3 represent the evaluation results of action names in parent-child relationships, respectively. Sibling and parental relationships were successfully extracted, but child behavior relationships were not.

4 Discussion

This section discusses the experimental results. The proposed method outperformed the baseline in the accuracy of discovering sibling and a parent action, but was limited in its ability to improve the accuracy of discovering child actions. It was confirmed that graph computation enabled the discovery of behaviors with the same objectives and identifying appropriate parent behaviors through a web document-based method. However, it was shown that there is room for improvement in identifying child actions.

The proposed method effectively resolved notational distortions such as "preventing cold" and "humidifying" and identified action names with the same purpose. However, in some cases, such as "build muscles" and "strengthen muscles," synonymous action names decreased accuracy. In particular, for "prevent dryness," the weighting tended to give "humidifying" a higher rating and general expressions a lower rating. This may be because weighting effectively penalized general expressions.

In the Web document-based method for finding parent and child actions, the accuracy of child actions was significantly lower than that of parent actions. This discrepancy can be attributed to the scarcity of search results for the "[*given action name*] for *" queries used for parent actions, in contrast to the abundance of results for "* for [*given action name*]" queries for child actions, which tend to include a larger proportion of irrelevant noise. Based solely on frequency, the ranking predisposes, including noisy results at the top, to a substantial disparity in accuracy between parent and child actions.

5 Conclusion

In this paper, we propose a method for discovering alternative actions that can achieve the original purpose using product review data and Web documents. The method using graph mining and linguistic patterns made constructing a graph of the names of actions in an achievement relationship possible.

There is room for improvement in our approach in many aspects: Finding child behavior did not work well, and action names are not uniquely defined. We will solve these problems to generate an overall achievements graph.

Acknowledgements. This work was supported by JSPS KAKENHI Grants Number 21H03775, 21H03774, and 22H03905. We used "Rakuten Dataset" provided by Rakuten Group, Inc. via IDR Dataset Service of National Institute of Informatics [3].

References

1. Linden, G., et al.: Amazon.com recommendations: item-to-item collaborative filtering. IEEE Internet Comput. 76–80 (2003)
2. Pothirattanachaiku, S., et al.: Mining alternative actions from community Q&A corpus for task-oriented web search. In: Proceedings of Web Intelligence 2017, pp. 607–614 (2017)

3. Rakuten Group, Inc.: Rakuten dataset (2020). https://doi.org/10.32130/idr.2.1. https://rit.rakuten.com/data_release/
4. Yang, Z., et al.: Leveraging procedural knowledge for task-oriented search. In: Proceedings of ACM SIGIR 2015, pp. 513–522 (2015)

Generating News Headline Containing Specific Person Name

Taiga Sasaki[1]($\boxtimes$), Takayuki Kuge[1], Yoshiyuki Shoji[2],
Takehiro Yamamoto[1], and Hiroaki Ohshima[1]

[1] University of Hyogo, Kobe, Hyogo 651-2197, Japan
ad241026@guh.u-hyogo.ac.jp, t.yamamoto@sis.u-hyogo.ac.jp,
ohshima@ai.u-hyogo.ac.jp
[2] Shizuoka University, Hamamatsu, Shizuoka 432-8011, Japan
shojiy@inf.shizuoka.ac.jp

Abstract. This paper proposes a method for generating news headlines that include a specific person's name from the news article. We adopt T5, a modern large language model (LLM), for news headline generation. Conventional news headline generation methods input the entire article into the LLM. Attempting to summarize the entire article in the limited number of characters of a news headline may generate a sentence with a meaning that is not factually correct. Therefore, we propose a method that only inputs meaningful sentences extracted from the news article. The proposed method extracts two types of sentences: a sentence represents the article's main topic, and a sentence contains facts about the given specific person's name. Through evaluation experiments, our method demonstrated better accuracy in generating news headlines focusing on a specific person than methods inputting the entire article.

Keywords: LLM · Headline generation · Personalize

1 Introduction

Most people must have favorite celebrities, such as idols, comedians, and athletes. They surely want to read all the news about those celebrities, but it is impossible to detect all the news since the celebrity's name may not appear in the headline of the news article. Therefore, if the news headline could be personalized, they would be more likely to access the article and read the content.

For example, fans of *Haaland*, a football player in *Manchester City Football Club*, likely click on the article with the headline "*Haaland's Goal Secures Victory for Manchester City*". On the other hand, the fans of *Rodri*, another player in the same club, would not click on that article because his name is not contained in this news headline. However, since the article is about the same team, this news often contains information relevant to *Rodri*. In this case, *Rodri* fans who did not click on the article would miss out on their favorite information.

In general, it is common that even if the headline does not contain the person's name, the body of the article often contains information about the person. In such cases, readers may overlook news about their favorite celebrity in the body of the article.

To address this issue, we propose a method for generating headlines focusing on a specified person from the articles. For example, suppose an article reports on the game results and content of a *Manchester City FC* match. The body of this article contains two pieces of information: *Haaland scored a goal*, and *Rodri assisted a goal*. When focusing on *Haaland*, our method generates the headline as *"Haaland's Goal Secures Victory for Manchester City"* from this article. Focusing on *Rodri* generates the headline *"Rodri's assist Secures Victory for Manchester City."*

Such headline generation method has two benefits. First, it prevents readers from missing information about their favorite celebrity. Second, it makes more people click on the article. It is beneficial to both the readers and the content providers.

We define the characteristics that the generated headline should have as

1. **Generated headline is based on facts,**
2. **It contains the name of the given specified celebrity,** and
3. **It contains the summary of the entire article.**

Nowadays, it is prevalent to adopt generative language models for news headline generation [1,4]. In this study, we also employ generative models to generate headlines. We adopt T5, a modern generative language model suitable for text-to-text tasks, because our method takes both input and output as text. T5 is usually employed to solve tasks such as question answering, translation, and summarization.

Generating headlines by inputting the entire article is a common practice in using a generative model. However, inputting the entire article often causes a problem, as the model fails to pick the correct part suitable for the headline. Let us explain this issue using the above-mentioned examples in more detail. When inputting the entire article to generate a headline focusing on *Haaland*, the model possibly generates a false headline such as *"Haaland's assist Secures Victory for Manchester City."*, contrary to the fact that who assisted a goal was *Rodri*. When generating a headline focusing on *Haaland*, information only related to other players is unnecessary. Thus, removing information about *Rodri*'s act from the input would prevent such errors. Our approach is to input only essential information into the model. To generate headlines focused on a specific person, it extracts the sentences only relevant to the person, and then generates headlines based only on these sentences. This approach reduces the likelihood that the model generates inaccurate headlines.

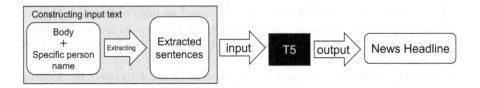

Fig. 1. Overview of proposed method

2 Proposed Method

This section explains the details of our approach to generating news headlines. Our proposed method aims to generate headlines that capture the article's essence and are focused on a specific person.

2.1 Overview

Our headline generation approach can be illustrated in outline as shown in Fig. 1. This method consists of two main components:

– Constructing input text, and
– Headline generation.

In the constructing input text phase, our method extracts essential sentences from the entire article. It extracts two types of sentences: one representing the article's main topic and the other containing information about the specific person. Our method uses these extracted two types of sentences as the input text for the T5. In the headline generation phase, the model fine-tuned with two types of extracted sentences generates actual personalized news headlines.

2.2 Constructing Input Text

Our method first extracts important sentences from the text of a news article that symbolize the news as a whole. We used two different existing important sentence extraction methods respectively, and used both as input.

The first important sentence extraction method is **Lead-1**, a technique for extracting a sentence at the beginning of a document. LEAD-1 is effective because crucial information frequently appears in the initial sentences of a document. Ishikawa et al. state that it is known that LEAD-1 is most effective when summarizing newspaper articles [2]. We added the first sentence of the article in the input text.

The second approach is extracting sentences highly similar to the genuine headlines. The news headlines written by professional journalists clearly represent the article's main topic and are easily understood by the reader. Therefore, sentences in the news text highly similar to the original headlines are likely to contain the article's main topic. These sentences are important information sources for generating news headlines that reflect the entire news content. In our method, the similarity is calculated based on the number of co-occurring morphemes between the original headlines and each sentence in the body of the article.

Next, we improved our method to include the given person's name in the generated news headlines. We established the criterion for extracting sentences containing information about the specified person to determine whether the person's name is contained in the sentence. For this purpose, our method extracts the sentences that contain information about the specified person. Articles often contain the person's name in sentences that mention information about that individual. We simply extract sentences containing the name of the specified person.

Finally, our method combined the extracted sentences mentioned above to complete the input text to the language model. In this process, the order of the sentences to be input into the language model is important. Our methods extracted sentences reflecting the entire news article or related to a specific person without considering the position in the news article. However, since news describes real events, the article expresses time-lapse and cause-and-effect relationships. In most cases, what happened first is written first. Therefore, our method re-sorted the extracted sentences in the order of their appearance in the original news article. Our method uses this extracted, sorted, and merged text as input for T5.

2.3 News Headline Generation with T5

We then generate actual personalized news headlines by fine-tuning a generative language model based on the input text prepared. Since we used Japanese news articles for the evaluation experiment, we used the pre-trained T5 model named "sonoisa/t5-base-japanese[1]".

We fine-tuned the model using the training set to generate headlines. As a training dataset, we collected Japanese sports news. We then processed them as described above to create input text. We treated the players' names included in the original news titles as given specific person names. This way, we can fine-tune the model with the task of generating the original news title for the given person's name and input sentence.

Once the fine-tuning was complete, we generated the news headline. T5 uses beam search and generates several news headlines in order of likelihood. As a post-process, our method eliminated those that did not contain the input specifiers' names from the output headlines.

3 Evaluation Experiment

We conducted evaluation experiments targeting game result articles of the *Hanshin Tigers*, a professional baseball team in Japan. The generated headlines

---
[1] Hugging Face: sonoisa/t5-base-japanese https://huggingface.co/sonoisa/t5-base-japanese.

to be evaluated are in Japanese. The evaluation experiment aims to confirm whether the accuracy of the generated headlines has improved due to the proposed method.

3.1 Evaluation Criteria

We evaluate the completeness of the headlines based on the following evaluation criteria:

1. **Factuality**: whether the headline is based on facts,
2. **Focusedness**: whether It contains the specified person name, and
3. **Entireness**: whether it contains the summary of the entire article.

These criteria are judged as 0 or 1 if they are applicable or not.

Three participants evaluated the generated news headers. They read the news content and news headers generated by the proposed and baseline methods, and evaluate them for three evaluation criteria.

3.2 Dataset for Evaluation

We created a news article dataset for evaluation. The training data consists of 2,738 articles related to the *Hanshin Tigers* from Daily Sports in 2021. We collected these data from Ceek.jp News[2]. This site is a news portal that compiles articles from multiple news publishers.

As a technique during crawling, we filtered news articles by their URL. This site organizes news articles by topic, which determines the URL. Therefore, we collected only articles with URLs containing *"Tigers"*.

We used this dataset for the fine-tuning. The number of epochs by early stopping with a patience of 10. The validation data consists of 152 articles related to the *Hanshin Tigers* from Daily Sports in 2021. We collected this data using the same method as the training data.

3.3 Comparative Method

We prepared a baseline method for comparison. The overview of the comparative method is shown in Fig. 2. The input text for the comparative method is the entire news article text. The generative language model used in the comparative method is also T5, the same as the proposed method. The purpose of using the same model for the proposed method and the comparative method is to avoid differences in the accuracy of generated headlines depending on the performance of the generative model.

We have constrained T5 to use a specific person's name as the token for the start position of the headline. The ability to specify a token for the starting position is provided in T5.

---

[2] Ceek.jp (in Japanese) https://news.ceek.jp/.

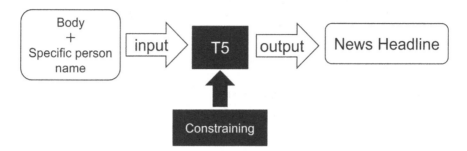

Fig. 2. Overview of comparative method

3.4 Evaluation Data

The dataset for the evaluation was made up of news from a different year than the training data set. The data for evaluation are articles about the *Hanshin Tigers* in the Daily Sports in the year 2022. We extracted 50 articles reporting the game results from the collected articles.

For each article, we generate a headline containing the names of two players. We randomly selected the target players from those appearing in the body of the article. We evaluated the top headline generated by the proposed method and the top headline generated by the comparative method.

3.5 Experimental Participants

The two factors are necessary to evaluate the accuracy of the headlines of articles on professional baseball game results: knowledge of professional baseball, and frequent exposure to headlines of articles on professional baseball. Therefore, we selected three participants who are Japanese Professional Baseball league enthusiasts.

3.6 Result

This section explains the results of our evaluation. Since individuals worked on this experiment, it is necessary to check the reliability of the evaluators. We calculated Fleiss' kappa for each criterion to check the degree of agreement among the evaluators. The right side of the Table 1 shows the actual result. We used the criterion proposed by Landis *et al.* [3] to evaluate the Fleiss' kappa. According to this criterion, Focusedness was evaluated as "Almost perfect." The criteria of Facutuality and Entireness were evaluated as "Substantial." Therefore, it can be said that sufficient agreement was obtained.

We then combined the evaluation scores for each participant for each criterion. Each evaluation score was labeled 0 or 1 for each news header. We determined by a majority vote of three evaluators whether a headline met the evaluation criteria. We calculated the percentage of headlines that met the evaluation

Table 1. Experimental results for accuracy and agreement. The % Good Header is the percentage of two or more participants responding 1 in a binary decision (*: $p < 0.05$)

	% Good Header		Agreement
	Proposed	Baseline	Fleiss' kappa
Factuality	*0.56	0.25	0.79
Focusedness	0.90	*0.99	0.89
Entireness	0.64	0.60	0.62

Table 2. Example where the proposed method prevents the generation of factually incorrect headlines (Specific person is *Sato*)

News Article	Method	Generated Headline
The Hanshin Tigers defeated the Giants. Fujinamis excellent pitching led the team to victory. Fujinami was the winning pitcher after 491 days. Satos impressive 18th home run played an important role in the teams victory	Original	Fujinamis Superb Pitching Propels Tigers to Victory
	Proposed	Satos Remarkable 18th Home Run Crucial in Tigers Victory
	Baseline	Satos Pitching Leads Tigers to Victory with Fujinami as Winning Pitcher

criteria. We calculated this percentage for each of the comparative and proposed methods.

The left part of Table 1 shows the overview of the experimental result. The proposed method is more accurate in terms of the criteria of Factuality. It is also slightly more accurate for the Entireness. On the other hand, for the Focusedness, our method was less accurate than the baseline.

We performed Welch's t test to determine whether the difference between the results of the proposed method and those of the baseline method is statistically significant. The significance level is 5%. A significant difference was observed in the criteria of Factuality and Focusedness.

4 Discussion

The experiments showed that the proposed method increases the likelihood of generating factual headlines. The difference between the proposed method and the compared methods in terms of the criteria of whether it is based on facts was statistically significant. This result is because the proposed method uses only the extracted key sentences as input, not the full text of the article. This approach reduces the risk that T5 generates headlines that are not factual. Table 2 is an example where the proposed method prevents the generation of factually incorrect headlines. The comparative method, which requires the full text of the article, generated a headline as if *Sato*'s pitching led the *Hanshin Tigers* to victory, even though it was *Fujinami*'s achievement in truth. On the other hand, the proposed method can generate headlines that are in accordance with the facts.

The results of the criteria of whether it contains the name of the specified celebrity showed that the comparative method was better. This result was statistically significant. A factor contributing to this result was the presence or absence of constraints. The comparative method places a constraint on T5, which specifies the token of the person's name of interest at the start of the headline. The proposed method does not apply this constraint.

The proposed method obtained slightly higher results for the accuracy of containing the summary of the entire article. The proposed method extracts sentences about the article's subject in advance and passes this information to T5. The comparative method has T5 find the subject of the article from the full text of the article. The evaluation results show that the proposed method is more accurate than the comparative method. However, the difference between the proposed method's and the comparative method's results is not statistically significant, so it is necessary to verify the proposed method's accuracy with more evaluation data in the future.

5 Conclusion

In this study, we addressed the generation of news headlines containing specific person names using T5. We proposed a method in which the input sentences are not the full article but extracts of important sentences. We found that this method increases the likelihood that the generative model will generate factual headlines.

Acknowledgements. This work was supported by JSPS KAKENHI Grant Numbers JP21H03775, JP22H03905, JP21H03554, JP24K03228.

References

1. Amin, O., Aijun, A.: Learning to generate popular headlines. IEEE Access **11** (2023)
2. Ishikawa, K., Ando, S., Okumura, A.: Hybrid text summarization method based on the TF method and the lead method. In: NTCIR Conference on Evaluation of Information Access Technologies (2001)
3. Landis, J.R., Koch, G.G.: The measurement of observer agreement for categorical data. Biometrics **33**(1), 159–174 (1977)
4. Vasilyev, O., Grek, T., Bohannon, J.: Headline generation: learning from decomposable document titles. arXiv preprint: arXiv:1904.08455 (2019)

Investigating Evidence in Sentence Similarity Using MASK in BERT

Kanako Nakai[✉], Yuka Kawada, Takehiro Yamamoto, and Hiroaki Ohshima

University of Hyogo, Kobe, Hyogo 651-2197, Japan
n.kanako0311@gmail.com, t.yamamoto@sis.u-hyogo.ac.jp,
ohshima@ai.u-hyogo.ac.jp

Abstract. This study tackled the problem of investigating the evidence for a model measuring the similarity between two related sentences. In each of the given sentences, there would be phrases that are semantically corresponding to each other. Some of them are phrases that increase the similarity of the sentences and, on the other hand, some of them are phrases that decrease the similarity of the sentences. We propose a method to extract such "similar" and "dissimilar" phrase pairs. In the proposed method, we first train a BERT model that can measure sentence similarity. Next, we find semantically corresponding phrases in a given pair of sentences. When we replace a certain corresponding phrase with a BERT [MASK] token if the trained BERT model determines that the sentences are less similar, we consider the phrase pair to be a "similar" phrase pair. On the other hand, the model determines that the sentences are more similar, we consider the phrase pair to be a "dissimilar" phrase pair. We have implemented the proposed method and conducted a case study.

Keywords: XAI · BERT · text classification

1 Introduction

Machine learning technology is advancing, and deep learning is used in many fields. For example, deep learning is used for medical image diagnosis and stock price prediction. There is a problem called "the black box problem" in using deep learning. The black box problem is the problem that we cannot understand the process by which a machine learning model predicts results. Explainable AI technologies have been developed to address the black box problem.

For example, let us assume we are conducting document classification using BERT. Several explainable AI techniques are proposed to investigate evidence and processes behind the classification results of documents. In the literature [1], a method is proposed to indicate which tokens are important in prediction using Attention weights. Yang et al. propose a method to measure the importance of sentences and tokens in a document by focusing on the Attention and the structure of the document [5]. We can get tokens that contribute positively or negatively to the resulting classification by LIME [3]. Ribeiro et al. propose a method to investigate the evidence for predictions by the rule called Anchors [4]. Kokalj et al. propose a method called TransSHAP, which adapts SHAP to the Transformer model including BERT [2].

In these methods, it can be said that we get the importance of each token. However, the tokens include prepositions such as "in" and "of". There are difficulties in interpreting evidence that we get by Attention or LIME.

Therefore, we tackle the problem of investigating evidence by **phrase** rather than token.

This study tackled the problem of investigating the evidence for a model measuring the similarity between two related sentences. An example of two related sentences would be *"The bus with two stories comes running on the road."* and *"The double-decker bus is stopped."* This pair refers to the same topic, but the two sentences have different meanings. *"There is a man jumping on a skateboard."* and *"A man is jumping on a skateboard."* are also a pair of related sentences. In this case, the two sentences have the same meanings.

In each of the given sentences, there would be phrases that are semantically corresponding to each other. For example, in the first pair of related sentences, *"The bus with two stories"* and *"The double-decker bus"* are semantically corresponding phrases. Furthermore, *"running"* and *"stopped"* are also semantically corresponding phrases. Some of them are phrases that increase the similarity of the sentences and, on the other hand, some of them are phrases that decrease the similarity of the sentences. The phrase pair *"The bus with two stories"* and *"The double-decker bus"* have the same meaning, and they may contribute to increasing the similarity of the sentences. We call such phrase pairs **"similar" phrase pairs**. On the other hand, the phrase pairs *"running"* and *"stopped"* have different meanings, and they may contribute to decreasing the similarity of the sentences. We call such phrase pairs **"dissimilar" phrase pairs**.

We propose a method to investigate "similar" and "dissimilar" phrase pairs. In the proposed method, we first train a BERT model that can measure sentence similarity. Next, we find semantically corresponding phrases in a given pair of sentences. When we replace a corresponding phrase pair with a BERT [MASK] token, if the trained BERT model determines that the sentences are less similar, we consider the phrase pair to be a "similar" phrase pair. On the other hand, the model determines that the sentences are more similar, we consider the phrase pair to be a "dissimilar" phrase pair. We have implemented such a method and conducted a case study.

2 Extracting Evidence for a Similarity Measure Model by Masking Corresponding Phrase Pairs

This section explains the method that investigating corresponding phrase pairs which provide evidence for similarity or dissimilarity. Sentence A: *"There is a man jumping on a skateboard."* and sentence B: *"A man is jumping on a skateboard."* We use these two sentences for explanations.

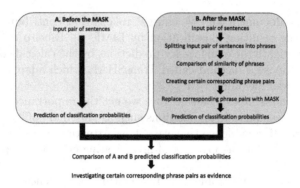

Fig. 1. Overview of the proposed method

2.1 Overview of the Proposed Method

We illustrate and outline our proposed method in Fig. 1.

First, we input a pair of sentences labeled as "similar" or "dissimilar" into the BERT model. The BERT model outputs the probability of being classified as similar or dissimilar. We define predicted classification probabilities as the probability of being classified as similar or dissimilar. Next, we create corresponding phrase pairs. Then, we replace a corresponding phrase in an input pair of sentences with a BERT [MASK] token and input them into the BERT model. Finally, we compare the predicted classification probabilities before and after replacing the corresponding phrase pairs with [MASK] tokens.

When the predicted classification probabilities change significantly by replacing a corresponding phrase pair with BERT [MASK] tokens, we investigate the corresponding phrase pair as evidence.

2.2 Training a Model to Measure the Similarity

We explain how to create a similarity measure model. We input a pair of sentences to BERT. Input pairs of sentences are labeled as "similar" or "dissimilar". These labels indicate the relationship between two sentences. We connect pairs of sentences by [SEP] token and input into BERT. The BERT model outputs the predicted classification probabilities. We convert a vector of BERT [CLS] token into predicted classification probabilities by sigmoid function.

2.3 Creating Corresponding Phrase Pairs

In this study, we use corresponding phrase pairs to investigate the evidence for the similarity between two related sentences. We explain how to create corresponding phrase pairs.

Splitting Input Pair of Sentences Into Phrases. We use a parser to split input pairs of sentences into phrases. For example, sentence A is divided into phrases such as "There is/a man/jumping/on a skateboard." sentence B is divided into phrases such as "A man/is jumping/on a skateboard."

Comparison of Similarity of Phrases. We input each phrase into Sentence-BERT to get the similarity between phrases. Sentence-BERT is a method used to determine the relationship between two sentences. For example, we input the first phrase of sentence A, "*on a skateboard*" and the first phrase of sentence B, "*A man*" into Sentence-BERT to calculate the cosine similarity. Then, we input the first phrase of sentence A, "*on a skateboard*" and the second phrase of sentence B, "*on a skateboard*" into Sentence-BERT. We calculate the cosine similarity for all combinations of phrases in the two sentences.

Creating Corresponding Phrase Pairs We create corresponding phrase pairs in order of high similarity between phrases. We don't duplicate phrases when creating corresponding phrase pairs.

For example, the top three phrase pairs with the highest similarity between phrases are "*on a skateboard*" and "*on a skateboard*", "*on a skateboard*" and "*is jumping*", and "*jumping*" and "*is jumping*". The first corresponding phrase pair is "*on a skateboard*" and "*on a skateboard*" with the highest similarity between the phrases. The next phrase pair with the highest similarity is "*on the skateboard*" and "*is jumping*". However, "*on a skateboard*" is included in the first corresponding phrase pair. Therefore, this phrase pair is not used as a corresponding phrase pair. The next phrase pair with the highest similarity is "*jumping*" and "*is jumping*". These are not included in corresponding phrase pairs created in the past. So we use this phrase pair as a corresponding phrase pair.

The number of corresponding phrase pairs is matched to the number of phrases in sentences with fewer phrases. If there are more phrases in one sentence and some phrases cannot create a corresponding phrase pair, we treat the phrases as **"not corresponding phrases"**. In the example sentence, sentence A is divided into four phrases, and sentence B into three phrases. To match the number of phrases in sentences with fewer phrases, we can create three corresponding phrase pairs for sentence A and sentence B. "*There is*" in the first sentence cannot create any corresponding phrase pair, so we treat it as **"not corresponding phrase"**.

2.4 Investigating Evidences

We replace a certain corresponding phrase in an input pair of sentences with a BERT [MASK] token and input them into the BERT model learned in Sect. 2.2. We compare the predicted classification probabilities for "Before the MASK" and "After the MASK". If the predicted classification probability decision is reversed, we investigate the corresponding phrase pair as evidence.

For example, in the example sentence, the corresponding phrase pair *"jumping"* and *"is jumping"* are replaced with BERT [MASK] tokens. The input into BERT is "[CLS]There is a man [MASK] on a skateboard. [SEP] A man [MASK] on a skateboard. [SEP]". The probability of being classified as similar is 0.8 and the probability of being classified as dissimilar is 0.2 before the MASK. The probability of being classified as similar is 0.3 and the probability of being classified as dissimilar is 0.7 after the MASK. In this case, the predicted classification probabilities are reversed. Therefore, we investigate the corresponding phrase pair *"jumping"* and *"is jumping"* as evidence.

In the case of **"not corresponding phrases"**, we replace only those phrases with [MASK] tokens and input them into the BERT model. If the predicted classification probability decision is reversed, we investigate the **"not corresponding phrase"** as evidence. *"There is"* in a sentence A is treated as **"not corresponding phrase"**. We replace only *"There is"* with a BERT [MASK] token and input them into the BERT model. The input into BERT is "[CLS] [MASK] a man jumping on a skateboard. [SEP] A man is jumping on a skateboard. [SEP]".

3 Experiment

Data. In this study, we use the JSTS dataset from JGLUE, which was published by Waseda University and Yahoo Japan Corporation. JGLUE is a benchmark for how well models understand Japanese. JSTS is constructed for the task of estimating the similarity of sentence pairs. Sentence pairs in the dataset are labeled with similarity values ranging from 0 to 5. The similarity of a sentence pair is 0 if the meanings are completely different and 5 if the meanings are identical.

Similar sentence pairs are relevant and semantically similar. We label similar sentence pairs with similarity between 3.8 and 5.0. Dissimilar sentence pairs are relevant and semantically dissimilar. We label dissimilar sentence pairs with a similarity between 2.0 and 2.6. We used a total of 4,000 sets of data in this experiment: 3,200 for training, 400 for validation, and 400 for test.

Experimental Setup. We train a similarity measure model using a trained BERT model from Tohoku University[1]. We set the learning rate to 2.0×10^{-5}, the batch size to 16, the number of epochs to 10, and the maximum input token length to 512. We set the patience to 3 and performed early termination.

We used a pre-trained Sentence-BERT model[2] for Japanese to calculate the similarity between phrases.

We used GiNZA as a parser.

---

[1] https://huggingface.co/tohoku-nlp/bert-base-japanese-v3.
[2] https://huggingface.co/sonoisa/sentence-bert-base-ja-mean-tokens-v2.

```
                              スケートボードでジャンプしている男性がいます。
              Phrase pair     There is a man jumping on a skateboard.
         (Proposed method)    [SEP]男性がスケートボードでジャンプしています。
                              [SEP] A man is jumping on a skateboard.
                              スケートボードでジャンプしている男性がいます。
                    LIME      There is  man jumping on a skateboard.
                (baseline)    [SEP]男性がスケートボードでジャンプしています。
                              [SEP] A man is jumping on a skateboard.
                                                  similar   dissimilar
```

Fig. 2. Comparison of evidence for similar sentences. We conducted our experiments in Japanese. The English sentences are translations of the Japanese sentences used in the experiments. The English shown as evidence is an example

Baseline. We use LIME as a baseline.

We investigate the top ten words that have a significant impact.

Evidence affecting similarity is visualized in blue. Evidence affecting dissimilarity is visualized in orange.

4 Experimental Results

Investigating Evidence for Similar Sentence Pairs. We explain the investigating **"similar" phrase pairs**.

For example, sentence A: *"There is a man jumping on a skateboard."* and sentence B: *"A man is jumping on a skateboard."* are "similar" in both prediction and correct labels. We investigate **"similar" phrase pairs** from the input pair of sentences.

Sentence A is divided into four phrases such as "There is/a man/jumping/ on a skateboard." Sentence B is divided into three phrases such as "A man/is jumping/on a skateboard." The number of corresponding phrase pairs is matched to the number of phrases in sentences with fewer phrases, so we create three pairs. Corresponding phrase pairs are "on a skateboard" and "on a skateboard", "a man" and "A man", and "jumping" and "is jumping". We treat "There is" in sentence A as **"not corresponding phrase"**.

Figure 2 shows the results of our investigating evidence by the proposed method and baseline. In the proposed method, the corresponding clause pairs "jumping" and "is jumping" is investigated as evidence of similarity. This corresponding phrase pair has the same meaning. We achieved our goal of extracting similar phrase pairs from similar sentences as evidence.

In LIME, "skateboard" which is common to the two sentences is investigated as evidence of similarity. On the other hand, "jumping" which is common to the two sentences is investigated as evidence of dissimilarity.

Investigating Evidence for Dissimilar Sentence Pair. We explain the investigating **"dissimilar" phrase pairs**.

For example, sentence A: *"The bus with two stories comes running on the road."* and sentence B: *"The double-decker bus is stopped."* are "dissimilar" in

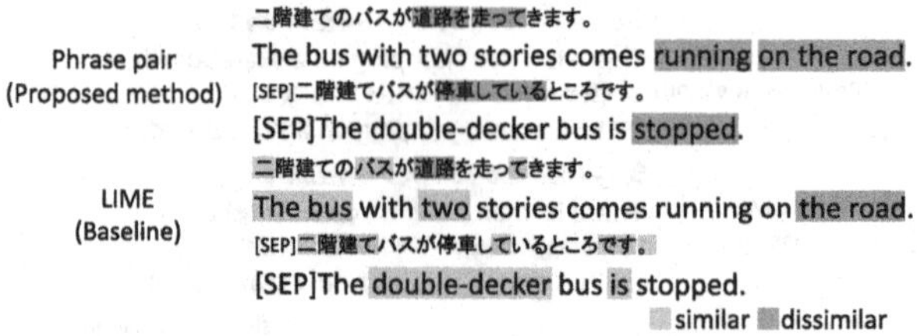

Phrase pair (Proposed method)	二階建てのバスが道路を走ってきます。 The bus with two stories comes ==running on the road==. [SEP]二階建てのバスが==停車している==ところです。 [SEP]The double-decker bus is ==stopped==.
LIME (Baseline)	二階建てのバスが道路を走ってきます。 The bus with ==two== stories comes running on ==the road==. [SEP]二階建てのバスが停車しているところです。 [SEP]The ==double-decker== bus ==is== stopped.

▪ similar ▪ dissimilar

Fig. 3. Comparison of evidence for dissimilar sentences. We conducted our experiments in Japanese. The English sentences are translations of the Japanese sentences used in the experiments. The English shown as evidence is an example

both prediction and correct labels. We investigate **"dissimilar" phrase pairs** from the input pair of sentences.

Sentence A is divided into five phrases such as "The bus/with two stories/comes/running/on the road." Sentence B is divided into three phrases such as "The double-decker bus/is/stopped." The number of corresponding phrase pairs is matched to the number of phrases in sentences with fewer phrases, so we create three pairs. Corresponding phrase pairs are "The bus" and "The double-decker bus", "comes" and "is", and "running" and "stopped". We treat "with two stories" and "on the road" in sentence A as **"not corresponding phrase"**.

Figure 3 shows the results of our investigating evidence by the proposed method and baseline. In the proposed method, the corresponding clause pairs "running" and "stopped" is investigated as evidence of dissimilarity. This corresponding phrase pair has a different meaning. We achieved our goal of extracting dissimilar phrase pairs from dissimilar sentences as evidence. In addition, "on the road" which appears only in sentence A is investigated as evidence of dissimilarity.

In LIME, "the road" is investigated as evidence of dissimilarity. On the other hand, "running" and "stopped" are not investigated as evidence of dissimilarity.

5 Conclusion

In this study, we proposed a method of investigating the evidence for a model measuring the similarity between two related sentences. We investigated the evidence for similarity measure by replacing corresponding phrase pairs with BERT [MASK] tokens and comparing the predicted classification probabilities before and after masking.

We investigate common phrase pairs as evidence from similar sentences by the proposed method. We investigate phrase pairs representing differences as evidence from dissimilar sentences by the proposed method. The proposed method can investigate the similarities and dissimilarities between the two sentences more accurately than LIME.

LIME investigates particles and auxiliary verbs as evidence. Auxiliaries and auxiliary verbs are difficult to understand as evidence. The proposed method improves comprehensibility by using phrases.

We describe the issues. We investigated only common phrase pairs from similar sentence pairs and only phrase pairs represent differences from dissimilar sentence pairs. This is because this study used predicted classification probabilities inversion to investigate evidence. LIME investigates not only common words but also words that represent differences from similar sentence pairs. We examine some methods for investigating words that represent differences in similar sentences and common words in dissimilar sentences.

Using phrases may not allow for semantic correspondence when we create corresponding pairs. For example, *"The bus with two stories"* and *"The double-decker bus"* have the same meaning. However, the former is divided into two phrases such as *"The bus/with two stories"* and the latter is one phrase. The corresponding phrase pair is *"The bus"* and *"The double-docker bus"*. We could not compare the two whole terms *"The bus with two stories"* and *"The double-decker bus"*.

To settle the issues above, it is necessary to create corresponding pairs using different methods instead of phrases. For example, we can use part-of-speech combinations to create corresponding pairs. Further investigation and experimentation should be undertaken to explore how to create corresponding pairs to investigate evidence.

Acknowledgment. This work was supported by JSPS KAKENHI Grant Numbers JP21H03775, JP22H03905, JP21H03554, JP24K03228.

References

1. Clark, K., Khandelwal, U., Levy, O., Manning, C.D.: What does BERT look at? An analysis of BERT's attention. In: Proceedings of the 2019 ACL Workshop BlackboxNLP: Analyzing and Interpreting Neural Networks for NLP, pp. 276–286 (2019)
2. Kokalj, E., Škrlj, B., Lavra, N., Pollak, S., Robnik-Šikonja, M.: BERT meets Shapley: extending SHAP explanations to transformer-based classifiers. In: Proceedings of the EACL Hackashop on News Media Content Analysis and Automated Report Generation, pp. 16–21 (2021)
3. Ribeiro, M.T., Singh, S., Guestrin, C.: "Why Should I Trust You": explaining the predictions of any classifier. In: Proceedings of the 2016 Conference of the North American Chapter of the Association for Computational Linguistics: Demonstrations, pp. 97–101 (2016)
4. Ribeiro, M.T., Singh, S., Guestrin, C.: Anchors: high-precision model-agnostic explanations. In: Proceedings of the Thirty-Second AAAI Conference on Artificial Intelligence and Thirtieth Innovative Applications of Artificial Intelligence Conference and Eighth AAAI Symposium on Educational Advances in Artificial Intelligence, pp. 1527–1535 (2018)
5. Yang, Z., Yang, D., Dyer, C., He, X., Smola, A., Hovy, E.: Hierarchical attention networks for document classification. In: Proceedings of the 2016 Conference of the North American Chapter of the Association for Computational Linguistics: Human Language Technologies, pp. 1480–1489 (2016)

Acceleration of Synopsis Construction for Bounded Approximate Query Processing

Tianjia Ni(✉), Kento Sugiura, Yoshiharu Ishikawa, and Kejing Lu

Graduate School of Informatics, Nagoya University, Nagoya, Aichi, Japan
ni.tianjia.w2@s.mail.nagoya-u.ac.jp

Abstract. *Approximate query processing (AQP)* has gained traction as an effective technique for executing queries on big data. *Bounded approximate query processing (BAQ)* is a recently proposed framework that stores a summary of an original table as a *synopsis* and ensures that its approximation errors remain below a user-specified threshold. Based on the BAQ framework, we proposed BAQ± to shrink a synopsis and improve query processing speed while guaranteeing strict error bounds. However, BAQ and BAQ± still have limitations in constructing synopses. They require time-consuming data sorting for each numerical attribute and cannot summarize high-cardinality categorical attributes, such as spatiotemporal data. To overcome these problems, we propose a novel framework called *Hierarchical BAQ (HBAQ)* and a synopsis construction method in this paper. HBAQ constructs multiple synopses based on the dimension tables of several categorical attributes and uses them to answer OLAP queries efficiently. We also introduce a new bucket definition to summarize numerical attributes effectively and support incremental updates for synopses. The experimental results show that HBAQ achieves half the construction time of BAQ with lower memory consumption. Furthermore, HBAQ can answer OLAP queries more efficiently than BAQ while ensuring strictly bounded errors.

Keywords: Approximate query processing · Query processing · Aggregate estimation

1 Introduction

The rapid increase in data volume and analytical complexity has made efficient query processing essential in databases. *Approximate query processing (AQP)* has emerged as a critical research area that expedites analytical operations within query processes by accepting some estimation errors in query results [4,10,14]. To navigate the challenges of extensive data analysis, AQP balances the precision of query results and the efficiency of query execution by leveraging various approaches [1,13]. A *synopsis* is one of the core concepts in AQP and condenses target data to make it feasible to handle large datasets.

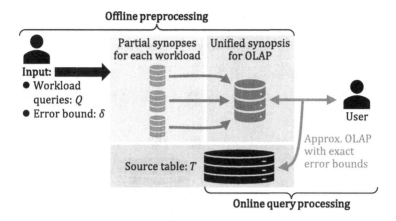

Fig. 1. Synopsis construction flow in BAQ±.

Bounded approximate query processing (BAQ) is a recently introduced framework based on synopses [11]. This method uses a user-defined error threshold (i.e., acceptable estimation errors) and expected workload queries to construct a synopsis that allows users to acquire OLAP query results with low latency. Unlike the previous approaches [3,7,9,12], BAQ strictly ensures accuracy for general aggregate functions (COUNT, SUM, AVG, MIN, and MAX). However, depending on target data and queries, BAQ may not guarantee error bounds accidentally and has space for constructing a more compact synopsis. To address these limitations, we have developed an enhanced work, BAQ± [15]. BAQ± improves the bucketing process for summarizing numerical attributes and revises the procedure for estimating OLAP query results in BAQ as shown in Fig. 1. The proposed approach allows BAQ± to construct a more compact synopsis, efficiently answer OLAP queries, and ensure strictly bounded estimation errors.

However, BAQ± faces two challenges: 1) significant synopsis construction time and 2) inefficient handling of high-cardinality categorical attributes. When constructing a synopsis, BAQ± generates partial synopses and then integrates them into a unified synopsis. Constructing partial synopses requires sorting numerical data for bucketing, and integrating them into a unified synopsis needs multiple full scans of an original table. Besides, BAQ± struggles to efficiently handle attributes with high cardinality, such as spatiotemporal data. Since BAQ±'s main contribution to data summarization is bucketing numerical attributes based on relative errors, they cannot summarize categorical attributes. Thus, many tuples may remain in a constructed synopsis if an original table includes high-cardinality categorical data.

To solve these problems, we propose a *hierarchical BAQ (HBAQ)* framework to process multi-level granularity in target data, as shown in Fig. 2. Compared to BAQ± as shown in Fig. 1, HBAQ constructs multiple synopses with a hierarchy based on dimension tables of categorical attributes and uses them to answer OLAP queries quickly. We also focus on improving synopsis management and propose a novel bucketing procedure to construct synopses efficiently

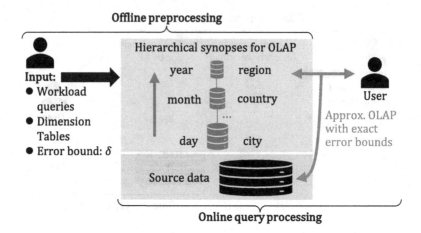

Fig. 2. Offline synopsis construction flow

and allow HBAQ to update them incrementally. Through experiments using simulated queries on multiple datasets, we compare our approach with several existing studies and demonstrate the effectiveness of the proposed method.

2 Preliminaries

In this section, we briefly introduce our previous study BAQ±. BAQ± introduces a novel bucket partitioning method and an improved query processing approach building upon BAQ's framework, as shown in Fig. 1.

A target OLAP query on a table T is constrained by the following conditions:

- It can have WHERE and GROUP BY clauses.
- Its SELECT clause can include aggregate functions on numerical attributes (i.e., MIN, MAX, SUM, and AVG), count functions (i.e., COUNT(*)), categorical attributes used in a GROUP BY clause.
- Its FROM clause has a single table T.
- Its WHERE clause can include basic comparison operators (i.e., =, ≠, >, ≥, <, and ≤) on some attributes and their concatenation based on logical operators (i.e., ∨ and ∧).
- Its GROUP BY clause can include categorical attributes.

BAQ± employs a *relative error* metric as its error criterion. The relative error between a true value $x \in \mathbb{R}$ and its estimated one $\hat{x} \in \mathbb{R}$ is defined as follows:

$$\text{err}(x, \hat{x}) = \begin{cases} \left|\frac{x-\hat{x}}{x}\right| & (|x| \geq \epsilon) \\ \frac{|x-\hat{x}|}{\epsilon} & \text{(otherwise)} \end{cases} \quad (1)$$

where $\epsilon \in \mathbb{R}$ represents a minimal positive value near zero.

Table 1. Table ''temperature.''

ID	region	date	temperature	status
1	Africa	2009-06-27	98.5	O
2	Africa	2009-08-01	103.5	O
3	Europe	2006-12-23	−13.1	O
4	Asia	2008-10-02	−10.8	F
5	Europe	2006-12-23	−12.7	F
6	Africa	2009-08-01	103.1	P
7	Asia	2008-10-02	−10.6	F
8	Europe	2006-12-01	-12.6	O
9	Africa	2009-08-10	107.1	P
10	Europe	2008-05-03	23.2	F

2.1 Numerical Attribute Bucketing

As in BAQ, BAQ± assumes the provision of a set of queries forming the workload, with the synopsis constructed based on these queries. Each workload query q is transformed into a *query column set* (QCS) π, from which a partial synopsis for each query is derived. Note that no partial synopsis is created for any QCS that is a subset of another QCS, as the queries can be answered using the superset QCS's synopsis.

For QCS π containing only categorical attributes, the partial synopsis groups tuples by the categorical values in π. For instance, with QCS $\pi_2 = $ {region, status}, a partial synopsis is constructed, as demonstrated in Table 3, from the table in Table 1. This synopsis maintains the count of tuples in each group as CNT.

When QCS π includes numerical attributes, BAQ± segments values into buckets $b \subset \mathcal{R}$, ensuring compliance with the following equation:

$$err(\min b, \overline{b}) \leq \delta \wedge err(\max b, \overline{b}) \leq \delta \qquad (2)$$

$\overline{b}$ represents the average value within the bucket, and δ is the error upper limit specified by the user. Firstly, if QCS π contains categorical attributes, tuples in the table are grouped accordingly. Then, within each group, numerical attribute values are sorted and allocated to the corresponding bucket b when they satisfy the above equation. For example, with QCS $\pi_1 = $ region, date, temperature, an error upper limit $\delta = 0.05$, and the table in Table 1, buckets and a partial synopsis are established as shown in Table 2.

Following the construction of partial synopses from the workload, BAQ± compiles a *unified synopsis* employing scan-based or greedy-based strategy. The details of each method are omitted here, but the basic idea involves extracting tuples from the original table that can serve as the entities for each partial synopsis. Given an input table T with N tuples and a total of M tuples in τ partial

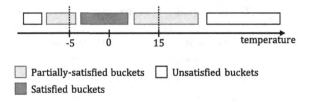

Fig. 3. Satisfied, partially-satisfied, and unsatisfied buckets for $-5 < $ temperature < 15.

Table 2. Synopsis in BAQ± with π_1 = {region, date, temperature}.

region	date	temp. (min., max., mean)	CNT_1
Africa	2009-06-27	(98.5, 98.5, 98.5)	1
Africa	2009-08-01	(103.1, 103.5, 103.3)	2
Europe	2006-12-23	$(-13.1, -12.7, -12.9)$	2
Asia	2008-10-02	$(-10.8, -10.6, -10.7)$	2
Europe	2006-12-01	$(-12.6, -12.6, -12.6)$	1
Africa	2009-08-10	(107.1, 107.1, 107.1)	1
Europe	2008-05-03	(23.2, 23.2, 23.2)	1

Table 3. Synopsis in BAQ± with π_2 = {region, status}.

region	status	CNT_2
Africa	O	2
Europe	O	2
Asia	F	2
Africa	P	2
Europe	F	2

synopses, the scan-based and greedy-based methods require time complexities of $O(MN)$ and $O(\tau MN)$, respectively.

2.2 Approximate Computation to OLAP Queries

Based on a constructed unified synopsis, BAQ± answers OLAP queries using query rewriting. For queries with selections on numerical attributes, BAQ± categorizes buckets as *satisfied*, *partially satisfied*, and *unsatisfied*. Figure 3 illustrates the idea by an example with the attribute "temperature" and its condition $-5 < $ temperature < 15. When selected rows include only satisfied buckets, BAQ± can compute their aggregated values without errors. When partially satisfied buckets exist, BAQ± verifies whether the estimation error on their aggregation is within a given error bound. If there is a possibility that its error exceeds the error bound, BAQ± reads the original tuples of partially satisfied buckets and computes the exact aggregated value. Due to limited space, we omit the detailed explanation of conditions for verifying aggregation errors here.

For MIN and MAX aggregations, BAQ± does not need query rewriting to answer queries. Since an attribute's minimum/maximum value is always in a specific bucket, BAQ± can use its representative (i.e., mean) value for approximation. As described above, since we build buckets while guaranteeing that all the values in a bucket satisfy bounded errors, the approximation errors of MIN/MAX aggregations are smaller than a user-specified threshold.

BAQ± rewrites COUNT, SUM, and AVG aggregation queries according to the buckets' states. If buckets fully satisfy selection conditions, BAQ± uses the following rules and can ensure no errors:

- COUNT($*$) → SUM(A_{cnt}),
- SUM(A) → SUM($A_{mean} \cdot A_{cnt}$), and
- AVG(A) → $\dfrac{\text{SUM}(A)}{\text{COUNT}(*)}$ → $\dfrac{\text{SUM}(A_{mean} \cdot A_{cnt})}{\text{SUM}(A_{cnt})}$,

where A, A_{mean}, and A_{cnt} represent a target numerical attribute, A's mean value in a bucket, and the number of tuples in a corresponding bucket, respectively. For partially satisfied buckets, BAQ± first computes the COUNT/SUM values of satisfied and partially satisfied buckets using the above rules. Let n and m denote satisfied buckets' total COUNT and SUM values. In addition, let t, a (>0), and b (<0) denote partially satisfied buckets' total COUNT, total positive SUM, and total negative SUM values, respectively. If the error verification for partially satisfied buckets passes, BAQ± can answer queries using the following rules:

- COUNT($*$) → $n + t/2$,
- SUM(A) → $m + (a+b)/2$, and
- AVG(A) → $\dfrac{m+(a+b)/2}{n}$.

3 Hierarchical Bounded Approximate Query Processing

Hierarchical bounded approximate query processing (HBAQ) is a new framework to summarize numerical and categorical attributes while ensuring strictly bounded approximation errors. As shown in Fig. 2, HBAQ contains multiple synopses according to *dimension tables* [8] of several categorical attributes, such as "place" and "date". Each synopsis follows the same concepts as BAQ and contains bucketed numerical attributes for AQP. Higher-level synopses contain more summarized (i.e., compact) data and allow HBAQ to answer some OLAP queries efficiently. HBAQ also uses lower-level synopses containing summarized but more concrete data to guarantee strict error bounds.

HBAQ constructs synopses hierarchically based on dimension tables input by a user. Although dimension tables are the same concept as in data warehouses [8], we represent the levels of each dimension as the total order. For example, the synopses in Fig. 2 have a temporal dimension comprising {day, month, year} with the following order:

$$\text{day} \prec \text{month} \prec \text{year}.$$

In this order relationship, the earlier elements represent finer granularity and convey more detailed information, while the latter exhibit coarser granularity and contain more aggregated data. As shown in Fig. 2, our framework supports multiple granular attributes, including temporal and spatial data.

The remainder of this section will discuss how to construct hierarchical synopses and how to use them to answer OLAP queries. It is important to note that although HBAQ introduces dimension tables, it does not cover generic analytical queries supported by traditional data warehouses, such as roll-up operations.

Table 4. Synopsis with {region, date, temperature} and {region, status}

region	date	temp. (ID, mean)	status	CNT$_1$	CNT$_2$
Africa	2009-06-27	(142, 98.5)	O	1	2
Africa	2009-08-01	(143, 103.3)	P	2	2
Europe	2006-12-23	(−100, −12.9)	O	2	2
Asia	2008-10-02	(−96, −10.7)	F	2	2
Europe	2006-12-01	(−100, −12.6)	F	1	2
Africa	2009-08-10	(143, 107.1)	−	1	−
Europe	2008-05-03	(112, 23.2)	−	1	−

Table 5. Synopsis with {region, month, temperature} and {region, status}

region	month	temp. (ID, mean)	status	CNT$_1$	CNT$_2$
Africa	2009-06	(142, 98.5)	O	1	2
Africa	2009-08	(143, 104.5)	P	3	2
Europe	2006-12	(−100, −12.8)	O	3	2
Asia	2008-10	(−96, −10.7)	F	2	2
Europe	2008-05	(112, 23.2)	F	1	2

3.1 Construction of Hierarchical Synopses

HBAQ constructs hierarchical synopses in a bottom-up manner. First, HBAQ constructs the synopsis of the finest granularity in dimension tables based on a novel bucket definition and a construction algorithm, as introduced in the next section. Then, based on the levels in the dimension tables, HBAQ constructs a hierarchical group of synopses. This construction process aggregates in sequence from the lower-level synopses to the upper-level synopses.

For example, consider the construction of synopses based on Table 1 and the QCSs {region, date, temperature} and {region, status}. This example does not have a dimension table for spatial data because the QCSs have only the "region" (i.e., the highest level) column. As shown in Table 4, the proposed method computes the synopsis at the "date" level as the minimum temporal granularity. Note that the IDs of the "temperature" column denote bucket IDs introduced in the next section. HBAQ then aggregates this synopsis based on the temporal dimension's upper level (i.e., "month", as shown in Table 5). Since the aggregated synopsis is more compact than the lower ones, HBAQ can reduce query latency, especially when handling OLAP queries that do not involve temporal attributes (e.g., queries on the QCS {region, status}).

3.2 Online Query Processing Based on Hierarchical Synopses

HBAQ processes OLAP queries in the same ways as BAQ± except for using hierarchical synopses. Although we have improved the bucketing approach from

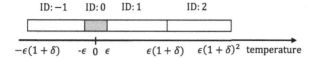

Fig. 4. Static Generation of Bucket Ranges and IDs

BAQ±, the bucket concept and properties remain the same. Thus, HBAQ can use the same procedures and query rewriting rules for handling OLAP queries, as described in the previous section.

HBAQ uses hierarchical synopses to answer OLAP queries efficiently. For example, consider a query with a selection on the "temperature" column in Fig. 2. During query processing, HBAQ first reads the synopsis of the "year" level and verifies the approximation error on partially satisfied buckets. If the error-bound verification fails in the "year" synopsis, HBAQ goes down to the "month" synopsis to read finer-grained data in the partially satisfied buckets. HBAQ repeats this process and finally reads the original table if the error-bound verification fails at the bottom level (i.e., the "day" synopsis). Although this process may need to read multiple synopses, the compactness of higher-level synopses mitigates its impact. Furthermore, we can reduce the number of read tuples in lower levels because their access ranges shrink as partially satisfied buckets are narrowed down.

4 Accelerating Synopsis Construction and Update

We propose new algorithms for synopsis construction and updating by modifying the definition of buckets. BAQ and BAQ± dynamically construct buckets by sorting input numerical attributes, but this approach becomes a critical bottleneck in synopsis construction. In particular, since BAQ and BAQ± require bucketing for each partial synopsis, they perform multiple full scans on an original table. Moreover, dynamically constructed buckets also make it challenging to update synopses because modifying (i.e., inserting, deleting, or updating) an original table may change the buckets' range and require reconstructing all synopses from scratch.

In the remainder of this section, we first introduce a new bucket definition based on static partitioning. Then, we explain a new synopsis construction algorithm and how to extend it to accept incremental updates.

4.1 Static Partitioning Into Buckets

We introduce the minimum unit ϵ_i of each numerical attribute A_i to divide its values into buckets and assign bucket IDs statically. Figure 4 illustrates our bucketing for the "temperature" attribute. First, we pack values where $|a_i| < \epsilon_i$ into a bucket with ID 0 because the minimum unit ϵ_i acts as a near-zero threshold, as defined in Eq. (1). Then, we use the following expressions to determine the ranges of bucket IDs j and $-j$:

$$[\epsilon_i(1+\delta)^{j-1}, \epsilon_i(1+\delta)^j) \text{ and} \quad (3)$$
$$(-\epsilon_i(1+\delta)^j, -\epsilon_i(1+\delta)^{j-1}]. \quad (4)$$

The advantage of this bucketing is that the following formula can compute a bucket ID j from the input value of a numerical attribute $a_i \in A_i$. Let sgn be the sign function that returns 0 for zero values and $+1$ or -1 for positive or negative values, respectively.

$$j = \text{sgn}\, a_i \left\lceil \log_{1+\delta} \frac{|a_i|}{\epsilon_i} \right\rceil \quad (5)$$

In other words, we can statically convert all numerical values in an original table into bucket IDs without the costly sorting used in BAQ/BAQ±.

Although we assume that a user specifies the minimum units of numerical attributes because their decision requires background knowledge, we also consider several approaches to automatically determining them. For example, if a target attribute represents integers, we can use $\epsilon = 1$ as a conservative setting. We can also use the significant digits of single/double-precision floating-point numbers for decimal attributes. From another perspective, the maximum number of maintainable buckets (i.e., representable by 64-bit integers) may derive minimum units.

4.2 Synopsis Construction Using Static Bucketing

We propose a new algorithm for constructing a synopsis using our bucketing approach. The algorithm requires only one scan of an original table compared to BAQ and BAQ±. Since our bucketing can statically convert numerical values into bucket IDs, the converted results of each tuple are independent of the other tuples in a table. Thus, we can convert each input tuple and insert it into a synopsis individually.

Algorithm 1 shows the proposed synopsis construction. The algorithm requires a source table T and query column sets Q. We first initialize a synopsis S as an empty table (line 1). Then, we convert each tuple $t \in T$ into synopsis records using static bucket IDs (line 3) and check every QCS to add the converted record to the synopsis (lines 4–18). If the synopsis already includes the same record regarding a query column set $\pi \in Q$, we update its content (lines 5–8). Otherwise, we check if the other QCSs share columns with π and have space for inserting new values to prevent the synopsis from including redundant records (lines 9–14). If there are no records for accepting the converted record, we insert a new record into the synopsis (lines 15–18). Note that we fill the new record by "NULL" except for the columns in π to allow the algorithm to insert new values in lines 9–14.

For instance, consider Table 1 and two QCSs $\pi_1 = \{\text{region}, \text{date}, \text{temperature}\}$ and $\pi_2 = \{\text{region}, \text{status}\}$ with a minimum unit $\epsilon = 0.1$ and an error bound

Algorithm 1: GenerateSynopsis(T, Q)

```
Input: T, Q                    // a source table and query column sets
Output: S                                                    // a synopsis
1  S ← ∅
2  foreach t ∈ T do
3     calculate the corresponding ID of numerical data.
4     foreach π ∈ Q do
          // check if the synopsis includes target QCS's values
5         if ∃r ∈ S, r.π = t.π then
6             mean(r.π)_A ← (mean(r.π)_A × SF_{r.π} + t.π_A) / (SF_{r.π} + 1)
7             increment CNT_{r.π} of r by 1
8             continue                             // go to the next QCS
          // the synopsis may include a part of QCS
9         foreach π' ∈ Q \ π do
              // check QCS in descending order of |π ∩ π'|
10            if ∃r ∈ S, (r.π' ≠ NULL ∧ r.π = NULL) ∨
                 (r.π ∩ π' = t.π ∩ π' ∧ ∀a ∈ r.π \ π', a = NULL) then
                  // update NULL to new values
11                update a ∈ r.π \ π' by t.π \ π'
12                mean(r.π)_A ← t.π_A
13                set CNT_π of r to 1
14                break
          // insert a new record if not exist
15        if ∄r ∈ S, r.π = t.π then
16            insert a new record r with t.π to S
17            mean(r.π)_A ← t.π_A
18            set CNT_π of r to 1
```

$\delta = 0.05$. Since the initial synopsis is empty, HBAQ converts the first tuple into ("Africa", "2009-06-27", (142, 98.5), "O") using Eq. (5) and adds it to the synopsis with $CNT_1 = CNT_2 = 1$. HBAQ then converts the second tuple into ("Africa", "2009-08-01", (143, 103.5), "O") and notices that the synopsis does not have acceptable records for π_1. Thus, HBAQ inserts a new record ("Africa", "2009-08-01", (143, 103.5), NULL) with $CNT_1 = 1$ and $CNT_2 = 0$. On the other hand, since the first record has the same π_2 values, HBAQ updates its CNT_2 to 2. When processing ("Africa", "2009-08-01", (143, 103.1), "P") converted from the sixth tuple, the synopsis does not have records with the same π_2 values, but the second record has the same value regarding {region} = $\pi_1 \cap \pi_2$. Therefore, HBAQ updates its "status" column from NULL to "P" with $CNT_2 = 1$. This procedure continues until HBAQ processes all the tuples and constructs the synopsis in Table 4.

4.3 Support for Incremental Updates

Since HBAQ introduces static bucket IDs, we can support incremental synopsis updates based on modifications to an original table. As described above, the conversion from source tuples into synopsis records is independent of the other contents of a table and an existing synopsis. Thus, we can safely compute a differential synopsis from the difference of a source table and apply it to a stored synopsis as incremental updates. This procedure allows HBAQ to avoid the reconstruction of a synopsis and maintain an up-to-date synopsis.

We can use our construction algorithm to create a differential synopsis. If a table modification only involves insert operations, we construct a differential synopsis from new tuples and merge it into an existing synopsis. If there is the deletion of tuples, we also use the proposed algorithm but decrease the tuple counts of corresponding records instead of incrementation. Since an update operation is a combination of delete and insert operations, we can construct a differential synopsis for update operations by combining the above insert and delete procedures.

5 Experiments

We compare HBAQ with BAQ and BAQ± using synthetic and real-world datasets. We used Python to implement every synopsis construction method and stored constructed synopses in PostgreSQL. Then, we submitted OLAP queries to PostgreSQL and measured query latency and relative errors.

5.1 Experimental Settings

We use the following three datasets in the experiments.

- **Orders:** This is the "ORDERS" table in the TPC-H benchmark [18] and contains 1.5 million records. We use five (four categorical and one numerical) columns with the dimension table of the "ORDERDATE" column. The dimension table has {day, month, quarter, year} levels.
- **Lineitem:** This is the "LINEITEM" table in TPC-H and contains 6 million records. We use six (five categorical and one numerical) columns.
- **Temperature:** This is an open dataset for machine learning in Kaggle [19]. It contains 2.8 million records with temperature measurements across 321 cities daily for over 36 years. We use three (two categorical and one numerical) columns.

We use synthetic queries in Table 6 to evaluate HBAQ, BAQ, and BAQ±. We prepare the workload and OLAP queries separately to facilitate a comprehensive evaluation. Each query has a single aggregate function, zero or one selection condition, and a grouping operation on a categorical attribute. We randomly select attributes for selection/grouping and then randomly generate selection conditions. For the Orders dataset, we added ten queries to evaluate the effect of dimension tables. Five OLAP queries have selections on the "ORDERDATE"

Table 6. Number of synthetic queries for lineitem and the temperature

Aggregate function	Selection condition					
	Workload query			OLAP query		
	None	Category	Numeric	None	Category	Numeric
COUNT	2	50	52	1	10	10
MIN, MAX	2	50	52	1	20	20
SUM, AVG	2	50	52	1	20	20

attribute with different levels. The other five OLAP queries use five different aggregate functions with no time selections.

We compare the competitors from four perspectives: 1) offline synopsis construction time [hours], 2) memory usage [MB], 3) OLAP query latency [milliseconds], and 4) the maximum relative error. We measure the total time required to construct all synopses (partial and unified synopses in BAQ and BAQ± or hierarchical synopses in HBAQ) for offline synopsis construction time. We use a unified synopsis for BAQ/BAQ± and all hierarchical synopses for HBAQ to measure memory usage. Since we submit each OLAP query thirty times, we use the averages of latency and the maximum relative errors (MRE) for comparison.

5.2 Experimental Results

We compare HBAQ with BAQ± and BAQ by setting the error bound δ to 0.05, 0.1, 0.15, 0.2, and 0.25.

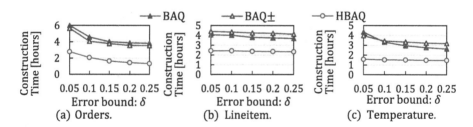

Fig. 5. Construction time in different datasets

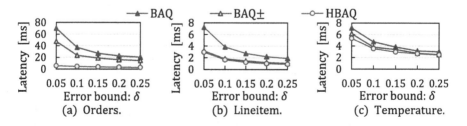

Fig. 6. Latencies of OLAP queries in different datasets

Construction Time. Figure 5 shows the synopsis construction times of the competitors for each dataset. The results show that HBAQ achieves about half the synopsis construction times of BAQ and BAQ± in all the datasets. As described above, BAQ and BAQ± require 1) sorting numerical values to construct partial synopses and 2) multiple full scans of a source table to construct a unified synopsis, which leads to longer construction times. In contrast, HBAQ can construct a synopsis with a single full scan of a source table, significantly reducing synopsis construction time. In addition, our construction algorithm avoids constructing redundant (i.e., partial) synopses and reduces temporal memory usage, resulting in shorter construction times.

Although HBAQ requires the construction of hierarchical synopses in the Orders dataset, the impact on the overall construction time is negligible. Since HBAQ constructs a higher-level synopsis by summarizing its lower-level synopsis, the sizes and computation times of hierarchical synopses decrease as the procedure goes to the upper levels. As a result, HBAQ can achieve shorter construction times than BAQ and BAQ±, regardless of constructing hierarchical synopses.

Latency. Figure 6 shows the average latency of the competitors for answering OLAP queries on each dataset. HBAQ can answer OLAP queries with less or comparable latency rather than BAQ and BAQ±. In particular, since HBAQ can use hierarchical synopses in the Orders dataset, HBAQ achieves about ten times faster latency than the existing methods. HBAQ does not have hierarchical synopses in the other datasets, but its latency is comparable to our previous work, BAQ±.

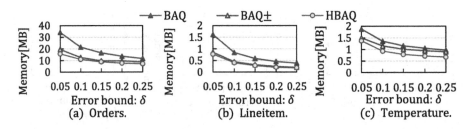

Fig. 7. Memory usage in different datasets

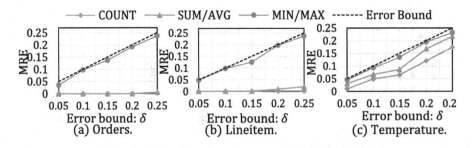

Fig. 8. Maximum relative error in different datasets

Memory Usage. Figure 7 shows the memory usage of the synopses constructed by the competitors. The results show that HBAQ can construct synopses with comparable or smaller sizes than the existing methods. In particular, although BAQ± achieves compact synopses using more giant buckets than BAQ and HBAQ, HBAQ achieves comparable sizes with the proposed static bucketing. These results demonstrate that the proposed construction algorithm effectively summarizes a source table into a synopsis while achieving efficient construction. Besides, the Orders dataset results show that hierarchical synopses do not significantly affect total memory usage.

Maximum Relative Error. Figure 8 shows the maximum relative errors of HBAQ on the results of OLAP queries in each dataset. The dashed lines are the upper bounds of relative errors, and bounded approximate processing must guarantee that its approximation errors are below these lines. The experimental results demonstrate that HBAQ successfully ensures strictly bounded errors while achieving efficient synopsis construction, fast online query processing, and synopsis size reduction.

6 Related Work

Approximate query processing (AQP) has become an crucial strategy in data management to address the challenges of delivering rapid query responses in growing data volumes [4,14]. Research such as BlinkDB [2] and VerdictDB [16] developed systems based on AQP to strike a balance between the speed of query execution and the accuracy of results. Fundamental to AQP are technologies like sampling, histograms, and synopses [5], which play crucial roles in summarizing and estimating large datasets. These technologies enable AQP to provide fast and highly accurate query results, which are critical for efficient data management in modern environments.

As Ioannidis emphasized [9], a histogram is one of the most critical components for summarizing data distribution and facilitating optimized query execution in large database systems. Histograms are available in various forms, such as equal-width, equal-depth, and V-optimal, to support different data and queries. Although histograms are helpful for efficient query planning, they face limitations in ensuring the accuracy of query results.

Sampling-based approximate query processing (SAQ) has been extensively studied within AQP [13]. SAQ extracts samples from source tables online and leverages these samples to estimate results. While SAQ offers a trade-off between performance and accuracy, it faces several challenges, such as precise error guarantees, acceptance of specific aggregations (e.g., MIN/MAX), and handling skewed data distributions.

Recent research has aimed to enhance AQP accuracy by integrating precomputation with online sampling techniques [6,17]. Sample+Seek [6] is a notable approach integrating precomputed measure-augmented inverted indexes to bolster accuracy. This approach underscores the importance of balancing enhanced precision with the practicalities of sample size and query latency. Despite its innovative strategy, Sample+Seek also grapples with the challenge of precise error quantification, a common hurdle in AQP methodologies.

Bounded approximate processing (BAQ) introduces a novel framework for achieving strict error guarantees [11]. BAQ summarizes a source table into a synopsis, guaranteeing the strict upper bound of errors on OLAP query results. However, BAQ cannot ensure the bounded errors according to specific input tables and queries and has space for improvement in synopsis size. Thus, we have extended BAQ into BAQ± to solve these problems [15]. BAQ± improves the bucketing and revises the query processing in BAQ to guarantee strict error bounds, but BAQ and BAQ± still face the challenge of long synopsis construction time.

7 Conclusion

This paper proposes the hierarchical bounded approximate query processing (HBAQ) framework. HBAQ uses hierarchical synopses to efficiently respond to OLAP queries with high-cardinality categorical data. HBAQ also accelerates synopsis construction while accepting differential updates. We implemented this framework and evaluated it on three datasets. The experimental results demonstrate that our method outperforms the existing methods regarding synopsis construction time, memory usage, and OLAP query latency while guaranteeing the strict upper bound on approximation errors. For future work, we aim to conduct a detailed examination of the framework to support broader types of queries.

Acknowledgments. This work was supported by JST CREST Grant Number JPMJCR22M2 and JSPS KAKENHI Grant Numbers JP20K19804, JP23K21726, and JP23K24850. This work also was supported by the "Nagoya University Interdisciplinary Frontier Fellowship" supported by Nagoya University and JST under university fellowships for the creation of science technology innovation, Grant Number JPMJFS2120.

References

1. Agarwal, S., et al.: Knowing when you're wrong: Building fast and reliable approximate query processing systems. In: Proceedings of SIGMOD, pp. 481–492 (2014)
2. Agarwal, S., Panda, A., Mozafari, B., Madden, S., Stoica, I.: BlinkDB: queries with bounded errors and bounded response times on very large data. In: Proceedings of EuroSys, pp. 29–42 (2013)
3. Braverman, V., Ostrovsky, R.: Generalizing the layering method of Indyk and Woodruf: recursive sketches for frequency based vectors on streams. In: Proceedings of SIGKDD, pp. 58–70 (2013)
4. Chaudhuri, S., Ding, B., Kandula, S.: Approximate query processing: no silver bullet. In: Proceedings of SIGMOD, pp. 511–519 (2017)
5. Cormode, G., Garofalakis, M., Haas, P.J., Jermaine, C.: Synopses for Massive Data: Samples, Histograms, Wavelets, Sketches, pp. 1–294. Now Publishers, Delft (2011)

6. Ding, B., Huang, S., Chaudhuri, S., Chakrabarti, K., Wang, C.: Sample+Seek: approximating aggregates with distribution precision guarantee. In: Proceedings of SIGMOD, pp. 679–694 (2016)
7. Guha, S., Harb, B.: Wavelet synopsis for data streams: minimizing non-Euclidean error. In: Proceedings of SIGKDD, pp. 88–97 (2005)
8. Han, J., Pei, J., Tong, H.: Data Mining: Concepts and Techniques, chap. 3, 4 edn. Morgan Kaufmann, Massachusetts (2022)
9. Ioannidis, Y.: The history of histograms (abridged). In: Proceedings of VLDB, pp. 19–30 (2003)
10. Li, K., Li, G.: Approximate query processing: what is new and where to go? A survey on approximate query processing. Data Sci. Eng. **3**, 379–397 (2018)
11. Li, K., Zhang, Y., Li, G., Tao, W., Yan, Y.: Bounded approximate query processing. IEEE TKDE **31**(12), 2262–2276 (2019)
12. Ma, Q., Triantafillou, P.: DBEst: revisiting approximate query processing engines with machine learning models. In: Proceedings of SIGMOD, pp. 1553–1570 (2019)
13. Ma, Q., et al.: Learned approximate query processing: make it light, accurate and fast. In: Proceedings of CIDR, pp. 1–11 (2021)
14. Mozafari, B., Niu, N.: A handbook for building an approximate query engine. IEEE Data Eng. Bull. **38**(3), 3–29 (2015)
15. Ni, T., Sugiura, K., Ishikawa, Y., Lu, K.: Approximate query processing with error guarantees. In: Proceedings of Big Data Analytics (BDA), vol. 13167, pp. 233–244 (2021)
16. Park, Y., Mozafari, B., Sorenson, J., Wang, J.: VerdictDB: universalizing approximate query processing. In: Proceedings of SIGMOD, pp. 1461–1476 (2018)
17. Peng, J., Zhang, D., Wang, J., Pei, J.: AQP++: connecting approximate query processing with aggregate precomputation for interactive analytics. In: Proceedings of SIGMOD, pp. 1477–1492 (2018)
18. Transaction Processing Performance Council: TPC-H homepage. https://www.tpc.org/tpch/ (2022)
19. University of Dayton, Environmental Protection Agency: Average daily temperature archive. https://ecommons.udayton.edu/cgi/viewcontent.cgi?article=10949&context=news_rls (2022)

Query Expansion in Food Review Search with Synonymous Phrase Generation by LLM

Arisa Ashizawa[✉], Ryota Mibayashi, and Hiroaki Ohshima

Graduate School of Information Science, University of Hyogo, Kobe, Hyogo, Japan
{ad23y002,af22h007}@guh.u-hyogo.ac.jp, ohshima@ai.u-hyogo.ac.jp

Abstract. In this study, we propose a method that considers the diversity of query expressions for searching food reviews. For example, if you search for reviews of udon noodles using the query *"firmness"*. In that case, it is easy to output reviews containing the expression *"firmness"*. On the other hand, we want to output reviews containing expressions like *"elasticity"* and *"chewiness,"* which do not match the exact query but have semantic similarity. Therefore, we propose two approaches using query expansion with ChatGPT, a large language model (LLM). Both approaches use the LLM to obtain several synonymous phrases for a query. The first method trains the BERT model for relevance judgment of reviews based on semantic similarity to the query. To create training data, collect candidate queries in advance and acquire their synonymous phrases. In the second approach, when a search query is input, the synonymous phrases of the query are obtained for on-demand query expansion. We implemented these approaches and compared them to methods such as BM25. The combined use of both proposal approaches showed better performance.

Keywords: Review Search · BERT · LLM · query expansion

1 Introduction

In online shopping, product reviews are important for purchase decisions. However, it is difficult to find the wanted information among the many reviews. It is usual for a product to have many reviews. In fact, there are more than 8,000 products with over 100 food reviews on Rakuten Ichiba.

When a user wants to find useful information from many reviews for a certain product, they use review search. A keyword query will be used for the review search. If a string match search is performed using this query, only reviews containing that query will be searched. However, the user also wants to search for reviews that contain phrases synonymous with the query.

In this study, we propose the following two approaches that can also search for reviews containing synonymous phrases.

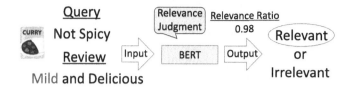

Fig. 1. This relevance judgment BERT outputs the relevance ratio of reviews to query. Based on this, the model judges whether a review is relevant or irrelevant.

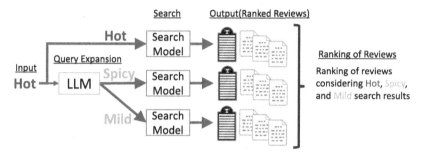

Fig. 2. Input query and on-demand query expansion is performed by LLM. The system searches for each query and outputs a ranking of reviews based on relevance ratio. Considering these rankings, the reviews are ordered by relevance.

- **Two approaches**
  - First Approach: **Relevance Judgment BERT**
  - Second Approach: **On-demand Query Expansion**

In the first approach, relevance judgment BERT is constructed to semantically judge whether review matches the query, as shown in Fig. 1. Constructing this model requires a large amount of training data, both relevant and irrelevant, considering semantic similarity. This training data is created in large amounts using synonymous phrases generated by the query expansion with LLM.

The second approach performs on-demand query expansion and searches with expressions relevant to the query, as shown in Fig. 2. It is necessary to get the synonymous phrases of the input query. These phrases are obtained by query expansion using LLM. In this way, each approach should be able to search for reviews that are semantically similar to the input query.

2 Related Work

When shopping online, some studies have been conducted to obtain information about products. For example, Yitzhak [2] and Liu [9] have studied effective ways of representing product information through faceted search and visual feature representation. Fang [6] and Hu [8] have also studied how to summarize reviews. This study uses review search to extract the reviews that users want.

This study uses BERT, a general-purpose language model, for review searches. BERT was proposed by Devlin [5] in 2018. BERT is used for a variety of tasks [1,3,10,14]. A sentiment analysis of the impact of the novel coronavirus on social life is being conducted by Chouikhi [4]. This is done using data collected from Twitter and the BERT model.

BERT is known to improve the performance of the relevant tasks by additional learning. Sushil [12] and Gururangan [7] have shown that additional learning improves the performance of tasks on the trained data. These previous studies have shown the usefulness of additional learning. In this study, additional learning is performed on the "Foods" and "Sweets and Snacks" review data.

This study proposes search methods using LLMs. ChatGPT, an LLM, is one of the natural language processing models developed by OpenAI. ChatGPT is capable of text generation and question answering. For example, how ChatGPT is used for debugging programming is discussed by Tinega [13]. In this study, we use ChatGPT to perform query expansion.

There are previous studies on query expansion using LLM. In Wang [15], the original query and the pseudo-document generated by LLM were concatenated to create a new query. In this study, multiple new queries are created by LLM from the original query, and each is used as one query.

3 Rakuten Dataset

We use the "Rakuten Dataset" [11]. The number of reviews for all categories is huge, about 70 million. Additionally, each category has different characteristics in the content of the reviews. We use data on about 6.5 million reviews, focusing on "Foods" and "Sweets and Snacks". This enables us to perform review searches that capture the characteristics of expressions in "Foods" and "Sweets and Snacks".

In this study, we refer to the combination of "review title" and "review content" as a "review". A review divided into one sentence, which is called a "review sentence".

The evaluation data and the training data are split in advance. First, we explain how we created the evaluation data. We obtained 28 genres with at least 20 products that each have at least 100 reviews. Five products were randomly selected from each of these genres, totaling 140 products for evaluation. Additionally, 100 of the latest reviews were obtained from each product, making a total of 14,000 reviews as the evaluation data. Next, the training data consisted of all 627,400 products not used in the evaluation, resulting in 6,435,083 reviews. A total of 25,051,124 review sentences were obtained.

4 Proposed Approaches

In this study, we propose two approaches to search for reviews containing semantically similar expressions to the search query, using LLM (gpt-35-turbo) with query expansion. The first approach, a relevance judgment BERT is constructed with training data created by query expansion. The second approach uses LLM to perform on-demand query expansion. These two approaches are independent. Therefore, they can be used in combination.

4.1 Approach 1: Relevance Judgment BERT

Training Data. We need to train what queries and what reviews are relevant/irrelevant in order to construct a relevance judgment BERT. Large amounts of relevant/irrelevant data for this study will be created by using the LLM. For example, relevant reviews for the query "*hot*" often include expressions such as "*spicy*" or "*pungent*". If these expressions "spicy" and "pungent" can be collected in advance, reviews containing those keywords can be made relevant data. We collected a large number of expressions semantically similar to pre-collected queries by using the LLM.

First, we collected the queries to be used for generation. We requested 200 lancers on Lancers[1] to collect keywords for review search. We asked them to think of specific products and provide keywords for searching reviews about those products. The task was performed for 10 products per lancer. Only appropriate queries were collected. In the end, we obtained 1,189 search keywords, which we refer to as "pre-collected queries". The queries obtained include queries such as "*delivery*" and "*sweet*".

Next, query expansion was performed by ChatGPT. Synonymous phrases for the "pre-collected queries" were obtained. When performing query expansion, we indicated which genre the queries targeted. We used a total of 38,283 pairs of queries and genres. Using these pairs of queries and genres, we performed query expansion with the following prompts. The example below shows a pair with the genre "foods/seasoning/spices/chili pepper" and the query "*spicy*". When generating, we specified temperature=0.0 to minimize variation in the results.

- **Example Prompt (This is an English translation of the Japanese prompt.)**
  There are reviews for "foods/seasoning/spices/chili pepper". What specific phrases would you consider a review about "spicy" if included here? Please give me 5 words each that are relevant to "spicy" and 5 words that contrast with "spicy". Please keep the words as short as possible. We only want words. Please do not duplicate words. Do not use parentheses. Do not include words that are not appropriate for the word "spicy" in "foods/seasoning/spices/chili pepper". Only 5 relevant words and 5 contrasting words should be stored in the following format.
  ## Format Relevant words: [a, b ,c] Contrasting words: [a, b ,c]
- **Example Generated (This is an English translation of the output generated in Japanese.)**
  Relevant words: [hot, pungent, piquant, fiery, peppery]
  Contrasting words: [mild, mellow, insipid, bland, flavorless]

---
[1] https://www.lancers.jp/.

If a review in the target genre did not contain the generated expression, we judged it inappropriate as a synonymous phrase. As a result, we obtained a total of 381,349 synonymous phrases.

Finally, we explain how to create relevant and irrelevant data. With this data, we train the relevance of queries and review sentences. The data consists of "genre, query, review sentence, relevant/irrelevant label".

First, we created a list for each genre containing both the "pre-collected queries" and synonymous phrases. This list, called the "query list" was created for all 38,283 genres.

Next, two queries were obtained from the "query list" each. If the same product has review sentences containing each query, two cases were randomly (random seed = 42) selected from the review sentences. Those were taken as relevant data, and 12,261,334 were obtained. We randomly selected two review sentences for the same product that did not contain any of the words from the "query list" as irrelevant data. They were obtained as 12,099,540 irrelevant data.

Relevance Judgment BERT. In this study, we constructed a review-trained BERT model by fine-tuning a Japanese pretrained BERT model with additional learning using reviews. The Japanese pre-trained BERT model is the BERT model (cl-tohoku/bert-base-japanese-v3) provided by Inui Laboratory at Tohoku University. In this study, all of the training data was created in the 3 section were used for additional learning. This trains the food review.

Fine-tuning was done on the review-trained BERT model using both relevant and irrelevant data. By including genre information, we trained the model to consider the differences in expressions for each product category. The input is genre, query, and review sentence, and it is possible to judge whether the review sentence is relevant to the query, as shown in Fig. 3.

The search results should also rank the reviews based on how well they match the query. However, the model is inputted with a review sentence, not a review. The relevance ratio through the sigmoid function is the score of the review sentence. we obtain the maximum score among the review sentences from the same user for the same product. This score is used to rank the reviews This searches for relevant reviews at the top.

4.2 Approach 2: On-Demand Query Expansion with LLM

We perform on-demand query expansion to find the top reviews related to the query input by the user. When a user inputs a query, ChatGPT is used to retrieve up to 10 expressions related to the query on-demand. Query expansion is done at the same prompt as the query expansion for constructing the relevance judgment BERT. Each expression is searched separately. A score is assigned to each review for each query. The largest score among them is the score of the review. We rank the reviews by their scores.

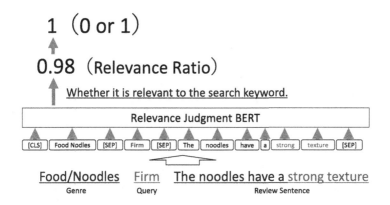

Fig. 3. When genre, query and review are input to the relevance judgment BERT, it outputs the relevance judgment of the review to the query considering the genre. If the value is high, it can judge as 1.

5 Evaluation and Discussion

5.1 Evaluation Data

Evaluation is evaluated by comparing the search results from the proposed method with the labeled correct data.

First, we explain the process of collecting queries for evaluation. We asked one student in our lab to create a query for a review search on 140 products we had collected in advance. For example, if a user searches for a review of a certain apple, the query is often *"sweet"*. However, they may search not only for *"sweet"* but also for queries such as *"mold"*. In this study, we collect queries users first think of, called "easy queries", and those they do not, called "difficult queries". We collected two queries per product, for a total of 280 queries obtained.

Next, we explain how the correct data was created by two students in the same laboratory. The data used were reviews that did not include a query for each product collected. The correct data is labeled on a 3-point scale indicating the relevance of the review to the query. The following is an example of the evaluation method in the case of the product *"curry"* and the query *"sweet"*.

– **Example of Presented Answer**

Review	Relevance Ratio	Case
I like spicy food, so I felt it was too sweet	2	Review relevant to obviously *"sweet"*
It tasted like something even a child could eat	1	Review inferred to be relevant to *"sweet"*
It was a thick curry. It is a repeat!	0	Review not relevant to obviously *"sweet"*

This resulted in collecting 26,988 correct data points.

5.2 Evaluation and Discussion

We evaluated whether we can find the top reviews relevant to our query. We used nDCG@10 for the evaluation. We compared methods such as TF-IDF, OkapiBM25, ChatGPT Embeddings (text-embedding-ada-002), and sentence-BERT (sonoisa/sentence-bert-base-ja-mean-tokens). In addition, on-demand query expansions to each of the methods were also compared. The evaluation results are shown in Table 1.

First, we evaluated with the queries collected in advance. In some cases, expressions with the same word stem, such as *"sweet"* or *"sweetly"* for *"sweetness"*, appeared frequently in the reviews. In this case, it was found that OkapiBM25 and others were highly evaluated. On-demand query expansion was found to be effective for relevance judgment BERT.

Next, we evaluated searches for each of the "easy queries" and "difficult queries". The evaluation results are shown in Tables 2 and 3. The results of the evaluation showed that OkapiBM25 was highly rated when review searches were performed on "easy queries". The method with on-demand query expansion for relevance judgment BERT was the second highest. We think that the value of OkapiBM25 is high for "easy queries" because there were many expressions that matched the stem in the review. When searching for "difficult queries", the combination of "relevance judgment BERT" and "on-demand query expansion" proposed by this study was the highest evaluated method. The second highest evaluated method was "relevance judgment BERT". In fact, when review searches were conducted on "difficult queries", relevance judgment BERT was able to search for reviews containing expressions that could not be obtained by the other methods. For example, when searching for *"cracked"* for *"chicken egg"* review, we were able to search review that included the expression *"damaged"*. Additionally, by on-demand query expansion, we were able to search review that actually included expressions relevant to *"cracks"* such as *"intact"* and *"broken"* at the top of the list. These results indicate that the proposed method is useful.

Table 1. nDCG@10

Method	No On-demand Query Expansion	On-demand Query Expansion
TF-IDF	0.298	0.418
OkapiBM25	0.512	0.448
ChatGPT Embeddings	0.464	0.406
SentenceBERT	0.311	0.253
Relevance Judgment BERT	0.450	0.536

Table 2. nDCG@10(easy queries)

Method	No On-demand Query Expansion	On-demand Query Expansion
TF-IDF	0.341	0.502
OkapiBM25	0.620	0.524
ChatGPT Embeddings	0.503	0.440
SentenceBERT	0.348	0.295
Relevance Judgment BERT	0.474	0.603

Table 3. nDCG@10(difficult queries)

Method	No On-demand Query Expansion	On-demand Query Expansion
TF-IDF	0.246	0.317
OkapiBM25	0.385	0.357
ChatGPT Embeddings	0.419	0.366
SentenceBERT	0.266	0.204
Relevance Judgment BERT	0.422	0.456

6 Conclusion

In this study, we proposed two approaches: a method using query expansion to construct relevance judgment BERT and on-demand query expansion method. In particular, we have shown that the approach that combines the two approaches is able to search for the most variety of expressions included.

In order to capture the characteristics of the data, we focused on "Foods" and "Sweets and Snacks" as the target data. However, users also purchase "Home Appliances" and "Clothes". In the future, we would like to expand the target genres and construct a general-purpose search system.

Acknowledgments. This work was supported by JSPS KAKENHI Grant Numbers JP21H03775, JP22H03905, JP21H03554, JP24K03228. In this paper, we used "Rakuten Dataset" (https://rit.rakuten.com/data_release/) provided by Rakuten Group, Inc. via IDR Dataset Service of National Institute of Informatics.

References

1. Bao, H., et al.: UNILMv2: pseudo-masked language models for unified language model pre-training. In: Proceedings of ICML 2020, pp. 642–652 (2020)
2. Ben-Yitzhak, O., et al.: Beyond basic faceted search. In: Proceedings of WSDM 2008, pp. 33–44 (2008)
3. Chen, T., Huang, S., Wei, F., Li, J.: Pseudo-label guided unsupervised domain adaptation of contextual embeddings. In: Proceedings of AdaptNLP 2021, pp. 9–15 (2021)
4. Chouikhi, H., Chniter, H., Jarray, F.: Arabic sentiment analysis using BERT model. In: Advances in Computational Collective Intelligence, pp. 621–632 (2021)

5. Devlin, J., Chang, M.W., Lee, K., Toutanova, K.: BERT: pre-training of deep bidirectional transformers for language understanding. In: Proceedings of NAACL 2019, pp. 4171–4186 (2019)
6. Fang, L., Qian, Q., Huang, M., Zhu, X.: Ranking sentiment explanations for review summarization using dual decomposition. In: Proceedings of CIKM 2014, pp. 1931–1934 (2014)
7. Gururangan, S., et al.: Don't stop pretraining: adapt language models to domains and tasks. In: Proceedings of ACL 2020, pp. 8342–8360 (2020)
8. Hu, M., Liu, B.: Mining opinion features in customer reviews. In: Proceedings of AAAI 2004, pp. 755–760 (2004)
9. Liu, B., Hu, M., Cheng, J.: Opinion observer: analyzing and comparing opinions on the web. In: Proceedings of WWW 2005, pp. 342–351 (2005)
10. Niwa, A., Nishiguchi, K., Okazaki, N.: Predicting antonyms in context using BERT. In: Proceedings of INLG 2021, pp. 48–54 (2021)
11. Rakuten Group, Inc.: Rakuten Ichiba data. Informatics Research Data Repository, National Institute of Informatics. (dataset) (2020). https://doi.org/10.32130/idr.2.1
12. Sushil, M., Suster, S., Daelemans, W.: Are we there yet? Exploring clinical domain knowledge of BERT models. In: Proceedings of BioNLP 2021, pp. 41–53 (2021)
13. Tinega, G.A., Mwangi, W., Rimiru, R.: Text mining in digital libraries using OKAPI BM25 model. In: Int. J. Comput. Appl. Technol. Res., 398–406 (2019)
14. Vaswani, A., et al.: Attention is all you need. In: Proceedings of NIPS 2017, pp. 6000–6010 (2017)
15. Wang, L., Yang, N., Wei, F.: Query2doc: query expansion with large language models. In: Proceedings of EMNLP 2023, pp. 9414–9423 (2023)

Question Answer Summary Generation from Unstructured Texts by Using LLMs

Yuuki Tachioka

Denso IT Laboratory, Shibuya, Tokyo 150-0002, Japan
tachioka.yuki@core.d-itlab.co.jp
https://www.d-itlab.co.jp/en/

Abstract. Recently, the importance of generating accurate question and answer summaries has increased because the number of documents has increased, but it is difficult to summarize them from unstructured documents where the answer is not associated with the question. To address this problem, first we make a group of answer sentences, which corresponds to one question based on the heuristic rules, and we find the corresponding answer from the summaries and subtopics of the question by using the BM25-based similarity calculation. Second, we use the large language model (LLM) to generate a summary of the answers from the answers found. Experimental results showed that our methods significantly outperformed LLM-based answer generation that inputs whole answer sentences including irrelevant parts, which correspond to another question, to LLM and that our methods were practical compared to human-generated gold summaries.

Keywords: BM25 · Gale-Shapley algorithm · LLM · RAG

1 Introduction

Recently, the number of documents has increased and it is hard to read through them. It is desirable to summarize the documents and release them in a concise form that stakeholders can easily understand. For summary generation, it is more difficult to deal with unstructured documents than with structured documents. We aim to develop technologies that generate summaries from unstructured documents.

NTCIR QA Lab-PoliInfo series [3–5, 7] extracts important political information from unstructured documents. This task deals with the unstructured government minutes of the Tokyo Metropolitan Assembly, which are composed of unseparated question and answer sentences. Question Answering 2 subtask of the NTCIR-17 QA Lab-PoliInfo-4 task [7] is to create a summary of questions and answers (QA) from these unstructured minutes. For preparation of Question Answering, it is necessary to convert unstructured texts into structured QA pairs by making a paragraph composed of each question and answer, and by associating a question with its answer. To make structured data, we use heuristic rules to make paragraphs composed of related sentences for the question and answer. Each sentence has a Q/A/O tag, where Q, A, and O are a question, an answer, and others, respectively. After the paragraphs are obtained, we use BM25 [9]-based similarities to align the summaries and subtopics of the questions and the corresponding answers.

Based on the QA alignment obtained, we use a large language model (LLM) to generate answer summaries. Retrieval-Augmented Generation (RAG) [6] is used to provide additional information to LLM. The associated answers are inputted into the text-to-text transfer transformer (T5) [8] model to generate the summary of the answers. Our paper aims to investigate whether the extraction of the corresponding part to LLM is needed or not in the case of RAG, based on the QA alignment to convert unstructured texts into structured data. In addition, to remove the influence of misalignments on model degradation, the training data were refined on the basis of ROUGE F-measures by comparing the generated answer summaries with human-generated gold summaries. The proposed approach can be applied to other summarization tasks of unstructured texts except for the heuristic rules to form paragraphs, which is specified in this task.

2 QA Alignment

We align the question summary and the answer summary in three steps without using the questions themselves, as shown in Fig. 1. The first step finds the corresponding part of the entire minutes by date and questioner ID. The second step combines multiple related sentences with "A" tags to form a paragraph by using heuristic rules in Sect. 2.1. The third step matches the summaries and subtopics of the question and the corresponding answer based on the similarities between them in Sect. 2.2.

We take the approach of generating an answer summary directly from the original answer in minutes [10]. In the PoliInfo-3 QA alignment subtask, as shown in Fig. 2 we have tried to find the original question asked in the Tokyo Metropolitan Assembly from a summary of questions as input, and then used the results of the QA alignment to find the original answer [11]. However, this approach requires linking twice, that is, linking between the question summary and the question and linking between the question and answer. If one of the two linking phases fails, the correct answer cannot be obtained. In addition, the specification of the corresponding questions is not necessarily essential for this task. Thus, we directly associated the summary and subtopics of the question with the answers to reduce the influence of misalignment between the question and the answer.

2.1 Heuristic Rules to Make Paragraphs

We can accurately combine sentences to make paragraphs using regular expressions that are optimized for minutes because questions and answers in the Diet have a fixed format. Heuristic rules use fixed phrases at the beginning or the end of the sentence that start paragraphs and a fixed phrase at the end of the

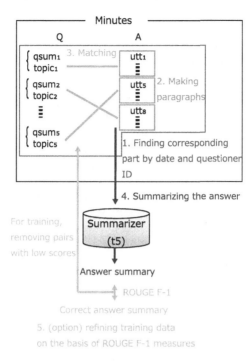

Fig. 1. Finding answers by direct linking with question summaries and topics and summarizing the answers for PoliInfo-4 Question and Answering 2 subtask.

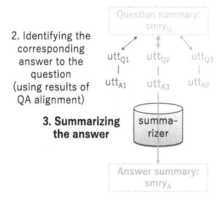

Fig. 2. Linking questions and answers and summarizing the answers for PoliInfo-3 Question and Answering subtask.

sentence that terminates paragraphs. These were originally developed for the QA Alignment subtask of the NTCIR-16 QA Lab-PoliInfo-3 task and the details of our heuristic rules are found in [11].

2.2 Feature for Matching

After we prepare word n-grams by morphological analysis, processed by MeCab[1], BM25 models [9] are constructed on the morphemes excluding tokens, auxiliary verbs, and post-positional particles. BM25 vectors are high-dimensional sparse vectors that only have values in the existing morpheme. BM25 vectors are calculated for both the i-th summaries of the questions and subtopics, $\bm{v}_Q^i$, and the j-th answer paragraphs, $\bm{v}_A^j$. For matching, cosine similarities between sparse vectors ($c_{ij} = (\bm{v}_Q^i)^\top \bm{v}_A^j / |\bm{v}_Q^i||\bm{v}_A^j|$) are used as a feature. Based on c_{ij}, we match the summaries of the questions and subtopics with the answer paragraphs by using a matching algorithm. The effectiveness of BM25 for this task was shown in our article [11], where BM25 outperformed Bidirectional Encoder Representations from Transformers (BERT) [1] and Wikipedia2Vec [12].

3 Question and Answering

After the answer parts are specified by the alignment in Sect. 2, they are inputted into LLM and the answer summaries are generated.

3.1 Summarizing the Answers by LLM

We utilized LLM for generating answer summaries by fine-tuning the pre-trained LLM on answer-answer summary pairs extracted from the training data. The LLM training task is the generation of answer summaries from the comma separated input composed of answer paragraph, the summary of the question, and the subtopics. The main difference between the standard RAG and our proposed method is that the former inputs all answer paragraphs including irrelevant answer paragraphs to T5 when generating answer summaries, but our method only inputs the corresponding answer paragraphs associated with the summary of the question. There are two advantages of our proposed method against the standard RAG. First, the length of input text can be smaller. In some cases, it is necessary for the standard RAG to truncate the input text because the answer paragraphs exceed the limit of prompt length of T5. Second, the removal of irrelevant parts can improve the performance of summary generation, which is validated on the content score of the generated answer summaries.

3.2 Refinement of Training Data

There are some misalignments between the summary of the question and the answer in the training data. These errors can degrade performance. To reduce this influence, we refine the training data based on the ROUGE F-measure scores. For training data, once the answer summaries are generated using the LLM that is fine-tuned on all training data, they are compared with the correct summaries in terms of the ROUGE F-1 scores. If the alignment is wrong, the generated answer summary is also wrong. Thus, pairs below the threshold can be removed from the training data, and the T5 models are fine-tuned on the refined data again.

---

[1] https://taku910.github.io/mecab.

4 Experiments

4.1 Experimental Setup

We evaluated performance using the "formal run" of the NTCIR-17 QA Lab-PoliInfo-4[2]. For matching, BM25 models were trained using distributed data[3] for "Himawari" derived from the minutes of the plenary session and the budget committee of the national diet, and we used the Gale-Shapley (hospital/resident) matching algorithm[4] [2]. Answer summaries were made by LLM, which was a pre-trained Japanese T5 model fine-tuned in the standard RAG style according to the baseline code provided by the task organizers and in the proposed methods. The pre-trained model was **sonoisa/t5-base-japanese**[5] trained with a 100 GB Japanese corpus.

The generated answer summaries were evaluated in terms of both objective scores and subjective scores. The objective scores were the ROUGE-1 F measure, which was calculated by comparing the generated summaries with the human generated gold summaries. The gold summaries were extracted from "TogikaiDayori", which was an official report published by the Tokyo Metropolitan Assembly. Subjective scores[6] were composed of correspondence[7], content[8], well-formed[9], and overall[10] scores, and were evaluated for 100 generated answer summaries at three levels: A (good), B (fair), and C (bad). The total scores were summed over them as

$$score = 2\,|A| + 1|B| + 0\,|C|, \qquad (1)$$

where $|A|$, $|B|$, and $|C|$ is the count of the answer summaries scored as A, B, and C, respectively.

4.2 Results and Discussion

Table 1 shows the ROUGE-1 F-measure. "Prop. 1" was our proposed method, and "Prop. 2" was our proposed method that was fine-tuned again on refined training data in Sect. 3.2. The comparison between the baseline and our methods shows that the extraction of the corresponding part significantly improved performance, and refinement did not improve performance in terms of the ROUGE-1 F-measure. Figure 3 shows the Rouge-1 F-measure for training data. The scores were concentrated around 1.0, but the distribution was long-tailed and there were some samples that had low scores.

---

[2] https://github.com/poliinfo4/PoliInfo4-FormalRun-Question-Answering-2.
[3] https://csd.ninjal.ac.jp/lrc/index.php.
[4] https://pypi.org/project/matching/.
[5] https://huggingface.co/sonoisa/t5-base-japanese.
[6] Subjective scores (human evaluation scores) were judged by the task participants.
[7] Is generated summary in accordance with the answer expression?
[8] Are the important points of an answer covered by the summary?
[9] Is summary natural for Japanese in terms of expression and grammar?
[10] Overall impression including length, expression, content, and grammar.

Table 1. ROUGE-1 F-measure results.

	ROUGE-1 F-measure
Baseline (organizer)	0.2736
QA alignment-based extraction (Prop. 1)	0.3246
+refinement of training data (Prop. 2)	0.3246

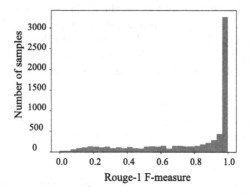

Fig. 3. Rouge-1 F-measure evaluated on training data for data refinement.

Table 2 shows the examples of the summaries generated by the baseline, two types of proposed methods, and the professional human (gold). These were translated from the original Japanese texts. Although the scores for the ROUGE-1 F measure were the same in Prop. 1 and 2, the generated summaries were different. The baseline summary was abstract and lacked important three underlined points. The summaries of our proposed methods covered important points and were concrete.

Table 3 shows the human evaluation scores. The proposed methods improved all scores from baseline. The baseline content score was worse than that of gold by 22 points, but our content score was equivalent to that of gold, showing that the use of structured data was important for generating summaries. With refinement, the correspondence and content scores improved, but the well-formed scores degraded. The overall scores were equivalent among both proposed systems. Our systems performed the best in terms of objective scores and subjective scores among task participants [7]. Although there was room for improvement of the overall scores, the performance of our methods was practical[11].

---

[11] We did not conduct significant test for human evaluation results because the evaluated scores for each answer summary are not open to the participants (only total scores are given), but the improvement of correspondence score (13, 14 points improvement), content score (21, 23points), and overall score (31, 30 points) is obvious.

Table 2. Examples of translated generated answer summaries, where 'SubTopic' is "Undesired loneliness (望まない孤独)" and 'QuestionSummary' is "Comprehensive promotion of related measures (施策の合的な推を)".

Gold	We realize a society in which no one is left behind by establishing a cross-disciplinary consultation support system in all municipalities, and by supporting municipalities that work to create a place to stay. (全区市町村で分野横断的な相支援体制の、居所の出に取りむ区市町村支援等、一人取り残さない社会をする。)
Baseline	We develop a bright future for Tokyo from a citizens-first perspective by focusing on people-focused measures and people-focused policies. (人に焦点を当てた施策や人に焦点を当てた施策を重点的に推し、都民ファストの点で京の明るい未来を切り拓く。)
Prop. 1	We realize a society where no one is left behind through measures such as building a cross-disciplinary consultation support system in the municipalities and supporting municipalities that create a place to stay by taking advantage of the strengths of both online and offline services. (区市町村での分野横断的な相支援体制のや、オンライン双方のみを生かした居所出に取りむ区市町村を支援する等、ひとり取り残さない社会を。)
Prop. 2	We realize a society in which no one is left behind by building a cross-disciplinary consultation support system in all municipalities and supporting municipalities that create places to stay by utilizing the strengths of both real and online services. (全区市町村での分野横断的な相支援体制のや、リアルとオンライン双方のみを生かした居所の出に取りむ区市町村を支援する等、もが安心して暮らせる社会をする。)

Table 3. Human evaluation results.

	Correspondence				Content				Well-formed				Overall			
	A	B	C	score	A	B	C	score	A	B	C	score	A	B	C	score
Baseline (organizer)	86	8	6	180	35	49	16	119	89	6	5	184	48	29	23	125
Prop. 1	94	5	1	193	46	48	6	140	92	8	0	**192**	67	22	11	**156**
Prop. 2	94	6	0	**194**	47	48	5	**142**	84	12	4	180	65	25	10	155
Gold	93	6	1	192	47	47	6	141	96	3	1	195	69	28	3	166

5 Conclusion

To summarize unstructured documents, we first convert unstructured documents into structured documents by making paragraphs of questions and answers and by associating question summaries with answers based on the BM25-based similarity calculation. Second, we train LLM to generate summaries of the answer from the corresponding answer and the summaries and subtopics of the questions. In addition, when training LLM, we refine training data on the basis of ROUGE F-measures. The experimental results showed that our methods significantly improved the performance from standard RAG baseline when inputting all answers including irrelevant parts into T5 and that human-rated scores of our methods were practical compared to human-generated gold summaries.

References

1. Devlin, J., Chang, M.W., Lee, K., Toutanova, K.: BERT: pre-training of deep bidirectional transformers for language understanding. In: Proceedings of the 2019 Conference of the North American Chapter of the Association for Computational Linguistics: Human Language Technologies, vol. 1, pp. 4171–4186 (2019)
2. Gale, D., Shapley, L.: College admissions and the stability of marriage. Am. Math. Monthly **92**, 261–268 (1962)
3. Kimura, Y., et al.: Overview of the NTCIR-15 QA lab-PoliInfo-2 task. In: Proceedings of the 15th NTCIR Conference on Evaluation of Information Access Technologies, pp. 101–112 (2020)
4. Kimura, Y., et al.: Overview of the NTCIR-16 QA lab-PoliInfo-3 task. In: Proceedings of The 16th NTCIR Conference on Evaluation of Information Access Technologies, pp. 156–174 (2022)
5. Kimura, Y., et al.: Overview of the NTCIR-14 QA lab-PoliInfo task. In: Proceedings of the 14th NTCIR Conference on Evaluation of Information Access Technologies, pp. 121–140 (2019)
6. Lewis, P., et al.: Retrieval-augmented generation for knowledge-intensive NLP tasks. In: Proceedings of the 34th International Conference on Neural Information Processing Systems. NIPS 2020, Curran Associates Inc., Red Hook, NY, USA (2020)
7. Ogawa, Y., et al.: Overview of the NTCIR-17 QA lab-PoliInfo-4 task. In: Proceedings of The 17th NTCIR Conference on Evaluation of Information Access Technologies, pp. 217–231 (2023)
8. Raffel, C., et al.: Exploring the limits of transfer learning with a unified text-to-text transformer. arXiv (2019)
9. Robertson, S., Zaragoza, H.: The probabilistic relevance framework: BM25 and beyond. J. Found. Trends Inf. Retrieval **3**(4), 333–389 (2009)
10. Tachioka, Y.: ditlab at the NTCIR-17 QA lab-PoliInfo-4 task. In: Proceedings of the 17th NTCIR Conference on Evaluation of Information Access Technologies, pp. 241–243 (2023)
11. Tachioka, Y., Keyaki, A.: ditlab at the NTCIR-16 QA lab-PoliInfo-3. In: Proceedings of the 16th NTCIR Conference on Evaluation of Information Access Technologies, pp. 207–210 (2022)
12. Yamada, I., et al.: Wikipedia2Vec: an efficient toolkit for learning and visualizing the embeddings of words and entities from Wikipedia. In: Proceedings of the 2020 Conference on Empirical Methods in Natural Language Processing: System Demonstrations, pp. 23–30. Association for Computational Linguistics (2020)

Real Estate Information Exploration in VR with LoD Control by Physical Distance

Yuki Nakayama[1](✉), Yuya Tsuda[1], Yoshiyuki Shoji[2], and Hiroaki Ohshima[1]

[1] University of Hyogo, Kobe, Hyogo 651-2197, Japan
ad24h044@guh.u-hyogo.ac.jp, ohshima@ai.u-hyogo.ac.jp
[2] Shizuoka University, Hamamatsu, Shizuoka 432-8011, Japan
shojiy@inf.shizuoka.ac.jp

Abstract. We propose a method for real estate information exploration in VR. It is difficult to find a real estate property that matches the user's preferences from many properties. Each real estate property has multiple attributes such as rent, layout, nearest station, and so on. In web services for real estate information exploration, search results are usually displayed in a list similar to a standard Web search. Listed search results are not suitable for an overview of the information. We have developed an interface for information search that enables the user to overview real estate information in VR with an unlimited display area. In the proposed system, the user can select a target attribute or narrow down the results by attribute values. Real estate information is displayed as a panel in which information on several real estate properties is gathered. When the user physically gets closer to the panel, more detailed information is presented on the panel. In other words, level of detail (LoD) control by physical distance is implemented. Thus, the proposed system can intuitively and seamlessly switch between an overview of overall information and a presentation of detailed information. Users evaluated the LoD controls to be easy to operate.

Keywords: VR · Search interface · Exploratory search

1 Introduction

When looking for real estate properties, we frequently use real estate information search systems. Real estate properties have many attributes such as rent, distance to the nearest station, and layouts. It is important to compare multiple attributes to find a real estate property that meets the user's preferences. Most real estate information retrieval systems on Web sites display search results in the form of a list. In a listed display, it is difficult to place and overview a large amount of information due to the screen size. In this study, we propose a real

Fig. 1. The VR interface for real estate information exploration

estate information exploration method in VR. Because VR has an unlimited display area, it is suitable for placing a large amount of information. Users can also move around in the area, allowing them to get an overview of the information.

The method proposed in this paper has three functions: placement by category, operation of level of detail (LoD) by physical distance, and narrowing down. When the user selects an attribute, the real estate information is divided into several categories based on the attribute value, and the real estate information is placed in VR by these categories. Figure 1 shows how the real estate information is placed. When the user moves away from the placed information, an exterior photo is displayed. When the user moves closer, detailed real estate information such as rents and layouts are displayed. When the user selects a category, the real estate information displayed is that of the selected category. In this way, the user narrows down the real estate information displayed.

There are many studies on information exploration in VR. Austin et al. [6,7] researched user actions and preferences for three types of search result display on a VR head-mounted display. They tested users with three types of display conditions (list, grid, and arc) and collected data on user actions and perceptions. The results showed that users completed the task faster in the arc display. However, users preferred the list and grid display more. They also propose the Information Interactions in Virtual Reality (IIVR) system to study user interaction with abstract information objects in VR from information retrieval. This system will help in future studies that would study the effects of information layout and types of interactions in VR. Giunchi et al. [3] proposed a method that uses sketches as queries to search a collection of 3D models in VR. This allows users to search for models while immersed in the virtual environment quickly. User experiments showed that this method allows users to perform searches quickly and naturally. Kaliciak et al. [4] proposed an image retrieval framework that helps in awareness and understanding in monitoring the marine environment. In this framework, a VR headset is used to monitor images from a 360-degree camera in real time. In addition, information is presented in an AR-like way by adding information supplements to the images. In this study, we propose a VR interface that can overview real estate information.

Studies exist that use LoD controls for the interface. Daskalogrigorakis et al. [2] proposed an AR interface, Holo-Box, which alters queries and results using LoD control in glanceable interfaces. Glanceable interfaces is an ARUI that searches for information on objects in the line of sight and displays the results on top of it. The amount of information displayed by this UI is managed by the LoD control to improve visibility. In this study, we propose a method to change the information to be displayed using LoD controls by physical distance.

2 VR Interface for Real Estate Information Exploration

2.1 System Overview

In this system, users explore real estate information in VR. Real estate information is displayed on a square panel. Real estate information is displayed in a single panel with multiple real estate properties displayed together. We call this panel the real estate information panel. Real estate information is categorized based on an attribute value. For example, if the attribute is rent, real estate properties with rents between 0 and JPY20,000 are classified into one category. Real estate information is displayed together by category. The user selects a category and changes the displayed real estate information to that of the selected category. In this way, the real estate information is narrowed down. The LoD of the displayed real estate information is changed according to the distance from the user. We implemented the system with these three functions. For the implementation, we extracted and used 84,222 real estate information from the LIFULL HOME'S Dataset[1] for Kobe City, Hyogo Prefecture.

2.2 Categorization and Placement of Real Estate Information

The user selects an attribute by pressing a button as shown in Fig. 2. The attributes that can be selected are rent, layout, nearest station, distance from

---
[1] LIFULL Co., Ltd. (2015): LIFULL HOME'S Snapshot Data of Rentals. Informatics Research Data Repository, National Institute of Informatics. (dataset). https://doi.org/10.32130/idr.6.1.

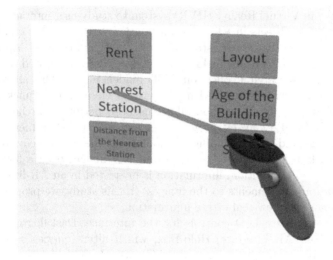

Fig. 2. Attribute selection

the nearest station, and age of the building. The system categorizes real estate information based on the selected attribute. The system places the real estate information grouped into categories. A category is a group of real estate information categorized by the value of the attribute. For example, if the attribute is rent, the categories are from 0 yen to 20,000 yen, from 20,000 yen to 40,000 yen, and so on. The maximum number of categories is eight. The placement of real estate information is as shown in Fig. 1. The percentage of real estate information in each category is represented by the height of the real estate information panel. This allows the user to view many real estate information from an overview perspective.

2.3 Narrowing Down the Search Results

The user narrows down the real estate information by selecting a category. At the bottom of the arranged real estate information, there is a button that represents the category information. By pressing this button, the user selects a category. Figure 3 shows how the user chooses a category. Only the real estate information contained in the selected category is displayed. If users change the attribute after this, they can place the real estate information after narrowing it down again. In this way, the user narrows down the real estate information displayed.

2.4 LoD Control

The LoD of real estate information is controlled by physical distance. When the user physically approaches a panel displaying real estate information, detailed real estate information is displayed on the panel. The information to be displayed

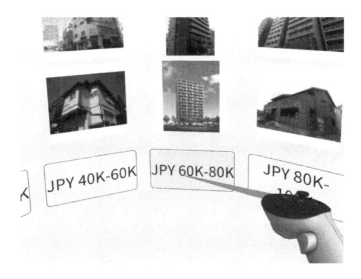

Fig. 3. Selecting a category

is the exterior image, layout, rent, nearest station and distance from the station, and age of the building. The detailed real estate information is shown in Fig. 4. This allows the user to switch between overview and detailed information.

3 Evaluation

3.1 Evaluation Experiment and Result

To evaluate the usefulness of the proposed method, we experimented with four students. All four subjects had little or no experience using VR. Most of the subjects had rarely done real estate information exploration. One of the subjects had done real estate information exploration three to five times. In the experiment, subjects were tasked with finding one property that met specified requirements. The specified requirements were the property's rent, layout, nearest station, and so on. The subjects were asked to take the tasks in the proposed method and real estate information exploration using the Web site. Users were asked to use the LIFULL HOME'S[2] Website. Here, the real estate information exploration on the Website is used as the conventional method.

We asked the subjects to answer several questionnaires to evaluate the proposed system. To evaluate usability, we used the System Usability Scale(SUS) [1]. In SUS, the subjects answer 10 questionnaires, and a score is calculated from the results. Based on the answers, a score from 0 to 100 is calculated.

In the questionnaire, subjects answered questions to evaluate both the proposed and the conventional methods, and questions to evaluate only the proposed methods. In both cases, subjects were asked to answer the questions on a 5-point

---

[2] https://www.homes.co.jp/.

Rent：JPY 63,000
Layout：Studio
Age of Building：54years
Address：Hyogo-ku, Kobe, Hyogo
Nearest Station：
Kobe City Subway Seishin-yamanote Line
Minatogawa-koen Station
Distance：3 minutes walk
Description：
Recommended for those who are not satisfied with a one-room apartment.
Bath and toilet are separated.

Fig. 4. View detailed real estate information

scale, with 1 indicating disagreement and 5 indicating agreement, and to give reasons for their answers.

Table 1. Questions to evaluate both the proposed and the conventional methods and scores (1: disagree, 5: agree; SD means standard deviation)

ID	Questions	Average		SD	
		VR	PC	VR	PC
1	I think the amount of information displayed is appropriate	3.00	2.75	1.00	0.83
2	I think the time it took to decide on the property was short	2.50	2.75	0.50	0.83
3	I think there were fewer operations to narrow down and display detailed information	4.25	2.50	0.43	0.50

This section describes the results of the questions to evaluate usability. The average SUS score for the proposed method was 31.2, showing that usability issues remain. Analysis of the answers shows that the average score for question 1 was 2.00 and for question 3 was 2.25. Subjects evaluated the system as difficult to use. The mean score for question 5 was 2.00, and the mean score for question 6 was 3.75. Subjects evaluated the functions as not integrated.

This section describes the results of questions evaluating each of the proposed and conventional methods. The questions and the average scores for each method

Table 2. Questions to evaluate only the proposed methods and scores (1: disagree, 5: agree; SD means standard deviation)

ID	Questions	Average	SD
1	Displaying only an overview of the exterior of the property was appropriate for real estate information exploration	3.0	1.2
2	The operation of approaching information to view detailed information was appropriate for information exploration	4.0	0.7
3	The operation of pressing buttons on the VR space to narrow down the information was appropriate for information exploration	3.5	1.1

are shown in Table 1. In question 3, subjects commented that the proposed method was able to display detailed information just by approaching. Subjects also commented that the conventional method required them to move between pages to see detailed information, which increased the number of clicks.

This section describes the results of the questions used to evaluate the proposed method. The questions and their average scores are shown in Table 2. As reasons for their answer to question 1, subjects commented that there were many similar exterior photographs and that they could not identify the places where the real estate properties were placed. As reasons for their answer to question 2, subjects commented that simply physically approaching the panel was intuitive and easy to understand and that they did not feel stressed by the operation.

3.2 Discussion

Based on the experimental results, we believe that the method of LoD control based on the physical distance from the information makes the operation intuitive and easy to understand. It was also shown to reduce the number of operations and the stress on real estate information exploration. On the other hand, it was shown that there are issues regarding usability and the amount of information to be displayed. We believe that usability should be improved by enhancing the functions necessary for real estate information exploration, improving the attributes that can be selected, and the ability to narrow down the exploration. In addition, we believe that the amount of information to be displayed needs to be such that the attributes can be understood even when the user is far away from the site.

4 Conclusion

In this paper, we propose and experiment with a real estate information exploration method in which information placed in a VR space is changed by LoD control. In the proposed method, real estate information is placed in the VR space based on one attribute. When the user moves away from the information placed on the VR space, only rough real estate information with exterior photographs is displayed, and when the user moves closer, detailed real estate information including rent and layouts is displayed. In this way, the user can narrow down the exploration while acquiring both general and detailed real estate information, enabling information exploration that meets his or her needs. In an experiment, we conducted real estate information exploration using the proposed method and the conventional method to confirm their usefulness. As a result, it was shown that the method of LoD control based on the physical distance from the information reduces the burden of operation in real estate information exploration in VR space. In the future, to enhance the functions necessary for real estate information exploration, we will add functions such as selecting a region by moving on a map in VR.

Acknowledgments. This work was supported by JSPS KAKENHI Grant Numbers JP21H03775, JP22H03905, JP21H03554, JP24K03228. In this paper, we used "LIFULL HOME'S Dataset" provided by LIFULL Co., Ltd. via IDR Dataset Service of National Institute of Informatics [5].

References

1. Brook, J.: SUS - a quick and dirty usability scale. Usabil. Eval. Ind. 189–194 (1996)
2. Daskalogrigorakis, G., McNamara, A., Mania, K.: Holo-box: level-of-detail glanceable interfaces for augmented reality. In: Proceedings of the 2021 Special Interest Groupe on Computer Graphics, pp. 1–2 (2021)
3. Giunchi, D., James, S., Steed, A.: 3D sketching for interactive model retrieval in virtual reality. In: Proceedings of the 2018 Joint Symposium on Computational Aesthetics and Sketch-Based Interfaces and Modeling and Non-Photorealistic Animation and Rendering, pp. 1–12 (2018)
4. Kaliciak, L., Myrhaug, H., Goker, A.: Content-based image retrieval in augmented reality. In: Proceedings of the 2017 International Symposium on Ambient Intelligence, pp. 95–103 (2017)
5. LIFULL Co., Ltd.: LIFULL HOME'S snapshot data of rentals (2015). https://doi.org/10.32130/idr.6.0
6. Ward, A.R., Capra, R.: Immersive search: using virtual reality to examine how a third dimension impacts the searching process. In: Proceedings of the 2020 International ACM SIGIR Conference on Research and Development in Information Retrieval, pp. 1621–1624 (2020)
7. Ward, A.R., Gu, Y., Avula, S., Chakravarthula, P.: Interacting with information in immersive virtual environments. In: Proceedings of the 2021 International ACM SIGIR Conference on Research and Development in Information Retrieval, pp. 2600–2604 (2021)

Voices of Asynchronous Learning Students: Revealing Learning Characteristics Through Vocabulary Analysis of Notes Tagged in Videos

Xiaonan Wang[1], Yancong Su[2], Yi Sun[3], Takeshi Nishida[1], Kazuhiro Ohtsuki[1], and Hidenari Kiyomitsu[1](✉)

[1] Graduate School of Intercultural Studies, Kobe University, Kobe, Japan
217c303c@stu.kobe-u.ac.jp, tnishida@people.kobe-u.ac.jp,
{ohtsuki,kiyomitu}@kobe-u.ac.jp
[2] School of Design and Art, Xiamen University of Technology, Xiamen, China
syc@xmut.edu.cn
[3] Graduate School of Information Technology, Kobe Institute of Computing, Kobe, Japan
sun@kic.ac.jp

Abstract. This paper describes a method for analyzing learning data from video courses. The data comes from an on-demand introductory programming course at a university. The study analyzes differences in note content characteristics across students by exploring patterns of student notes in a video learning environment, combining quantitative analysis with qualitative analysis centered on the vocabulary used in student notes. By deploying a custom designed learning support system, we facilitated the straightforward expression of students' questions and notes, thereby collecting rich data reflecting their learning process. Initial findings suggest that the frequency and distribution of specific vocabulary used in students' notes provide insights into the range of learners' interests and depth of understanding. For example, although different students recorded notes on the same topic at a specific point in a video, there were significant differences in their focus and note content. Some students tended to record how-to's and specific features, while others focused more on theoretical concepts and programming principles. These findings show that individual student differences significantly influence their learning focus and information-processing style in a video learning environment. This analysis method allows educators to target instructional interventions based on learners' unique characteristics of engagement and comprehension as demonstrated through vocabulary use. The preliminary results of this study provide educators with insights to help them understand learner diversity in digital learning environments, providing an essential basis for personalized lesson design and improved learning outcomes.

Keywords: Asynchronous Learning · Learning Strategy · Learner Characteristic

1 Introduction

The rapid and transformative development of online learning platforms and the widespread adoption of video-based learning have significantly reshaped the educational landscape in recent years. The global e-learning market has experienced substantial growth and is expected to continue its upward trajectory in the coming years [1]. This growth has been particularly evident in the field of computer science education. The asynchronous nature of video-based learning, its flexibility, accessibility, and ability to cater to diverse learning styles, makes it a highly attractive option for those seeking to acquire programming skills and knowledge [2, 3].

With the rise of online learning, an unprecedented amount of data has been generated, capturing a broad spectrum of learner interactions and behaviors. This data, ranging from quantitative metrics such as video view counts and interactive quiz performances [4], to qualitative insights, such as text-based interactions and learner-generated content [5], presents a rich resource for understanding and enhancing learning experiences. The notes stand out as a vibrant source of insight among the qualitative data. By meticulously examining such notes, researchers and educators can gain a deeper understanding of how learners interact with and digest educational content [6], opening new avenues for enhancing teaching strategies and learning outcomes.

This study explores the characteristics of students' notes taken during video-based learning and their impact on learning outcomes through qualitative analysis of the vocabulary used in the notes. Specifically, we seek to address the following research questions:

- What are the key characteristics of the students' notes in programming video learning?
- How do these vocabulary characteristics vary among students with different learning outcomes?
- How do we target instructional interventions based on these characteristics to personalize learning support?

By exploring these questions, this study seeks to provide valuable insights into the learning processes and strategies employed by students in video-based programming education. Understanding the characteristics of student notes and their relationship to learning outcomes can inform the design of more effective video learning platforms and support systems that cater to learners' diverse needs and preferences. Moreover, the findings of this study can contribute to the broader field of educational technology by shedding light on the role of note-taking in video-based learning and its potential impact on student engagement and achievement.

The novelty and relevance of this research lie in its unique approach to analyzing asynchronous video-based learning through the lens of student's notes, particularly in the domain of programming education. Examining the vocabulary and content of notes taken during video lectures illuminates the varied learning strategies employed by students and their impact on educational outcomes. This nuanced understanding can inform the development of more effective instructional designs, learning support systems, and personalized feedback mechanisms that adapt to individual learners' needs and preferences [7]. Furthermore, by focusing on the specific domain of computer science education, this study contributes to the growing body of research on effective strategies for teaching and learning programming skills in online environments.

2 Related Work

The rapid growth of online learning platforms and video instruction has generated a wealth of data capturing learner interactions and behaviors. These data include both quantitative metrics, such as video views and quiz scores, as well as text-based qualitative data generated by learners during learning activities [8]. Qualitative data, often manifested through student notes, forum posts, and reflective essays, offers a nuanced view of the cognitive and affective processes underlying learning [9]. It stands as a critical resource for educators and researchers aiming to glean insights into student learning behaviors and strategies.

Among the qualitative data, student notes offer a vibrant window into the learner's cognitive process during their educational journey. Examining these notes allows researchers to better understand how learners interact with and process educational content. Kiewra et al. [10] found that note-taking can serve both encoding and external storage functions, facilitating learners' processing and retention of information. Nakayama et al. [6]. Explore the impact of self-efficacy and note-taking activities on student learning and reflection in blended learning environments, revealing significant correlations between note-taking behaviors, self-efficacy, and students' engagement in their studies. These findings underscore the potential of analyzing student notes to gain insights into their learning processes and strategies.

In computer science education, researchers have investigated using video-based instruction to enhance student engagement and learning outcomes. For example, Nariman [11] reported positive student perceptions and increased motivation when using video lessons in a programming course. In programming education, students' learning processes are often reflected in their code. Researchers have developed various learning support systems and analytical approaches to capture and analyze these learning processes. For example, Shrestha et al. [12] developed a tool to visualize students' coding processes, allowing educators to identify common coding patterns and difficulties.

Another valuable source of learning data in video-based programming courses is students' notes while watching video lectures. These notes can provide a rich understanding of student's thought processes, questions, and reflections as they engage with the course content. To facilitate the collection of this distinctive learning data, we developed a system that allows students to take notes directly linked to specific frames in the video lecture [13]. This approach enables students to capture their thoughts, questions, and reflections in real-time, creating a rich learning data repository. By employing a combination of qualitative analysis, based on the content of students' notes, and quantitative analysis, involving feature classification and statistical methods, we aim to investigate the specific characteristics of these notes and their relationship to learning outcomes in a video-based programming course.

Qualitative analysis methods, such as content analysis, have been increasingly applied in educational research to understand learners' experiences and processes. Content analysis involves systematically examining and interpreting textual data to identify patterns, themes, and meanings [14]. In the context of educational data, content analysis has been used to explore students' self-regulated learning strategies; for instance, Kauffman et al. [15] used text-mining techniques to analyze students' self-regulated learning strategies in an online learning environment. In our study, we employ a similar approach to analyze the qualitative learning data, which requires a blend of methodologies that can both categorize and interpret textual information's rich, complex nature. For example, the thematic analysis of student notes can highlight prevalent misunderstandings or misconceptions in a given subject area. In addition, Machine learning and natural language processing (NLP) tools have been increasingly employed to automatically categorize and analyze text data, facilitating the handling of large datasets. An exemplary study by Crossley et al. [16] utilized NLP to automatically classify student reflections in an online learning module, revealing insights into students' self-regulated learning strategies and their correlation with academic performance.

The related work section underscores the critical role of qualitative learning data in enhancing our understanding of student learning processes and outcomes. Building upon these methodological approaches, this study applies vocabulary analysis to the specific domain of video-based programming education. It aims to uncover students' unique learning characteristics and strategies in this context. The significance of this research lies in its potential to provide a more nuanced understanding of students' learning experiences in video-based programming courses. By combining quantitative and qualitative analysis methods, we can uncover what students focus on in their notes and how they process and engage with the course content. These insights can inform the design of more effective instructional interventions and personalized learning support systems tailored to learners' diverse needs and preferences.

3 Methodology

3.1 Data Collection

For this study, we utilized an internally developed system to enhance asynchronous learning experiences in online courses, particularly in computer programming. This comprehensive system, developed with HTML and JavaScript, enables students to interact with video content through a web application accessible via computers or smartphones. This system enabled students to link their queries directly to specific video frames, thereby reducing the need for de-tailed descriptions to convey their questions or comments. This integration facilitated the collection of sequential data linked to video frames, which, through visualization techniques, transformed into actionable insights for educators. Instructors can review these annotations and engage with the learners' queries or notes via an instructor's interface. The system's robust data collection capabilities also capture detailed click-stream data, offering comprehensive insights into student engagement patterns with the video material.

For example, when students select a course and go to the course's video page, there will be a note toolbar on the right side of the video, as shown in Fig. 1. Notes are recorded by marking the time points in the video and presented as markers in the timeline. Students can click the "Add" button at the top of the tool to generate a note below the toolbar, displaying the corresponding video time point. Multiple notes are displayed on the right-hand side, and students can click on them to enter the editing mode.

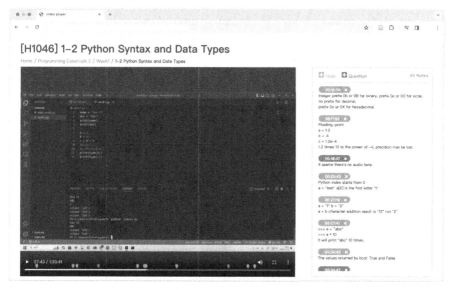

Fig. 1. Student Video Learning Interface, Consisting of Two Columns with the Video Area on the Left and Notes Area on the Right.

The study participants were 38 students enrolled in an introductory Python programming course offered asynchronous online format at a university. The course is conducted as an On-demand video course. We collected data from three lecture videos totaling approximately 90 min for this analysis. Out of the 38 enrolled students, 26 generated data, including 56 questions and 303 learning notes. We also obtained the student's scores on the chapter test. Informed consent was obtained from all participants, and the data was anonymized for analysis.

Preliminary results show a significant increase in student feedback (notes and questions) compared to previous asynchronous video courses, and student comments indicate that the system's note-taking feature is particularly beneficial.

3.2 Analysis Method

First, we analyzed the frequency of vocabulary usage and the number of users. By analyzing high-frequency words, we aimed to understand the common themes or concepts students focused on. By analyzing words with high frequency but low user numbers, we sought to gain insights into unique or individual note-taking practices.

Second, based on the content features discovered through the frequency analysis, we identified common patterns or features in the notes, such as code-related vocabulary (high frequency, relatively fewer users) and language-related vocabulary (many users, reflecting shared focus points). This analysis revealed students' tendencies to record examples or summarize definitions. We then proposed a method for categorizing notes and conducted a correlation analysis between different types of notes and learning outcomes (Score).

The correlation analysis between learning activities (notes) and learning outcomes (Score) enabled us to identify specific student groups and unique individuals (outliers). Based on this finding, we analyzed individual differences in students by correlating notes with videos and comparing the content of notes from the same period.

4 Analysis and Results

4.1 Student Engagement Analysis

First, we counted the frequency of words and the number of users in all the student notes collected, as detailed in Table 1. These notes provided a deep insight into the student's engagement with the course material.

Table 1. Vocabulary frequency and number of users in student notes.

Word	Frequency	No. of students	Word	Frequency	No. of students
int	150	18	value	26	17
list	76	14	result	25	11
float	73	17	variable	25	10
type	55	15	number	24	12
print	52	9	change	22	12
python	51	18	code	20	18
tuple	44	17	format	17	13
string	37	15	data	16	10
element	32	13	indentation	15	11
age	27	8	access	14	9

High-frequency words such as "int", "list", "float", "tuple", and "string" are used by many students, aligning closely with the course content. This correlation underscores these terms' pivotal role in Python programming, highlighting students' active engagement with fundamental data types discussed in the videos. On the other hand, words like "element", "value", "code", and "format", while have a relatively lower frequency, nevertheless used by many students. The widespread use of these words indicates that students value these concepts, and actively engaging with the core concepts of the Python programming language.

Interestingly, certain high-frequency words such as "print" (Frequency 52, Students 9) and "type" (Frequency 55, Students 15) have a relatively lower user count despite their high occurrence. These words are Python's built in functions. Similarly, "age" (Frequency 27 times, Students 8) is a term used in the code examples provided by the instructor during the course. The high frequency but low user count of these program-specific vocabularies suggests that some students extensively record code examples in their notes. The same Python's built in functions seem to have words like "int", "float" and "list".

The frequency analysis shows that these notes exhibit a di-verse range of patterns and characteristics. This diversity in note-taking strategies, such as focusing on code examples or organizing and summarizing knowledge, reflects the students' different preferences and approaches to capturing and processing the course material.

Based on the vocabulary frequency and user count analysis, we observed that the notes exhibit different characteristics and patterns. Some notes tend towards code examples, while others tend towards summaries and definitions. To clarify these features, we individually examined the notes' contents and proposed a way to categorize them. As shown in Table 2, each category reflects a different aspect of student engagement with the course material.

Table 2. Examples of the four types of notes.

Note Type	Example
Code-Note	list1 = [9, 1, −4, 3, 7, 11, 3] \n len(list1) \n max(list1) \n min(list1) \n list.count(3)
Summary-Note	Difference between list and tuple: 1. The first is that tuples use (), while lists use []. 2. Lists can have elements added, deleted...
Tag-Note	Starting python and saving files List of Uses and Modifications
Cryptic-Note	Mutability subscript order

Code-Note primarily consists of code snippets along with brief explanations or comments. They serve as a reference for the student and help reinforce the understanding of programming concepts through practical examples. Code-Note demonstrate the students' focus on practical application and their effort to reinforce programming concepts through concrete examples.

Summary-Note is characterized by students' attempts to summarize the content in their own words. They demonstrate the student's understanding and interpretation of the material. The creation of Summary-Note reflects a more profound cognitive process of assimilation and comprehension. This strategy reflects the students' active engagement with the material and attempts to build a personal understanding of the concepts.

Tag-Note is characterized by their direct reference to specific points within the video content, often containing brief descriptions or keywords. This category arises from a unique behavior observed in the dataset, where students utilize notes to mark specific video frames for later reference. Marking specific points for future reference or emphasis indicates a strategic approach to learning, where students are keen on pinpointing and revisiting critical content segments.

Cryptic-Note is a brief note. Sometimes, it comprises only one word or phrase that may take time to be comprehensible or out of context. These notes could represent personal reminders or very condensed summaries of concepts. While Cryptic-Note may lack clarity for an external reader, they likely hold significant meaning for the individual student, as a mnemonic strategy or as a quick reference point that supports their unique learning journey.

These categories offer a structured way to examine students' diverse note-taking strategies. We can further explore how each category may influence or reflect students' learning processes and outcomes by categorizing notes into distinct types. In the following sections, we will further explore the distribution of these note types among the students and investigate their potential impact on academic performance.

4.2 Relationship Between Note Types and Learning Outcomes

Our investigation into the impact of note-taking activities on learning outcomes in asynchronous video learning environments has yielded significant findings. A polynomial regression analysis was conducted to delve into the non-linear relation between the quantity of four distinct types and students' performance scores, revealing crucial insights.

Figure 2 illustrates the relationship between the number of note-taking activities and student's scores for each type of note. The horizontal axis represents the students, denoted by their name codes (A–Z), arranged in descending order based on their scores on the chapter test. The vertical axis represents the number of note-taking activities for each type of note. The backgrounds A, B, and C delineate the score intervals. The scatter points in the graph indicate the actual number of note-taking activities for each student and each type of note. The dashed lines represent the polynomial regression curves for each type of note. These curves are fitted to the scatter points using a polynomial regression model, which captures the non-linear relation between the number of note-taking activities and students' scores.

The Summary-Note and Code-Note activities exhibit significant correlations with students' performance scores. The Summary Note (red) shows a positive correlation with learning outcomes, suggesting that students who engage more in summarizing content tend to achieve higher scores. This trend might reflect the cognitive processing benefits of summarization, which requires understanding and consolidating information, facilitating deeper learning. On the other hand, the Code-Note (green) trend also indicates a positive correlation but with a distinct pattern that suggests nuances in how coding activities relate to learning outcomes.

The Tag-Note (blue) and Cryptic-Note (yellow) activities show less pronounced trends in the analysis, indicating a more complex or indirect relationship with learning outcomes. This complexity suggests that while these forms of note-taking are part of students' learning strategies, their direct impact on performance might be influenced by other factors, or they might serve more as supplementary or personal preference styles of note-taking.

The detailed exploration of the polynomial regression analysis identified the correlations between note types and learning outcomes. Based on these correlations, we

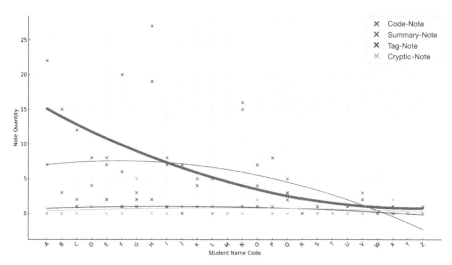

Fig. 2. Relationship between four note-taking activities and learning outcomes (scores). The scatter points represent the data, while the dashed lines show the polynomial regression curves.

can even identify specific student groups from the graph's relative distribution of scatter points and the associated curves. Students whose note-taking activity scatter points lie above the regression curves for "Summary-Note" and "Code-Note" appear to have put in more effort to achieve their current scores. Conversely, suppose a student's note-taking activity falls below the predicted curve. In that case, they may benefit from increased engagement in these beneficial note-taking strategies, potentially leading to better grades with a little more effort. In this way, educators can more accurately encourage and guide students. These findings allow educators to encourage and guide students more accurately and offer valuable clues for educational systems to develop personalized learning plans.

4.3 Individual Differences Analysis

The comprehensive analysis of the relationship between the four types of note-taking activities and academic performance (scores) not only highlighted the general trends and specific student groups but also pinpointed outliers within the dataset. Notably, students H and N were identified as such outliers, engaging in a significantly high volume of Summary-Note activities yet achieving relatively lower scores compared to their peers. In this section, we will analyze these individual differences, contrasting these students' note-taking content with that of students A and B, who boasted markedly better academic outcomes despite having a similar quantity of summary notes. This stark contrast in note-taking strategies and academic performance is a fascinating area for further exploration.

We correlated the notes of these four students with the timing of the video. Figure 3 shows the distribution of the notes of these four students in the video course. Each dot represents a note taken by a student at a specific point in the video content. The x-axis represents the video time in seconds; the y-axis lists the students. They are segregating their note-taking activities. The notes are color-coded, with students with high scores (A, B) and lower scores (H, N) indicated in blue and red, respectively, where all code notes are uniformly indicated in bright yellow.

The difference can be seen in the graph. Students A and B have a concentrated distribution of notes, predominantly in summary notes. Conversely, students H and N exhibit a more dispersed note-taking pattern with a relatively higher frequency of code notes.

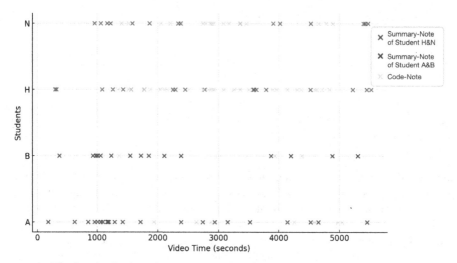

Fig. 3. Distribution of Notes Over Video Time for Students A, B, H, and N.

Focus on Summary Notes. For students A and B, the summary notes are more concentrated at specific intervals along the video timeline. This concentration suggests that these students may have identified and focused on particular segments of the course content as being more critical or challenging, dedicating their efforts to summarizing and understanding these key areas. Identifying and concentrating on crucial content could contribute to their higher performance, as it indicates efficient study strategies and perhaps a better grasp of foundational concepts.

Contrast the note-taking Strategies of Students H and N. Students H and N, in contrast to A and B, have their summary notes evenly distributed across the entire span of the video content. This distribution pattern may imply a broader, albeit less focused, engagement with the material. While this shows a recognized effort to engage with the entirety of the course content, the lack of concentration in specific areas might suggest a less strategic approach to note-taking and learning, which could be one of the reasons behind their relatively lower performance.

Uncover the Implications of Summary Note Density. The depth of engagement, as inferred from the density of summary notes, varies significantly between the two groups. High-performing students' notes are less dense but appear more strategically placed, potentially indicating a more analytical approach to note-taking, focusing on understanding rather than mere documentation. In contrast, the denser, more uniform spread of summary notes by H and N might reflect an exhaustive, but possibly surface-level, engagement with the material, focusing on breadth over depth.

Figure 3 shows many notes taken at almost the same point in time. To further investigate the individual differences in students' understanding and processing of information, we extracted and compared the notes taken by Student A and Student H at nearly identical time points. Table 3 shows Student A and Student H recorded notes at similar time points.

Table 3. Notes from Student A and Student H, at similar points in time.

Video Time(s) A	Notes from A	Video Time(s) H	Notes from H
1292.8	Use spaces to make code more readable	1260.6	After "#" is a comment (it's normal to put a space)
1715.1	Automatically format in PyCharm with Ctrl + Alt + L	1774.3	Python can output integers of any size and handle negative numbers. Integer define a = 125, no declaration needed
1933.6	.4 is 0.4 1.2e−4 is 1.2 times 10 to the power of −4	2035.6	a = 1 type(a) print data type
2386.1	Integers and floating-point numbers can also be calculated. The result is a floating-point number	2250.2	The result of 4 / 2 is a floating-point number. 4 // 2 is an integer
2640.5	Backslash plus n means newline	2519.8	String a = 'test' a = "Chinese" a = "'a'"
2743.2	Adding a backslash / "before" makes it lose its function and become a regular symbol	2749.2	Symbol after\will display
2872.8	len() gets the length of the string	2878.6	len(a) gets length
2934.7	Addition of strings is possible, multiplication is not. Get the length of a string with len(a)	2909.2	a = "1" string b = "2" a + b '12' result is character addition
3522.4	String formatting. Add f in front of the string and directly reference variables inside	3502.7	name1 = 'abc' age1 = 30 new_str_3 = f "I am {name1}, this year {age1} years old" print(new_str_3)

(*continued*)

Table 3. (*continued*)

Video Time(s) A	Notes from A	Video Time(s) H	Notes from H
4143.7	sep inserts a custom symbol between characters, default is space. End adds a custom symbol at the end, default is newline	4140.9	print('a[0], a[1], sep = '-', end = '**')
4524.1	The entire value of a string can be changed, but its internal characters, etc., cannot be changed	4523.8	Elements of a string cannot be changed; strings are immutable data types
4588.2	List reversal add.reverse() can reverse the order in the list	4594.7	List reversal list3 = [1 2 3] list3.reverse()
4649.6	Adding.sort() sorts the values in a list in ascending order. Specify reverse = True in parentheses (to sort in descending order)	4625.3	List sorting list4 = [9, 1, −4] list4.sort() small to large list4.sort(reverse = True) large to small
…	…	…	…

Through these examples, we observe several distinct differences in their note-taking approaches:

Content Selection and Emphasis. Student A frequently chooses to make separate notes on closely related topics, such as the note on string operations and numerical data types, demonstrating an inclination towards categorizing information into distinct concepts. Notably, when encountering points related to coding, Student A tends to describe these points using descriptive language, while Student H leans towards directly recording code examples. This distinction illuminates a fundamental difference in their engagement with the material; Student A prefers to process and translate coding concepts into verbal descriptions, possibly to reinforce understanding or for easier recall. In contrast, while practical, Student H's preference for directly noting down code examples might not engage the same cognitive processes required for a deeper understanding of the underlying principles.

Detail Orientation. Student A's notes exhibit more detail, providing specific examples and steps (e.g., "Automatically format in PyCharm with Ctrl + Alt + L"). This suggests a focus on actionable insights and practical application, which is crucial for programming. This approach likely helps Student A understand and apply coding principles effectively, which is reflected in higher academic performance. Student H's approach, while demonstrating engagement with the material, may not foster the same level of applicability due to a focus on examples over the conceptual framework that supports their use.

Expression and Clarity. The notes from Student A are structured in a way that directly addresses the 'how-to' aspect, offering clear, actionable advice. He demonstrates his propensity for critical thinking and problem-solving skills essential in programming. Student H's notes, while informative, tend to be more declarative, stating facts or observations, such as "After '#' is a comment", which points to a difference in how they might intend to use these notes for future reference.

In summary, while the quantity of notes taken by both students suggests a similar level of engagement with the course material, the qualitative differences in their notetaking and study strategies point to deeper underlying factors influencing their academic performance. Student A's approach, focusing on depth of understanding, application of knowledge, critical thinking, and personalization, appears to contribute to higher academic success. In contrast, although effective in specific contexts, Student H's strategy might not provide the same cognitive engagement or application skills level, which could explain the lower scores.

5 Discussion

5.1 Note Content Characteristics

The observed vocabulary frequency and user count among students' notes offer a window into the diverse strategies learners employ in an asynchronous video programming course. The classification into four note types, Code-Note, Summary-Note, Tag-Note, and Cryptic Note, reveals students' multifaceted strategies to navigate and internalize asynchronous video content. With their practical focus on re-cording code snippets and brief explanations, Code-Note demonstrates students' emphasis on hands-on application and reinforcement of programming concepts. On the other hand, Summary notes reflect a more reflective and synthesizing approach, where students actively process and internalize the course material by summarizing content in their own words. Tag-Note represent a strategic form of engagement where students leverage the flexibility of video-based learning to mark specific points for later reference or emphasis. This behavior highlights the importance of providing learners with tools to customize their educational experiences in asynchronous settings. Lastly, with their personalized shorthand and concise expressions, Cryptic-Note reveals the idiosyncratic ways students navigate and make sense of the learning material. Moreover, this categorization is a foundation for further exploring the relationship between note-taking practices and learning outcomes.

5.2 Note Types and Learning Outcomes

The polynomial regression analysis illustrating the relationship between the quantity of note types and academic performance emphasizes the significance of summarization and practical application in learning. The positive correlation between summary notes and learning outcomes suggests that the cognitive process of summarizing content aids in more profound understanding and retention. Similarly, the engagement in Code-Note reflects practical application. However, the pattern of the curves suggests that increasing Code-Note alone does not directly lead to an increase in scores after a certain point. A plausible explanation is that interpreting code is a fundamental skill in programming learning. However, as you get deeper into the program, the interpretation of algorithms and logic becomes more and more independent of the code, and the trend of the Code-Note curve in the graph seems to confirm this.

These findings provide educators valuable opportunities to guide and support students' learning journeys. Instructors can offer personalized recommendations and feedback by examining the position of a student's note-taking activity relative to the regression curves. For students whose engagement falls above the curve, acknowledging their effort and encouraging them to maintain effective note-taking strategies can help sustain their motivation and performance. Conversely, for students below the curve, educators can suggest increasing their participation in the specific note-taking activities associated with higher learning outcomes, such as Summary-Note and Code-Note.

These insights significantly impact educational systems, offering a basis for developing personalized learning plans and adaptive curricula. By understanding the correlations between note-taking activities and learning outcomes, institutions can tailor their support systems to cater to individual student needs. For example, students consistently falling below the regression curves for beneficial note-taking activities could be provided with additional resources, such as tutorials or study groups, to help them develop more effective learning strategies. On the other hand, students demonstrating strong engagement in these activities could be given more advanced or challenging material to enhance their learning experience.

5.3 Difference Among Individuals

The analysis of individual differences, particularly the case study of students H and N, offers a nuanced understanding of how similar levels of engagement, as measured by the quantity of notes, can result in varied academic performance. This discrepancy underscores the importance of quality and strategy in note-taking over sheer quantity. Students A and B's ability to discern and focus on crucial content segments, as evidenced by their targeted note-taking, suggests a strategic approach that likely contributes to their higher performance. Conversely, the broader but less focused engagement of students H and N may reflect a scattergun approach to learning, indicating that a more directed and analytical note-taking strategy could enhance learning efficiency and outcomes.

This differentiation between students offers a valuable diagnostic tool for educators, allowing for the identification of students who could benefit from increased engagement in beneficial learning activity practices. Tailoring educational interventions to encourage effective learning strategies could help optimize learning outcomes. This approach highlights the need for adaptive and student-centered teaching approaches. Such strategies recognize learners' individual efforts and challenges and provide pathways for every student to achieve their academic potential, thereby enriching the educational experience for all participants in the asynchronous video learning ecosystem.

5.4 Research Limitations and Future Directions

The current study's findings are based on data collected from a single introductory programming course at one university, which may limit the generalizability of the results.

Therefore, future research should aim to recruit a more diverse sample of participants from various courses and institutions. This approach would provide a more comprehensive understanding of note-taking strategies and their effectiveness across different learning contexts, such as advanced programming courses or other disciplines. Furthermore, the study focuses on the immediate impact of note-taking strategies on learning outcomes without examining their long-term effects on knowledge retention and application. Future studies could employ longitudinal designs to investigate how different note-taking approaches influence students' ability to retain and apply programming concepts over time. It could involve assessing students' performance on programming tasks or projects at various intervals after finishing the course and analyzing the relationship between their note-taking strategies and long-term learning outcomes.

Another avenue for future research is conducting in-depth analyses of student notes' content and quality. While the current study categorizes notes based on their general characteristics (e.g., Code-Note, Summary-Note), a more detailed examination of the specific information captured in the notes could reveal additional insights into effective note-taking practices. For example, researchers could explore the relationship between the depth and accuracy of the content in Summary-Note and students' performance on conceptual programming questions. Additionally, future studies could investigate how different note-taking strategies influence students' problem-solving abilities in programming tasks. It could involve designing experiments that compare students' performance using different note-taking approaches (e.g., Code-Note vs. Summary-Note) on authentic programming problems. Such research could shed light on how note-taking strategies support the development of practical programming skills.

6 Conclusion

This study delves into the nuanced world of asynchronous video-based learning in programming education, uncovering students' diverse note-taking strategies and their impact on learning outcomes. Through an elaborate analysis of vocabulary frequency and user engagement within student notes, distinct patterns emerged, revealing students' varied and nuanced strategies for navigating their learning environments. These findings highlight the necessity for educational technologies to be adaptable, catering to the myriad learning preferences students exhibit.

The relationship analysis between note types and learning outcomes further elucidates the positive impact of summarization and application on academic performance, highlighting the cognitive benefits of actively engaging with and processing course content. This analysis affirmed the value of specific note-taking strategies and revealed the potential for educators to tailor interventions that encourage these effective practices, thereby enhancing student learning experiences.

Future research can further extend this study's findings by increasing the sample's representative size, conducting in depth analyses of note-taking content, and exploring the long-term effects of note-taking strategies. This study enhances our understanding of student learning characteristics in asynchronous video programming courses. It provides practical implications for optimizing instructional design and personalized learning support, contributing to creating more effective and adaptive digital learning experiences.

Acknowledgment. This work was partly supported by JSPS KAKENHI (Grant Number 19K03000).

References

1. Global Market Insights: E-learning Market Size - By Technology (Online E-learning, LMS, Mobile E-learning, Rapid E-learning, Virtual Classroom, Others), Pro-vider (Service, Content), Application (Corporate, Academic, Government) & Forecast, 2023–2032. https://www.gminsights.com/industry-analysis/elearning-market-size, 2024/04/02
2. Bates, A.T.: Technology, e-Learning and Distance Education. Routledge (2005)
3. Dahalan, N., Hassan, H., Atan, H.: Student engagement in online learning: learner's attitude toward e-mentoring. Procedia Soc. Behav. Sci. **67**, 464–475 (2012)
4. Kim, J., Guo, P.J., Seaton, D.T., Mitros, P., Gajos, K.Z., Miller, R.C.: Understanding in-video dropouts and interaction peaks in online lecture videos. In: Proceedings of the First ACM Conference on Learning@ Scale Conference, pp. 31–40. (2014)
5. Andresen, M.A.: Asynchronous discussion forums: success factors, outcomes, assessments, and limitations. J. Educ. Technol. Soc. **12**(1), 249–257 (2009)
6. Nakayama, M., Mutsuura, K., Yamamoto, H.: Student's reflections on their learning and note-taking activities in a blended learning course. Electron. J. e-Learn. **14**(1), 43–53 (2016)
7. Chen, C.M.: Intelligent web-based learning system with personalized learning path guidance. Comput. Educ. **51**(2), 787–814 (2008)
8. He, W.: Examining students' online interaction in a live video streaming environment using data mining and text mining. Comput. Hum. Behav. **29**(1), 90–102 (2013)
9. Onan, A.: Sentiment analysis on massive open online course evaluations: a text mining and deep learning approach. Comput. Appl. Eng. Educ. **29**(3), 572–589 (2021)
10. Kiewra, K.A., DuBois, N.F., Christian, D., McShane, A., Meyerhoffer, M., Roskelley, D.: Note-taking functions and techniques. J. Educ. Psychol. **83**(2), 240 (1991)
11. Nariman, D.: Impact of the interactive e-learning instructions on effectiveness of a programming course. In: Complex, Intelligent and Software Intensive Systems: Proceedings of the 14th International Conference on Complex, Intelligent and Software Intensive Systems (CISIS-2020), pp. 588–597 (2021)
12. Shrestha, R., Leinonen, J., Hellas, A., Ihantola, P., Edwards, J.: Codeprocess charts: visualizing the process of writing code. In: Proceedings of the 24th Australasian Computing Education Conference, pp. 46–55 (2022)
13. Wang, X., Sun, Y., Su, Y.: A study of interaction design for E-learning system based on distributed asynchronous communication. In: Proceedings of the Tenth International Symposium of Chinese CHI, pp. 249–255 (2022)
14. Vaismoradi, M., Jones, J., Turunen, H., Snelgrove, S.: Theme development in qualitative content analysis and thematic analysis (2016)
15. Kauffman, D.F., Zhao, R., Yang, Y.S.: Effects of online note taking formats and self-monitoring prompts on learning from online text: Using technology to enhance self-regulated learning. Contemp. Educ. Psychol. **36**(4), 313–322 (2011)
16. Crossley, S., Paquette, L., Dascalu, M., McNamara, D.S., Baker, R.S.: Combining clickstream data with NLP tools to better understand MOOC completion. In: Proceedings of the Sixth International Conference on Learning Analytics & Knowledge, pp. 6–14 (2016)

Review Search Interface Based on Search Result Summarization Using Large Language Model

Marino Fujii[(✉)], Yuka Kawada, Takehiro Yamamoto, and Takayuki Yumoto

Graduate School of Information Science, University of Hyogo, Kobe, Hyogo, Japan
924norimaki@gmail.com, {t.yamamoto,yumoto}@sis.u-hyogo.ac.jp

Abstract. For users of e–commerce sites, it is important but difficult to find the desired information from large numbers of reviews efficiently. In this paper, we propose a review search interface that provides users with summaries of reviews as search results. This interface uses large language models (LLMs) to vectorize queries and review sentences and to generate summaries. In this interface, when a user selects a phrase in a review as a query, the query is vectorized using a LLM. Then, the interface searches for reviews based on their similarity to the query vector. Finally, it summarizes the searched reviews using a LLM, and presents the summary as the search results to the user. By using a LLM, the proposed interface can be used for reviews in any domain. We conducted user experiments to quantitatively compare the proposed method with string matching methods. We evaluated the proposed method by taking a questionnaire with seven items, including comprehensiveness of viewpoints and opinions, and readability of search results. As a result, the proposed method was superior in terms of the readability of search results.

Keywords: Product review · LLM · Search interface

1 Introduction

In recent years, more and more people are purchasing products on e–commerce websites. According to a survey by the Ministry of Economy, Trade and Industry (METI) of Japan in 2022, the BtoC–EC market size in the goods sales field has increased year by year from 5.993 trillion yen in 2013 to 13.9997 trillion yen in 2022. When purchasing products on EC websites, users use product reviews as one of the factors in their purchasing decisions. Users can get more information about the actual usability and durability of the product from product reviews.

While reading a review, users may come up with something new that they want to know. In this case, they take two actions. The first action is to reread the reviews from the beginning. In this case, when the number of reviews is very large, it takes time for them to find the information they want to know. The

---

This work was supported by JSPS KAKENHI Grant Numbers JP24K03228, JP21H03775, JP22H03905.

solution is to extract only the reviews that contain the information they want to know, or to summarize the reviews. The second action is to search by keywords. Typical review search systems use string matching. However, these systems can not find reviews written using synonyms such as "appearance" and "design". Furthermore, although users express what they want to know as a keyword, but the keyword is not always contained in the reviews. Suppose that there are the two reviews of the hair dryer, "No more damaged hair" and "I was surprised at how moist and settled (my hair) was." Although both of these are written about "hair quality", they do not appear in the search results when a user searches with the query "hair quality". The solution is to display similar words and sentences of the query in the search results.

There are many studies on product reviews. Kurihara et al. extracted movie reviews containing similar expressions to queries using Doc2Vec [2]. Spatharioti et al. compared an interactive search interface using ChatGPT with a traditional keyword search interface for retrieving product features [4]. Ichimura found that an interface that displayed summary sentences using dependency parsing was higher satisfaction than an interface that displayed all reviews in the search results [1]. Liu et al. confirmed the usefulness of review summarization using ChatGPT by human evaluation [3]. In this paper, we use a GPT-based large language model (LLM) as well as ChatGPT for summarization.

We propose a review search interface based on search result summarization using LLM. The proposed interface is outlined as follows. When users select a phrase in the review as a query, the system vectorizes the query and review by using a LLM. The interface finds reviews similar to the query based on the cosine similarity with the query vector. The interface summarizes similar reviews using a LLM and presents them to the users. LLMs have been trained on a large amount of data in various domains. Therefore, the proposed interface with LLM-based summarization can be used for reviews in any domain. Furthermore, this interface allows users to search for similar sentences and to grasp the contents of the search results at once.

2 Proposed Review Search Interface

2.1 Interface Usage

The usage of the proposed interface is shown in Fig. 1. This interface displays review articles of the target product to users. They select the description they are interested in by dragging it from the displayed review article. For example, a user is interested in the "the cold air is weak" in a review article of a hair dryer. Then, the user selects the description, and the selected sentence is automatically entered into the search window. When user clicks the "Search Similar Sentences" button, summaries of review sentences similar to the query "the cold air is weak" are displayed. The number of summarized reviews is also displayed next to each summary. When the user clicks on the summary result, full text of the review sentences are displayed and the user can check the details.

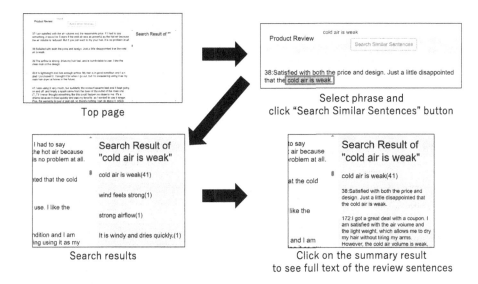

Fig. 1. The usage of the proposed interface.

2.2 Process Overview of the Proposed Interface

The process overview of the proposed interface is shown in Fig. 2. Preliminarily, each review article is divided into sentences. For example, the review article "Satisfied with both the price and design. Just a little disappointed that the cold air is weak." is divided into the review sentences "Satisfied with both the price and design." and "Just a little disappointed that the cold air is weak." Then, each sentence is vectorized. We use OpenAI's embedding model, text–embedding–ada–002[1] . By this process, each review sentence is represented by a 1,536–dimensional vector. When a user input a query, the query is vectorized in the same way. We calculate the cosine similarity between the query and the review sentences to find the review sentences similar to the query. We consider the review sentences whose cosine similarity to the query is 0.85 or higher to be review sentences similar to the query. By using sentence vectors for calculating similarity, we can find reviews that are semantically similar to the query. The found review sentences are summarized using a LLM. Finally, summaries with exactly the same wording are aggregated and presented to the user.

2.3 Summarization of Review

Review sentences with high similarity to a query are summarized using OpenAI's gpt–35–turbo–16k[2]. This model is a LLM that responds in natural language to a natural language instruction called a prompt. The reason for using a summary in

---

[1] https://openai.com/blog/new-and-improved-embedding-model.
[2] https://platform.openai.com/docs/models/gpt-3-5.

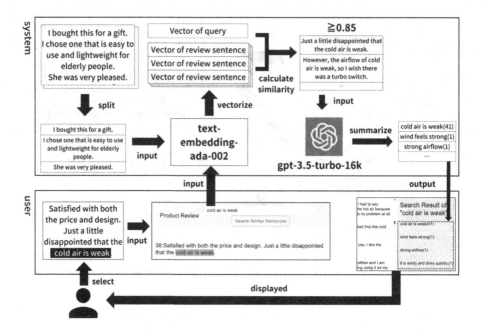

Fig. 2. The process overview of the proposed interface.

the proposed interface is to enable users to easily grasp the reputation described in the review sentences. Because LLMs have been trained on a large amount of data in various domains, the proposed interface with LLM-based summarization can be used for reviews in any domain.

We use a few-shot prompt for summarization. Few-shot prompt is a prompt that improves the accuracy of the response by providing several examples of input and output in the prompt. The prompt template used for summarization in the proposed interface is shown in Fig. 3. In the prompt, we use the original review sentence of the query as "Example of input", and the query as "Example of output".

The same result should be output for each query each time. There is a parameter called temperature in gpt-35-turbo-16k. As the value of this parameter decreases, the probability of a change in response decreases. We set this parameter to 0.

2.4 Use Case

We describe an actual use case of the proposed interface. Note that this use case is originally in Japanese. Suppose that a user selects the description "Not for damaged hair" from the review sentence "The finish of my hair was quite dry, so it is good for children's healthy hair, but not for damaged hair". Then, this description is entered as a query to the system, and the system finds five review sentences whose similarity to the query is 0.85 or higher. To summarize one of

```
#Explanatory text
Please summarize the input text in about 10 characters according to the output
format.
#Input
(Review sentence to be summarized)
#Example of input
(Review sentence containing a query)
#Example of output
(Query)
#Output format
Summary results of about 10 characters
```

Fig. 3. The prompt template used for summarization (originally in Japanese).

the five review sentences, "The quality of the hair when dried is not so nice", the proposed interface uses the prompt with the following substitutions for the prompt template of Fig. 3.

- *Input* section: The quality of the hair when dried is not so nice.
- *Example of input* section: The finish of my hair was quite dry, so it is good for children's clean hair, but not for damaged hair.
- *Example of output* section: Not for damaged hair

The summary "Hair quality is not so nice" is generated using this prompt.

This process is repeated for other four review sentences similar to the query. Review summaries with exactly the same wording are grouped and presented to the user.

3 User Experiment

3.1 Overview of Experiment

To evaluate the usefulness of the proposed interface, we conducted the user experiment. The experiment was originally conducted in Japanese.

We used review articles of two dryer products collected from Rakuten Ichiba[3]. Dryer reviews include perspectives expressed in multiple synonyms and not in direct keyword phrases. As of December 11, 2023, both were priced at approximately 6,000 yen. We used only the text of the review article, not the review title and the rating score. We also collected review articles of one toaster oven product for the training task. We collected 200 most recent reviews as of July 26, 2023 for all products.

The experimental participants were eight students (four males and four females) from University of Hyogo. We conducted the experiment from January 9, 2024 to February 15, 2024. The participants were paid approximately 1,500 yen.

---
[3] https://www.rakuten.co.jp/.

In the experiment, the participants were given the following instructions and asked to perform a search task using the proposed interface or the baseline interface.

This experiment was conducted in within–subjects design.

As a baseline, we used an interface that searches by string matching and displays the full text of the review sentences as search results. When a user enters words in the search window of the baseline system, the unsummarized results of the string matching search are displayed. There are three differences from the proposed interface.

- A user enters a word as a query.
- Only review sentences that contains the exact same words as the query are displayed in the search results.
- The unsummarized reviews are always displayed as the search results.

3.2 Experimental Procedure

The experimental procedure is conducted in the following four steps.

1. Practice task
2. Search task using the proposed interface (or baseline interface) and questionnaires for it
3. Search task using another interface and questionnaires for it
4. Interview

Preliminarily, we explained to the participants the search interface and the data to be collected in the experiments. The data we collect in the experiments are query and search time logs, and audio data from interviews.

First, we conducted a practice task. The goal of the practice task is to get participants familiar with the search interface. The participants used the proposed interface to search for review sentences of one toaster oven product using three queries. We also explained the baseline interface after this task was completed. We allowed time for a free search using each of the two interfaces.

We then performed the search task. The participants searched for review sentences of two dryer products to see how these products were evaluated. They searched one of the two products with the baseline interface and the other with the proposed interface. The time limit was 10 min per each product. When the participants felt that they had searched sufficiently, they could finish the task even within the time limit.

We conducted a questionnaire at the end of each search task. The questions of the questionnaire are shown in Table 1. We used a 5–point Likert scale for all questions.

Finally, we interviewed the participants to learn more about their reasons for answering the questionnaire.

Table 1. Questionnaire items (originally in Japanese).

Evaluation items	Question
1. Readability and accuracy of summary	You could easily find out the opinions contained in the review
2. Comprehensiveness to user interests	You could find what you wanted to search
3. Comprehensiveness of viewpoints	You could search from different perspective
4. Comprehensiveness of opinions	You could find opinions without omissions within the same point of view
5. Readability of search results	You think the way the results were displayed was easy to understand
6. Difficulty of using the system	You think the system was easy to use
7. Overall satisfaction	You are satisfied with the function of the system

Table 2. Questionnaire results: Mean (SD) of ratings.

Evaluation items	Baseline interface	Proposed interface
1. Readability and accuracy of summary	4.00 (1.31)	4.80 (0.48)
2. Comprehensiveness to user interests	3.63 (1.19)	4.20 (0.52)
3. Comprehensiveness of viewpoints	3.88 (1.55)	4.40 (0.52)
4. Comprehensiveness of opinions	2.88 (1.55)	3.60 (1.14)
5. Readability of search results	3.25 (1.28)	4.20 (0.84)
6. Difficulty of using the system	4.63 (0.74)	4.40 (0.71)
7. Overall satisfaction	3.38 (1.51)	4.40 (0.67)

3.3 Questionnaire Results and Discussions

We analyzed the results of the questionnaires and interviews. First, we show the results of the questionnaire in Table 2. The mean and standard deviation (SD) of the questionnaire ratings were calculated for each interface and question. The proposed interface outperformed the baseline using string matching for all six items except "difficulty of using the system". In particular, "readability of search results" was found to be significantly different by Welch's t–test at the 5% significance level. In the interviews, many subjects stated that the area in which the proposed interface outperformed the baseline interface was "readability of search results".

Regarding "readability of search results", the participants stated as follows.

- In the second half (proposed method). I could see how many cases were written about it ...
- When there was a large number of search results, I thought this one (the proposed interface) was really easy to see.

The proposed interface received high ratings in the item "readability of search result" because the search results by the proposed interface concisely described what users want to know. In the baseline interface, on the other hand, a large number of characters are displayed in the search results. Review search using the baseline interface is easier than reading all the reviews, but it is still hard to find what users want to know in the reviews. The proposed interface provides a short summary of the search results and displays the number of reviews corresponding to each summary. This reduces the burden on users to read the reviews. Therefore, significant differences were found in the questionnaire ratings, and many participants mentioned them in the interviews.

4 Conclusion

In this paper, we proposed a review search interface that provides users with summaries of reviews as search results. The proposed interface vectorizes a description in the review selected by the user and searches for the similar review sentences. It summarizes each searched review sentence using a LLM. It aggregates reviews with the same summary and display the summaries along with the number of the aggregated reviews. In the experiment, we compared the proposed interface with the baseline interface using conventional string matching. We conducted a user experiment and surveyed the users on the seven items such as comprehensiveness of viewpoints and opinions, and readability of search results. The proposed interface was superior to the baseline interface, especially in terms of readability of search results.

As future work, we need to improve the prompt for summarization. In the prompt used in the proposed interface, the query is used as the output example and the original review text is used as the input example. However, there were many cases where the output was similar to the query. Furthermore, some cases could not be summarized in 10 characters. Therefore, we need to consider summarization methods other than using few-shot prompt.

References

1. Ichimura, S.: Searching for delicious restaurants using review. J. Inf. Process. Soc. Jpn. **61**(11), 1748–1756 (2020). (in Japanese)
2. Kurihara, K., Shoji, Y., Fujita, S., Dürst, M.J.: Target–topic aware Doc2Vec for short sentence retrieval from user generated content. In: Proceedings of iiWAS2019, pp. 463–467. ACM, New York (2020). 10.1145/3366030.3366126

3. Liu, J., Liu, C., Lv, R., Zhou, K., Zhang, Y.: Is ChatGPT a good recommender? A preliminary study. arXiv preprint arXiv:2304.10149 (2023)
4. Spatharioti, S.E., Rothschild, D.M., Goldstein, D.G., Hofman, J.M.: Comparing traditional and LLM-based search for consumer choice: a randomized experiment. arXiv preprint arXiv:2307.03744 (2023)

Yes-No Flowchart Generation for Interactive Exploration of Personalized Health Improvement Actions

Naoya Oda[1(✉)], Yoshiyuki Shoji[2], Jinhyuk Kim[2], and Yusuke Yamamoto[3(✉)]

[1] SIOS Technology, Inc., Tokyo, Japan
n-oda@sios.com
[2] Shizuoka University, Hamamatsu, Japan
{shojiy,kimj}@inf.shizuoka.ac.jp
[3] Nagoya City University, Nagoya, Japan
yusuke_yamamoto@acm.org

Abstract. This paper proposes a system to facilitate efficient interactive exploration of personalized health improvement actions. With the proposed system, the users issue a lifestyle concern they wish to address to improve their overall health, e.g., a lack exercise or irregular sleep. The system then utilizes a large language model (LLM) to generate a list of candidate improvement actions. Then, the system calculates the semantic similarity among the actions and employs hierarchical clustering to construct a binary tree. Finally, the LLM is utilized again to transform the non-leaf nodes of the binary tree into a series of Yes/No questions, resulting in a dynamic Yes/No flowchart. By navigating this interactive flowchart, the users can identify relevant health improvement actions tailored to their needs in an efficient manner. Experimental results demonstrate the system's effectiveness in terms of facilitating efficient search for health improvement actions. However, user satisfaction with the proposed system was found to be lower than that for the traditional web search and ChatGPT methods.

Keywords: LLM · Health · Interactive system

1 Introduction

Lifestyle-related diseases, in which various lifestyle factors, e.g., diet and exercise, are the main causes of disease onset, have a significant impact on people's health. For example, unhealthy dietary habits are widespread in developed countries, and the mortality rate due to lifestyle-related diseases, e.g., diabetes, is increasing [4,23]. Lifestyle-related diseases are also a serious problem in Japan, and many studies have investigated the impact of unhealthy lifestyles on the risk of lifestyle-related diseases [1,5,13]. To prevent such diseases, it is necessary to improve lifestyle habits, e.g., increasing exercise levels and improving diet (hereafter referred to as "health improvement actions") [14]. Previous studies have developed systems that enable users to perform these actions [7,11].

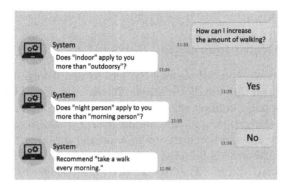

Fig. 1. Screenshot of the proposed system.

Information on health improvement actions is readily available online, and web-based search engines make it easy to learn about health improvement actions [3,19]. However, it is difficult to identify personalized health improvement actions that are suitable for individuals from the vast amount of available information because the optimal health improvement actions differ depending on individual tastes and habits. Thus, the top health improvement actions listed in search engine results may not necessarily be the most appropriate. In addition, it is difficult to search for the best health improvement actions with limited resources, and examining a large amount of information to identify the most effective action is time consuming. Further, the user must clarify relevant constraints to identify an effective health-improving behavior. However, it is also difficult for inexperienced and uninformed users to organize their constraints to perform an effective search for relevant information.

In this paper, we propose a system that interactively searches for optimal health improvement actions by generating questions to guide the search for personalized health improvement actions (Fig. 1). After the user inputs a lifestyle topic they wish to improve, e.g., a lack of exercise or irregular sleep, the proposed system generates a list of relevant actions to improve the identified topic using a large language model (LLM). For example, if the user wants to increase the amount of exercise, the proposed system generates a list of health improvement actions, e.g., "walking with friends", "walking to work instead of driving to work", and "doing yoga before going to bed at night". The proposed system then performs hierarchical clustering by calculating the semantic similarity of these health improvement actions and generates a binary tree in which the health improvement actions are positioned at the leaf nodes. Then, based on the health improvement actions, e.g., "walking with friends" and "walking to work instead of driving to work", the following question is asked: "Do you prefer to do activities alone rather than with others? using the LLM. Here, the nodes other than the leaf nodes of the binary tree are converted into question sentences. Finally, a YES/NO chart is generated with the improvement behavior as a leaf node, and questions are presented based on this chart.

2 Related Work

2.1 Issues in Health Information Retrieval

Previous studies have reported various barriers to the average user's search for health information on the web. For example, Zhang et al. found that older adults evaluate the web page quality when searching for health-related information based on limited indicators, e.g., the administrator of the page and the number of advertisements, and some misunderstand the meaning of the indicators [21]. In addition, Chi et al. conducted a survey to investigate how people search for health-related web pages. They found that most people browse specialized web pages if the search topic is serious; however, they also consult sources that are not corroborated by experts for topics they consider to be less serious [2]. Suzuki et al. investigated the effect of confirmation bias on web search behavior for health topics [17,18]. The results demonstrated that participants who were given confirmation bias prior to the experiment frequently viewed only the top search results and completed the search task more quickly. Yamamoto proposed a conversational search strategy called *suggestive answer strategy* to encourage searchers to make critical decision making on various topics, including health issues [20]. The goal of the current study is to enable users with limited information retrieval skills to search for relevant health improvement actions.

2.2 LLMs in the Health Sector

Previous studies have also investigated the utilization of LLMs to support health. For example Khaokaew et al. designed a framework that combines wearable sensors and LLMs to provide personalized advice sleep [8], and Lu et al. proposed a system that interprets pathology images and documents using LLMs [9]. In addition, Zhou et al. stated that the lack of explainability and insufficient data are important challenges when utilizing LLMs in actual medical practice [22]. In the current study, a LLM is used to generate remedial actions and questions to enable users to explore health information interactively and effectively.

2.3 Interactive Information Retrieval

Many previous studies have attempted to employ interactive information retrieval methods in a wide range of non-medical fields. For example, Nguyen et al. proposed a system that allows users to search for a desired video interactively by selecting the content of the video, the date the video was shot, and similar videos [10]. In addition, Schedl et al. proposed a system that enables users to search for songs interactively by representing songs as three-dimensional objects in a three-dimensional space [15]. This system facilitates visual understanding of song information by changing the color of blocks according to the genre and the distribution of acoustic features, as well as by positioning songs that the users do not listen to frequently at a greater distance. Hiramoto et al. proposed a system that enables users to obtain map information interactively [6]. In contrast, we

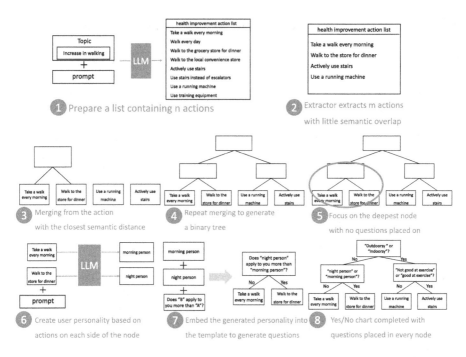

Fig. 2. Flowchart of the proposed system.

propose a system to search for appropriate health improvement actions interactively by presenting questions to the user using a YES/NO chart generated by a LLM.

3 Proposed System

The proposed system searches for relevant health improvement actions for each user. As shown in Fig. 1, the proposed system first receives a text description of a lifestyle topic the user wants to improve. Then, the system presents the user with several questions and recommends health improvement actions according to the user's answers. Internally, the proposed system generates a YES/NO chart to realize the presentation of questions and health improvement action recommendations. The process used to generate the YES/NO chart is described below with reference to Fig. 2.

3.1 Generation of Health Improvement Action List

To generate a YES/NO chart, it is necessary to prepare a list of health improvement actions that are diverse and have little overlap with the topics. Thus, the list of health improvement actions is generated in two steps: generating candidate health improvement actions and extracting health improvement actions with little semantic overlap.

Algorithm 1. Algorithm to extract A_{in} actions from the list of A_{in} candidate health improvement actions (excluding redundant actions).

Input: $A_{in} = \{a_1, ..., a_n\}, m \in \mathbb{N}$
Output: A_{out}
1: Select action a_i from A_{in} randomly.
2: $A_{out} \leftarrow \{a_i\}, A_{in} \leftarrow A_{in} \setminus \{a_i\}$
3: **while** $|A_{out}| < m$ **do**
4: $\quad a \leftarrow \underset{a \in A_{in}}{\arg\min} \text{ LIST-SIM}(a, A_{out})$
5: $\quad A_{out} \leftarrow A_{out} \cup \{a\}$
6: $\quad A_{in} \leftarrow A_{in} \setminus \{a\}$
7: **end while**

In the first step (i.e., generating candidate improvement actions), n candidate health improvement actions corresponding to the input health topics are generated using ChatGPT[1] (refer to (1) and (2) in Fig. 2). The prompt for the candidate generation process is "I am interested in improving my lifestyle. Please output n methods for <topic>.

In the second step, semantic duplicates are omitted from the candidates generated in the first step. Note that the health improvement action list generated by ChatGPT may contain semantic duplicates. For example, as shown in Fig. 2, "actively use stairs" and "use stairs instead of escalators" are similar actions. Here, if a YES/NO chart is generated using all of these behaviors as they are, the system will only present similar actions. Therefore, the semantic similarity between the actions in the list should be as low as possible. Thus, we generate a list A_{out} with $m(m < n)$ elements by omitting redundant actions from the list of health improvement actions A_{in} generated by ChatGPT.

The health improvement action extraction algorithm is shown in Algorithm 1. First, A_{in} is a list containing n improvement actions. Then, an empty list A_{out} is generated, and a_{in} extracted randomly from A_{in} is moved to A_{out}. Then, the semantic similarity between all elements a_{in} and A_{out} in A_{in} is calculated, and a_{in} with the lowest semantic similarity is moved into A_{out} repeatedly until the number of elements in A_{out} reaches m. Given an improvement action x and a list of improvement actions Y, the following formula is employed to calculate the similarity LIST-SIM(x, Y) between x and Y.

$$\text{LIST-SIM}(x, Y) = max(\{sim_{cos}(x, y) \mid y \in Y\}) \qquad (1)$$

Here, $sim_{cos}(x, y)$ is the cosine similarity of the vectors $\mathbf{v}(x)$ and $\mathbf{v}(y)$ of x and y. In addition, a Japanese SentenceBERT model is employed to vectorize each health improvement action [12].

3.2 Binary Tree Generation

Then, the extracted m actions (Sect. 3.1) are summarized in a binary tree structure to generate the frame of a YES/NO chart. Note that the YES/NO chart

---
[1] https://openai.com/blog/chatgpt.

is used to present questions in sequence from the root node to the leaf nodes to identify the user's requirements in a step-by-step manner and present optimal improvement actions. Thus, it is desirable that questions that roughly divide improvement actions are asked in the shallow part of the generated binary tree, and those that divide similar health improvement actions are placed in the deep part of the binary tree.

In this study, hierarchical clustering using the Ward method is applied to the list of health improvement actions to generate a dichotomous tree in which similar health improvement actions are placed at the closest leaves. The vector of health improvement actions in the list is converted into a 768-dimensional vector using the Japanese SentenceBERT model, and the Euclidean distance is employed to measure the distance between the clusters. For example, as shown in Fig. 2 (3), "Take a walk every morning" and "Walk to the store for dinner," which are the closest in distance, are merged. Then, the merging of actions with the close distances is repeated, and a binary tree is generated with the health improvement actions as leaves, as shown in Fig. 2 (4).

3.3 Question Generation

Finally, ChatGPT is employed to generate the YES/NO questions and position them at the nodes of the binary tree (Sect. 3.1). The generated questions are designed to clarify the user's attributes and requirements. Here, it is desirable to narrow down the range of improvement actions according to the answers to the questions. Thus, ChatGPT is employed to estimate the user's attributes and requirements based on the improvement actions in the YES/NO chart and generate relevant questions using the estimated attributes and the prepared question templates.

The flow of the question generation process is described as follows. First, we focus on the deepest clause in the binary tree where no questions are placed. Then, we list the improvement behaviors in the subtrees to the left and right of the target clause. The listed health improvement actions are embedded in the prompts shown in Fig. 3 and input to ChatGPT to estimate the user attributes corresponding to each behavior. In Fig. 2 (6), "Take a walk every morning" is listed as an action on the left subtree of the node in focus, "Walk to the store for dinner" is listed as an action on the right subtree, and the corresponding user attributes are estimated for each action. After the user's attributes are estimated, the estimated attributes are embedded in a question template to generate questions that correspond to the clauses. The template is "Does <attribute B> apply to you more than <attribute A>?" Here, < attribute A> and <attribute B> are generated from the actions on the left and right partials, respectively. For example, in Fig. 2 (7), the questions are generated and placed by applying "morning person" and "night person" to the template. The above operations are repeated until the questions are placed in all the clauses, thus completing the YES/NO chart.

```
#Instruction:
  listA contains actions for a healthy lifestyle. ListB contains my favorite actions
  from listA. ListC contains my friend's favorite behaviors from listA.Please
  analogize what kind of personalities my friend and I have each.Please note
  that there should be 5 outputs for each of them.

# Constraints :
  · The output format should be a list like the following.
listB2=[
"Want to exercise every day",
"Plan ahead",
"Health conscious"
]
  · Please output my personality to listB2 and my friend's personality to listC2.
  · The elements of listB2 and listC2 should be semantically contrastive.
```

Fig. 3. Prompt for attribute generation.

4 Experiment

We conducted an online user experiment to verify whether the proposed system can be used to efficiently search for improvement actions that are suitable for users. This experiment was conducted on GoogleMeet using a PC or Mac from January 10–16, 2024. The experimental participants were recruited from Shizuoka University students and graduate students. A total of 12 students participated in the experiment (11 male and one female).

4.1 Hypotheses and Evaluation Items

The hypothesis tested in this experiment and the two evaluation items for its validation are described as follows.

- H1: The users can search for health improvement actions efficiently using an interactive search system with YES/NO charts.
  Evaluation items: time taken to search and ease of conducting search.
- H2: The users can find suitable health improvement actions using the proposed interactive search system with the YES/NO chart.
  Evaluation items: degree of optimization of health improvement actions, feasibility of health improvement actions, degree of acceptance of health improvement actions, and motivational contribution of health improvement actions.

Fig. 4. Screenshot of the proposed system. After questions are presented (1), a health improvement action is presented (2).

Here, the time taken to search was evaluated by measuring the time taken by the participants to search for improvement actions, and the degree of optimization of health improvement actions was evaluated by assigning a score to the explored improvement actions and the improvement actions for comparison. These scores were then ranked. When ties occurred, the overlapping rankings were accepted, and the next ranking was skipped. For example, if four actions were scored "5, 4, 4, and 3," the corresponding ranking was given as "1st, 2nd, 2nd, and 4th," respectively. In addition, the ease of conducting the search, the feasibility of health improvement actions, the degree of acceptance of health improvement actions, and the contribution of motivation to health improvement actions were evaluated by asking the participants to answer a questionnaire on a five-point Likert scale.

4.2 Compared Methods

In the experiment, the proposed system and two existing methods were used by the participants to search for health improvement actions. A screenshot of the proposed system is shown in Fig. 4.

In this experiment, the following two general information retrieval methods were used for comparison.

- Web search
  Here, search queries were freely entered into a web browser to explore health improvement actions using the Google search engine.

- Search in ChatGPT
  Health improvement actions were also search by freely entering queries into ChatGPT (version GPT3.5).

4.3 Experimental Design and Procedures

This experiment was conducted according to a within-subjects design. In other words, the experimenter performed the search for the health improvement action using all three search methods, i.e., web search, ChatGPT, and the proposed system. The experiment was conducted as follows.

1. Respond to a preliminary questionnaire
2. Search for health improvement actions using web search
3. Respond to a post-task questionnaire
4. Search for health improvement actions using ChatGPT
5. Respond to a post-task questionnaire
6. Explore health improvement actions using the proposed system
7. Respond to a post-task questionnaire

In the preliminary questionnaire, the participants were asked to rate their level of interest in each search topic on a five-point scale (from "not at all" to "very much").

The topics to be explored using each search method were two randomly assigned topics from the following six candidate topics.

- Increase in walking
- Increase in metabolism
- Increase in vegetable intake
- Reduce salt intake
- Reduce stress
- Improve sleep quality

To avoid combinations of search method and topic affecting the search, the combination of search method and topic was randomized using Latin square design[2], as shown in Table 1. However, the order in which the search methods were implemented was not randomized, and all participants performed the search task in the following order: web search, ChatGPT, and the proposed system.

The measurement method used for each evaluation item is described as follows. The time taken to search was measured by the author using a stopwatch application installed on a Mac computer. Here, the measurement began as soon as the participant initiated the search task and ended when they reported the task was completed.

The feasibility of health improvement actions, the degree of acceptance of health improvement actions, and the contribution of motivation to health improvement actions were evaluated using the post-task questionnaire, where each question was rated on a five-point scale (from "very applicable" to "not applicable at all"). Here, the participants were asked the following questions.

---
[2] https://mathworld.wolfram.com/LatinSquare.html.

Table 1. Combination of randomized search methods and topics by Latin square design.

Collaborators	Web search	ChatGPT	Proposed system
S1, S2	Reduce stress	Reduce salt intake	Improve sleep quality
	Increase in metabolism	Increase in walking	Increase in vegetable intake
S3, S4	Increase in vegetable intake	Increase in metabolism	Reduce salt intake
	Reduce stress reduction	Improving sleep quality	Increase in walking
S5, S6	Increase in walking	Reduce stress	Increase in metabolism
	Increase in vegetable intake	Reduce salt intake	Improving sleep quality
S7, S8	Improve sleep quality	Increase in vegetable intake	Reduce stress
	Increase in walking	Increase in metabolism	Reduce salt intake
S9, S10	Reduce salt intake	Increase in walking	Increase in vegetable intake
	Improving sleep quality	Reduce stress	Increase in metabolism
S11, S12	Increase in metabolism	Improve sleep quality	Increase in walking
	Reduce salt intake	Increase in vegetable intake	Reduce stress

- Ease of conducting search: "How easy was the search for health improvement actions?"
- Feasibility of health improvement actions: "How easy do you think it is for you to implement the action you finally adopted?"
- Degree of acceptance of health improvement actions: "How satisfactory do you think the action you finally adopted are for you?"
- Contribution of motivation to health improvement actions: "How much do you think the action you finally adopted would increase your motivation to improve your lifestyle?"

The degree of optimization of health improvement actions was evaluated by asking the participants to score the 16 improvement and comparison actions for each topic in post-task questionnaire. Here, the questions were answered using a five-point scale. For example, a participant who specified "increasing vegetable intake" as a search topic was given a list of 16 system-generated improvement actions, and the participant scored each action on a five-point scale. In addition, participants who searched for "increasing vegetable intake" using web search or ChatGPT were also asked to score the behavior they searched for themselves out of five. By generating a ranking of the score improvement actions, we investigated the degree to which the explored improvement actions were preferred over the compared improvement actions.

5 Results

We analyzed the subjective and behavioral indicators of the 12 experimental participants to analyze the impact of the proposed system on their search for health improvement actions. Here, the collected data did not follow a normal distribution; thus, the Kruskal–Wallis test of nonparametric methods was employed

Table 2. Mean and standard deviation of search time, response scores to questions, and rank order of improvement actions explored, as well as the statistical significance in the analysis of variance (*: 0.05, ***: 0.001). ^{W,C,S} denotes the significance (p < 0.05) of the posterior test in two-pair conditioning for web search, ChatGPT, and the proposed system, respectively.

Evaluation items	Condition			p-value
	Web search	ChatGPT	Proposed system	
Time taken to search (s)	164.13 (112.92)^S	139.79 (103.86)^S	26.58 (5.04)^{W,C}	***
Q1. Ease of conducting search	4.29 (0.65)^{C,S}	4.75 (0.34)^{W,S}	5.00 (0.0)^{W,C}	***
Q2. Feasibility of health improvement actions	4.37 (0.57)	4.29 (0.58)	3.88 (0.83)	0.22
Q3. Degree of acceptance of health improvement actions	4.50 (0.56)	4.33 (0.78)	4.17 (0.58)	0.35
Q4. Contribution of motivation to health improvement actions	3.91 (0.73)	3.79 (0.81)	3.71 (1.14)	0.98
Ranking of the selected action	2.96 (3.61)^S	3.13 (1.51)	5.08 (2.50)^W	*

for all three search methods. In addition, we used the Steele–DeWeese test for multiple comparisons in the posterior analysis [16].

Table 2 shows the mean values, standard deviations, statistical significance of the analysis of variance, and the results of multiple comparisons for the compared search methods for the time taken to search, the answers to the four questions, and the ranking of the selected action. Here, the time taken to search is measured in seconds, and a higher mean value indicates that more time was required for the search. Q1 to Q4 show that the higher the mean value, the more positive the results for the search method and the improvement actions explored. The ranking of the selected action indicates that the lower the mean value, the more the improvement actions searched for by the collaborators were preferred compared to the other actions.

Table 2 shows that a significant difference was observed between both the web search and ChatGPT method in multiple comparisons with the proposed system in terms of the time taken to search (web search: $p < 0.05$; ChatGPT: $p < 0.05$). In addition, the proposed system tended to have a significantly lower mean value than the web search and ChatGPT methods (search time: 164.13 s vs 139.79 s vs 26.58 s; $p < 0.001$). These results demonstrate that the proposed system allowed the participants to complete the search for improvement actions in less time than the web search and ChatGPT methods.

For Q1, significant differences were confirmed between both the web search and ChatGPT methods compared to the proposed system (web search: $p < 0.05$; ChatGPT: $p < 0.05$). Significant differences were also confirmed between the web search and ChatGPT methods ($p < 0.05$). In addition, the mean tended to be significantly higher for the proposed system, ChatGPT, and web search, in that order (Q1: 4.29 vs. 4.75 vs. 5.00; $p < 0.001$). These results demonstrate that the participants found that searching with the proposed system was easier than searching with the ChatGPT and web search methods.

For Q2–Q4, the mean tended to be higher in the order of web search, ChatGPT, and the proposed system; however, no significant difference was observed (Q2: 4.37 vs. 4.29 vs. 3.88; $p > 0.05$). These results indicate that the participants felt the same degree of feasibility and satisfaction with the improvement behaviors explored by all search methods, and that they were inspired to improve their lifestyles.

Regarding the ranking of the explored improvement actions, a significant difference was observed between the proposed system and the web search method in multiple comparisons (web search: $p < 0.05$). In addition, the average ranking of the web search tended to be higher than that of the proposed system (ranking of the explored improvement actions: 2.96 vs. 5.08; $p < 0.05$). These results demonstrate that the participants preferred the improvement actions explored using the web search method compared to those explored using the proposed system.

6 Discussion

The experimental results demonstrated that the proposed system enabled the participants to complete the search for health improvement actions faster than using the web search and ChatGPT methods. We also found that using the proposed system to search for information was easier than the general information retrieval methods, i.e., the web search and ChatGPT methods.

When using the web search or ChatGPT methods to search for health improvement actions, the participant must scrutinize the information obtained from the search results. In contrast, with the proposed system, the participant only needs to respond to the proposed questions using yes or no answers. Thus, the participants completed the search in a short time and found it easy to search for health improvement actions using the proposed system. From the above results, we confirm that H1 ("The users can search for health improvement actions efficiently using an interactive search system with YES/NO charts.") is supported.

However, we found that the average values obtained for the proposed system were less than those of the compared methods in terms of the feasibility of improvement actions, the degree of acceptance of improvement actions, and the contribution of motivation to improvement actions. The health improvement actions explored by the compared method received higher evaluations than those explored using the proposed method because the web search and ChatGPT methods allowed the participants to search until they identified actions they felt were satisfactory or feasible. In addition, in terms of the degree of optimization of the improvement actions, we found that the web search method search significantly outperformed the proposed system.

The proposed system presented less information than the compared method, which may explain the difference in the evaluation of the search results. With the proposed system, only the name of the health improvement action is presented to the user during the search. In contrast, when searching using the web search method, additional information is presented to the user, e.g., relevant cautions and advice for effective implementation of the corresponding action. We consider that the presence or absence of such useful information resulted in the differences in the participants' evaluations of the search results.

From the above results, we confirm that H2 ("The users can find suitable health improvement actions using the proposed interactive search system with the YES/NO chart.") is not supported. However, significant differences could not be confirmed for three out of the four evaluation items, and the ranking of the health improvement actions recommended by the proposed system was relatively high, ranking fifth out of the 16 cases on average. Thus, the health improvement actions recommended by the proposed system were not completely inappropriate.

From the results for H1 and H2, the proposed system improved the search efficiency compared to the conventional search methods without losing much quality in terms of the acquired search results. However, in this study, the order in which the search methods were implemented was not randomized, and the influence of the order effects on the experimental results cannot be denied. Thus, to eliminate the order effect, we will consider increasing the number of participants and conducting a more extensive experiment in which the order of the compared search methods is randomized using the Graeco-Latin design.

There are several limitations to this study, which are summarized as follows. First, the health effects of the health improvement actions generated by the ChatGPT method are not guaranteed; thus, we cannot guarantee that the actions explored by the proposed system will in fact lead to lifestyle improvement when executed. To present actions that will contribute effectively to lifestyle improvement, we believe that LLMs that incorporate more health information as training data are required.

Second, the time complexity of the search, i.e., $O(n)$, cannot be avoided. As discussed in Sect. 3.2, we generated the binary trees using the Ward method. In this case, depending on the shape of the generated tree, the worst-case complexity for the search process may become $O(n)$. However, with self-balancing binary trees, the time complexity of the search process is $\log_2 n$; thus, it would be desirable to generate self-balancing binary trees to improve search efficiency.

7 Conclusion

This paper has proposed a system that realizes an interactive search for the best health improvement actions by summarizing the health improvement actions using a YES/NO chart format. The proposed system utilizes ChatGPT to generate health improvement action lists and questions that refine the user's attributes and requirements. In addition, an extraction algorithm was designed and implemented to improve the diversity of the health improvement action lists generated by ChatGPT.

To investigate the impact of the proposed system on the users' search for health improvement actions, we conducted a user experiment with student participants from Shizuoka University. The experimental results demonstrated that the proposed system allowed the participants to search for health improvement actions more efficiently than the compared methods without considerable deterioration in terms of the quality of the searched health improvement actions.

In the future, we plan to implement a method to generate the YES/NO chart using a self-balancing binary tree to realize a more efficient search. In addition, we plan to investigate a method to reliably generate relevant questions that are consistent with the health improvement actions to enable search for health improvement actions that are more suitable for each user.

Acknowledgements. The author(s) declare that financial support was received for the research, authorship, and/or publication of this article. This work was supported in part by the Grants-in-Aid for Scientific Research (21H03554, 21H03775, and 22H03905) from the MEXT of Japan and Daiko Foundation.

References

1. Adachi, H., et al.: Trends in nutritional intake and coronary risk factors over 60 years among Japanese men in Tanushimaru. Heart Vessels **35**, 901–908 (2020)
2. Chi, Y., He, D., Han, S., Jiang, J.: What sources to rely on: laypeople's source selection in online health information seeking. In: Proceedings of the 2018 Conference on Human Information Interaction & Retrieval, CHIIR 2018, pp. 233–236 (2018)
3. Eng, T.R., Maxfield, A., Patrick, K., Deering, M.J., Ratzan, S.C., Gustafson, D.H.: Access to health information and support: a public highway or a private road? JAMA **280**(15), 1371–1375 (1998)
4. Ford, E.S., Bergmann, M.M., Boeing, H., Li, C., Capewell, S.: Healthy lifestyle behaviors and all-cause mortality among adults in the united states. Prev. Med. **55**(1), 23–27 (2012)
5. Fukumoto, Y.: Nutrition and cardiovascular diseases. Nutrients **14**(94) (2022)
6. Hiramoto, R., Sumiya, K.: A web search method using user operation on digital maps. In: Proceedings of the 7th International Conference on Mobile Data Management, MDM 2006, p. 106 (2006)
7. Kao, H.T., Yan, S., Hosseinmardi, H., Narayanan, S., Lerman, K., Ferrara, E.: User-based collaborative filtering mobile health system. In: Proceedings of the ACM Interactive, Mobile, Wearable and Ubiquitous Technologies, vol. 4, no. 4, pp. 1–17 (2020)
8. Khaokaew, Y., Nguyen, T.H., Ji, K., Kegalle, H., Alaofi, M.: ZzzGPT: an interactive GPT approach to enhance sleep quality. arXiv preprint arXiv:2310.16242, abs/2310.16242(1) (2023)
9. Lu, M.Y., et al.: A foundational multimodal vision language AI assistant for human pathology (2023)
10. Nguyen, T.N., Puangthamawathanakun, B., Arpnikanondt, C., Gurrin, C., Caputo, A., Healy, G.: Efficient search with an interactive video retrieval system for novice users in IVR4B. In: Proceedings of the 20th International Conference on Content-Based Multimedia Indexing, CBMI 2023, pp. 168–172 (2023)
11. Rabbi, M., Aung, M.H., Zhang, M., Choudhury, T.: MyBehavior: automatic personalized health feedback from user behaviors and preferences using smartphones. In: Proceedings of the 2015 ACM International Joint Conference on Pervasive and Ubiquitous Computing, UbiComp 2015, pp. 707–718 (2015)
12. Reimers, N., Gurevych, I.: Sentence-BERT: sentence embeddings using siamese BERT-networks (2019)

13. Sakaue, A., et al.: Association between physical activity, occupational sitting time and mortality in a general population: an 18-year prospective survey in Tanushimaru. Japan. Eur. J. Prev. Cardiol. **27**(7), 758–766 (2020)
14. Sato, Y., Nagasaki, M., Nakai, N., Fushimi, T.: Physical exercise improves glucose metabolism in lifestyle-related diseases. Exp. Biol. Med. **228**(10), 1208–1212 (2003)
15. Schedl, M., Mayr, M., Knees, P.: Music tower blocks: multi-faceted exploration interface for web-scale music access. In: Proceedings of the 2020 International Conference on Multimedia Retrieval, ICMR 2020, pp. 388–392 (2020)
16. Steel, R.G.: A rank sum test for comparing all pairs of treatments. Technometrics **2**(2), 197–207 (1960)
17. Suzuki, M., Yamamoto, Y.: Analysis of relationship between confirmation bias and web search behavior. In: Proceedings of the 22nd International Conference on Information Integration and Web-Based Applications & Services, iiWAS 2020, pp. 184–191 (2021)
18. Suzuki, M., Yamamoto, Y.: Characterizing the influence of confirmation bias on web search behavior. Front. Psychol. **12**, 1–11 (2021). https://doi.org/10.3389/fpsyg.2021.771948
19. Tonsaker, T., Bartlett, G., Trpkov, C.: Health information on the internet: gold mine or minefield? Can. Fam. Physician **60**(5), 407–408 (2014)
20. Yamamoto, Y.: Suggestive answers strategy in human-chatbot interaction: a route to engaged critical decision making. Front. Psychol. **15**, 1–16 (2024). https://doi.org/10.3389/fpsyg.2024.1382234, https://www.frontiersin.org/journals/psychology/articles/10.3389/fpsyg.2024.1382234
21. Zhang, Y., Song, S.: Older adults' evaluation of the credibility of online health information. In: Proceedings of the 2020 Conference on Human Information Interaction and Retrieval, CHIIR 2020, pp. 358–362 (2020)
22. Zhou, H., et al.: A survey of large language models in medicine: progress, application, and challenge. arXiv preprint arXiv:2311.05112 (2023)
23. Zimmet, P., Alberti, K.G., Magliano, D.J., Bennett, P.H.: Diabetes mellitus statistics on prevalence and mortality: facts and fallacies. Nat. Rev. Endocrinol. **12**(10), 616–622 (2016)

GDMA

Enhancing Link Prediction Based on Simple Path Graphs

Zhiren Li, Yuzheng Cai, and Hongwei Feng[✉]

School of Data Science, Fudan University, Shanghai, China
{lizr19,hwfeng}@fudan.edu.cn, yuzhengcai21@m.fudan.edu.cn

Abstract. Link prediction is one of the prominent tasks in the field of graph machine learning, aiming to fill in missing edges in graph data or predict potential connections between nodes in the graph. Early link prediction methods primarily utilized traditional machine learning models, leveraging heuristic rules, embedded representations of nodes, and node features for prediction. However, graph neural networks possess powerful learning capabilities, effectively capturing the topological structure and node features of graphs, thus exhibiting increasingly superior performance in link prediction tasks. Methods for link prediction tasks based on graph neural networks primarily fall into two categories: node-based methods and subgraph-based methods. Among them, subgraph-based methods excel in link prediction tasks due to their ability to capture the graph topological structure between nodes and effectively address the issue of node isomorphism. Currently, subgraph-based methods typically employ enclosing subgraphs formed by the first or second-order neighboring nodes around the target nodes. Despite the effectiveness of enclosing subgraph-based methods in link prediction tasks, they face challenges such as the scalability issue arising from the large size of enclosing subgraphs and susceptibility to influence from hub nodes in the graph. To address this challenge, this study proposes a novel subgraph pattern, namely, utilizing target nodes to construct the simple path graph and conducting link prediction on these simple path graphs. We demonstrate in this study that, under certain order constraints, simple path graph is a subgraph of enclosing subgraph, thus alleviating the issue of excessive subgraph size encountered by subgraph-based methods. Furthermore, through experimental validation, we show that even with relaxed order constraints, the size of the simple path graph remains smaller than that of the enclosing subgraph. Experimental results demonstrate that the proposed link prediction method based on the simple path graph outperforms other methods, both on datasets with node features and datasets without node features.

Keywords: Graph neural network · Link prediction · Simple path graph

1 Introduction

In recent years, with the continuous research and development of graph neural networks [7,9,14], the task of link prediction [4,16] has garnered increasing attention as an important task within graph neural networks. Link prediction has diverse applications across different types of graph data, such as friend recommendation in social networks, recommendation systems on user-item bipartite graphs, and knowledge graphs completion [12].

The methods of link prediction can be divided into four main categories: rule-based methods, statistical methods, latent feature methods, and graph neural network-based methods. Rule-based methods [1,8,11] primarily rely on the structure and attributes of the graph, employing heuristic rules to predict the connections between nodes based on local or global features of the graph. Statistical methods [6,15], on the other hand, focus on utilizing probabilistic graphical models within probability models, performing link prediction through complex learning and inference processes. Latent feature methods [5,13] map nodes to low-dimensional feature spaces and predict connections between nodes based on their positions in this space. Methods based on graph neural networks learn node embeddings from the structure of the graph and node features in an end-to-end manner, enabling better capture of the complex structure of the graph and relationships between nodes, thereby improving the accuracy and generalization capability of link prediction.

Link prediction methods based on graph neural networks are typically divided into node-based methods [10] and subgraph-based methods [17,18]. In node-based link prediction methods, graph neural networks learn embeddings of the entire graph and predict links by learning embeddings of target nodes. For isomorphic nodes in the graph, graph neural networks learn the same embedding representation, implying that they occupy the same position in the embedding space. Consequently, the link relationships with other nodes in the network are also the same for these two isomorphic nodes. However, in real-world graph structures, isomorphic nodes often exhibit different connectivity patterns and local neighborhoods. For example, some nodes may be more inclined to connect to one isomorphic node than the other, with no direct connection to the latter. Therefore, node-based link prediction methods have limitations in addressing the issue of node isomorphism in graphs.

Subgraph-based link prediction methods address the problem of node isomorphism by extracting subgraphs around two target nodes, transforming the link prediction task into a subgraph classification task. Subgraph-based link prediction methods typically involve three steps [18]. Firstly, subgraphs around target nodes are extracted. In previous studies, these subgraphs consist of the first or second-order neighborhood of target nodes and the edges between their neighborhood, also known as enclosing subgraphs. After extracting the enclosing subgraph, the node information on the subgraphs needs to be integrated. This information includes node features, graph embedding representations, and node labels. The purpose of node labels is to differentiate nodes based on their distance or positional relationships with target nodes in the subgraph. Finally, the integrated information is input into a graph neural network to train a graph classification model, which completes the link prediction task based on the model's predictions.

In summary, methods based on graph neural networks, especially subgraph-based methods, outperform other methods in link prediction tasks. However, subgraph-based methods still face some challenges. Since subgraph-based methods typically extract enclosing the subgraph around the pair of target nodes, the enclosing subgraph tends to be large, with second or third-order enclosing subgraph potentially being the size of the original graph. Therefore, subgraph-based methods face the challenge of large subgraph sizes and susceptibility to the influence of hub nodes. We start our investigation by examining the paths between target nodes and find that the scale of the simple path graph [3] is smaller than that of the enclosing subgraph. Based on this observation, we explore a more efficient subgraph method to accomplish the link prediction task. We denote the method based on the simple path graph as SPG4LP. The contributions of the paper can be summarized as follows:

- **Smaller subgraph size:** The method proposed in this paper extracts relatively small subgraphs, indicating that the approach based on simple path graphs can reduce the demand for large-scale computing resources.
- **Insensitivity to hub nodes:** Simple path graph focuses on the paths between target nodes and the neighbors of hub nodes are not heavily incorporated into the simple path graphs.
- **Providing a wider range of subgraph selection:** Simple path graph offers a more refined subgraph partitioning, capable of meeting the requirements of different datasets.
- **Better performance:** Extensive experiments demonstrate the effectiveness and efficiency of the proposed approach.

2 Preliminaries

Definition 1. *Simple Path:* *Given an undirected graph $G = (V, E)$, a simple path in graph G refers to a sequence of nodes $v_1, v_2, \ldots, v_k$, where $v_i \in V$ and $v_i \neq v_j$ (for all $1 \leq i \leq j \leq k$), and adjacent nodes in the sequence are connected by edges, i.e., $(v_i, v_{i+1}) \in E$ (for all $1 \leq i < k$). The length of a simple path is the number of edges it contains. A simple path of length k between nodes s and t is denoted as $SP_{s,t}^{(k)}$.*

With the concept of simple path, we can then provide the definition of a simple path graph.

Definition 2. *Simple Path Graph:* *Given an undirected graph $G = (V, E)$, for any two nodes $s, t \in V$ and an integer k, the set of all simple paths with lengths less than or equal to k between nodes s and t is termed the simple path graph under the constraint of integer k, denoted as $SPG_{s,t}^{(k)}$. $SPG_{s,t}^{(k)} \subseteq G$. Formally, $SPG_{s,t}^{(k)} = \bigcup_{l \leq k} SP_{s,t}^{(k)}$.*

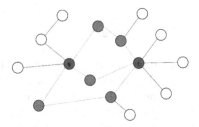

Fig. 1. Given a graph and the nodes s, t on the graph, the subgraph composed of all red and blue nodes as well as green edges on the graph represents the 3rd-order simple path graph between s and t, denoted as $SPG_{s,t}^{(3)}$.

Figure 1 depicts an example of a 3rd-order simple path graph in a given graph. In this paper, the minimum value of the order of a simple path graph is 2, because there are no edges between target nodes in link prediction tasks, making simple path graphs with orders less than 2 non-existent. To facilitate comparison with other subgraph-based link prediction methods, we provide the definition of the subgraph used by SEAL [18], namely the enclosing subgraph.

Definition 3. Enclosing Subraph: Given an undirected graph $G = (V, E)$ and an integer k, for any two nodes $s, t \in V$ in graph G, a kth-order enclosing subgraph $G_{s,t}^{(k)}$ is a subgraph of G containing nodes s and t and all their neighbors within a distance of at most k orders. Formally, the kth-order enclosing subgraph can be recursively defined as $ESG_{s,t}^{(k)} = \left(V_{s,t}^{(k)}, E_{s,t}^{(k)}\right)$, where the node set $V_{s,t}^{(k)} = \left\{u \mid \forall v \in V_{s,t}^{(k-1)}, u \in \Gamma(v)\right\} \cup V_{s,t}^{(k-1)}$, and the edge set $E_{s,t}^{(k)} = \left\{(u, v) \mid \forall u, v \in V_{s,t}^{(k)}, (u, v) \in E\right\}$.

Figure 2 presents an example of a 1st-order enclosing subgraph in a given graph. By carefully observing Figs. 1 and 2, it is evident that for any given pair of target nodes s, t, the 3rd-order simple path graph is a subset of the 1st-order enclosing subgraph, i.e., $SPG_{s,t}^{(3)} \subseteq ESG_{s,t}^{(1)}$. More generally, $SPG_{s,t}^{(2k+1)} \subseteq ESG_{s,t}^{(k)}$, meaning that the size of $SPG_{s,t}^{(2k+1)}$ is smaller than $ESG_{s,t}^{(k)}$, and $SPG_{s,t}^{(2k+1)}$ is a subgraph of $ESG_{s,t}^{(k)}$.

Corollary 1. For any undirected graph $G = (V, E)$, any pair of nodes s, t, and any positive integer k, we have $SPG_{s,t}^{(2k+1)} \subseteq ESG_{s,t}^{(k)}$.

Proof. For any node v in $SPG_{s,t}^{(2k+1)}$, assuming v is not in $ESG_{s,t}^{(k)}$, then v must not be a k-order neighbor of nodes s, t. Therefore, the shortest path length from v to nodes s, t must be greater than or equal to $k + 1$, contradicting the fact that v is a node in $SPG_{s,t}^{(2k+1)}$. Thus, v is also in $ESG_{s,t}^{(k)}$. For any edge (u, w) in $SPG_{s,t}^{(2k+1)}$, it is evident that $(u, w) \in E$, as both u and w are nodes

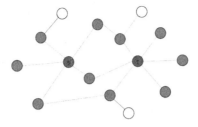

Fig. 2. Given a graph and the nodes s, t on the graph, the subgraph composed of all red and blue nodes as well as green edges on the graph represents the 1st-order enclosing subgraph between s and t, denoted as $ESG_{s,t}^{(1)}$.

in $SPG_{s,t}^{(2k+1)}$. Hence, (u, w) is an edge in $ESG_{s,t}^{(k)}$. Therefore, $SPG_{s,t}^{(2k+1)} \subseteq ESG_{s,t}^{(k)}$.

According to Corollary 1, we can conclude that the $(2k + 1)$th-order simple path graph between two nodes is a subset of the k-order enclosing subgraph. Naturally, a question arises "when these two types of subgraphs are equivalent?". To answer this question, we find an example in a complete graph.

Corollary 2. *For any pair of nodes s, t and any positive integer k on a complete graph $K_n (n \geq 4)$, we have $SPG_{s,t}^{(2k+1)} = ESG_{s,t}^{(k)}$.*

Proof. *Since all nodes in a complete graph K_n are connected pairwise, for any nodes s, t, their first-order neighbors are all nodes in the complete graph K_n, thus $ESG_{s,t}^{(k)} = ESG_{s,t}^{(1)} = K_n$. For any edge (u, v) in the complete graph K_n, if only one node of this edge coincides with s, t, suppose $u = s$, it is clear that (s, v, t) forms a simple path of length 2 in the complete graph K_n, so (u, v) is an edge in $SPG_{s,t}^{(3)}$; if u, v are distinct from s, t, it is evident that (s, u, v, t) forms a simple path of length 3 in the complete graph K_n, so (u, v) is an edge in $SPG_{s,t}^{(3)}$. Hence, $SPG_{s,t}^{(2k+1)} = SPG_{s,t}^{(3)} = K_n$. Consequently, $SPG_{s,t}^{(2k+1)} = ESG_{s,t}^{(k)}$.*

3 SPG4LP

Methods based on subgraphs often encounter the challenge of large subgraph sizes. However, as demonstrated by Corollary 1, compared to the enclosing subgraph, simple path graph exhibits smaller scales. In other words, simple path graph contains relatively fewer nodes and edges, whereas the enclosing subgraph tends to be larger. Furthermore, according to Corollary 1, given an undirected graph and a pair of target nodes, there exist more orders of simple path graphs compared to enclosing subgraphs. Simple path graph offers higher granularity within subgraphs. For instance, while traditional link prediction methods based on subgraphs typically favor using first or second-order enclosing subgraph, according to the conclusion of Corollary 1, we know that second and third-order simple path graphs are subgraphs of the first-order enclosing subgraph, and fourth and fifth-order simple path graph is a subgraph of the second-order enclosing subgraph. Consequently, when employing methods based on simple path graphs, there are four types of subgraphs to choose from. Moreover, with increasing orders, the scale of simple path graphs expands at a slower rate compared to enclosing subgraph. Enclosing subgraphs beyond the third order are often not considered, whereas simple path graphs can continue to increase in order. Thus, methods based on simple path graphs offer more than twice the number of subgraphs that can be selected compared to methods based on enclosing subgraphs, rendering them more flexible and versatile.

Additionally, methods based on enclosing subgraphs exhibit a notable drawback: susceptibility to the influence of nodes with high degrees (referred to as hub nodes in this paper). When hub nodes are present in the enclosing subgraph, neighborhoods of these hub nodes are typically introduced, leading to a rapid increase in the size of the enclosing subgraph, potentially including many irrelevant nodes for prediction. In contrast, link prediction methods based on simple path graphs are not influenced by hub nodes in the graph. Since simple path graphs focus more on the paths between target nodes, even if hub nodes are present in the simple path graph, they do not significantly affect the scale of the simple path graph.

A fundamental assumption of this paper is that link prediction tasks rely more on the connectivity between two target nodes, whereby the path information between target nodes determines whether two nodes are related. In the rule-based link prediction methods, except for Preferential Attachment (PA) [2], other methods focus on the path information between target nodes. Taking the simplest common neighbors (CN) as an example, the common neighbors between two nodes are precisely the nodes that connect them. However, the PA heuristic, which represents the product of the degrees of two nodes, does not directly reflect the association between two nodes but instead includes some unrelated nodes, introducing noise. In fact, the PA heuristic performs the worst among all heuristic rules, consistently ranking lowest across all datasets. However, PA still demonstrates effectiveness on certain datasets, indicating that it indirectly reflects the association between nodes. After all, nodes with higher degrees have more opportunities to connect with other nodes. According to our fundamental assumption, if the path information between target nodes can be extracted to identify how target nodes are connected, link prediction tasks can be better accomplished. Simple path graphs happen to fulfill this requirement. Compared to enclosing subgraphs, simple path graphs contain crucial information for completing link prediction tasks while being free from additional noise. Taking Figs. 1 and 2 as examples, a third-order simple path graph contains all simple paths between target nodes with a length less than or equal to 3, while excluding nodes that may interfere with prediction, which are included in the first-order enclosing subgraph. In fact, methods based on simple path graphs also outperform those based on enclosing subgraphs in experimental results. Figure 3 illustrates the framework of our method.

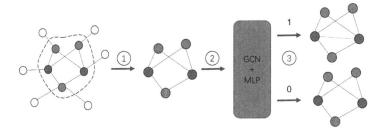

Fig. 3. Framework of link prediction method based on simple path graph. ①: Extraction of simple path graphs; ②: Integration of node information; ③: Model training and prediction

3.1 Key Steps

Our method based on simple path graphs primarily involves four key steps: generating positive and negative samples, generating simple path graphs, node labeling, and model training. These steps are detailed sequentially below.

(1) Firstly, in the task of link prediction, we need to prepare training data. To achieve this, we randomly generate positive and negative samples, where positive samples represent existing connections in the graph, while negative samples represent pairs of nodes without connections. Generally, we maintain a balanced ratio of positive to negative samples, typically 1:1, and divide the dataset into training and testing sets, commonly at a ratio of 9:1, for evaluating the model's performance during training.

(2) Secondly, to generate simple path graphs, we utilize the EVE-for-SPG [3]. For each pair of nodes in the training data (or each edge in positive samples), we generate a simple path graph. Thus, for each sample, we obtain a simple path graph, thereby constructing a collection of simple path graph data for the samples. After extracting simple path graphs, the edges between the two target nodes in positive samples, which are not present in actual prediction scenarios, need to be removed.

(3) Next, we need to label the nodes in the generated simple path graphs. Node labeling is performed to distinguish target node pairs from other nodes, assigning each node a specific label or feature based on its positional relationship with the target node pairs, facilitating subsequent understanding and processing by graph classification models. For comparability with SEAL, the same node labeling method as SEAL is employed in this paper.

(4) Finally, we employ the generated simple path graphs and node labels for model training. The model input includes the adjacency matrix A and feature matrix X of the graph, where for datasets without features for nodes, the feature matrix X is replaced by the identity matrix I. After the data passes through two layers of graph convolutional networks (GCN), each node is represented as a vector. Subsequently, we perform element-wise multiplication (Hadamard product) on the vectors representing the target node pairs

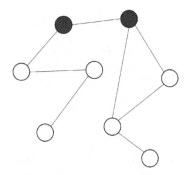

Fig. 4. When selecting the purple nodes as positive samples, there are no simple paths between the two nodes except for the directly connected edges.

and input the results into a multi-layer perceptron to complete the link prediction process. The model is trained by minimizing the cross-entropy loss function:

$$L = - \sum_{(s,t) \in D} \left(y_{(s,t)} \log(\hat{y}_{(s,t)}) + (1 - y_{(s,t)}) \log(1 - \hat{y}_{(s,t)}) \right) \quad (1)$$

where D is the sample set obtained after data processing in step one, $y_{(s,t)}$ represents the true connection between nodes s and t, and $\hat{y}_{(s,t)}$ represents the link prediction result between nodes s and t.

3.2 Dealing with Sparse Graphs

The density of a graph typically refers to the ratio between the actual number of edges in the graph and the total number of possible edges. For a graph with n nodes, the total number of possible edges is $\frac{n(n-1)}{2}$. Graphs with relatively low density are referred to as sparse graphs. Generally, sparse graphs also have low average node degrees. During the extraction of simple path graphs, some graph data may be relatively sparse, resulting in a lack of simple paths or only directly connected edges between pairs of target nodes, thus making it impossible to generate simple path graphs. To address this issue, this study adopts a hybrid strategy that combines simple path graphs and enclosing subgraphs.

This hybrid strategy mainly alters the subgraph extraction process while keeping other steps unchanged. During the subgraph extraction process, if there are no simple path graphs of length $k(2 \leq k \leq 10)$ between the pairs of target nodes, an enclosing subgraph of length $l(1 \leq l \leq 2)$ is extracted. Here, k and l are learnable hyperparameters, and by adjusting these parameters, the strategy can flexibly adapt to different datasets. This hybrid strategy introduces enclosing subgraphs as supplements, allowing for the provision of effective information to support link prediction tasks even in the absence of simple path graphs.

4 Experimental Results

Our study employs the same evaluation criteria as other link prediction methods, namely Area Under the Curve (AUC) of the Receiver Operating Characteristic (ROC) curve. To validate the effectiveness of the method based on simple path graph, we select several well-performing methods as baselines, including CN, PA, Adamic-Adar (AA), Node2Vec, Variational Graph Autoencoder (VGAE), and SEAL. These baselines encompass a variety of link prediction approaches. We select datasets from which both node feature-inclusive and node feature-exclusive datasets are derived. The dataset without node features consists of eight datasets: USAir, NS, PB, Yeast, C.ele, Power, Router, and E.coli. For datasets with node features, we employed three academic paper citation networks: Cora, Citeseer, and Pubmed.

To facilitate comparison with other methods, particularly SEAL, constraints on the order of subgraph were incorporated into the experimental design. Firstly, constraints were imposed on the order k of the simple path graph. When the performance using first-order enclosing subgraph on the dataset was superior, the order of simple path graph was ensured to be $k \leq 3$; when the performance using second-order enclosing subgraph was superior, the order was limited to $k \leq 5$. This was done because SEAL uses either first-order or second-order enclosing subgraph. According to the conclusion drawn from corollary 1, when $k \leq 3$ and $k \leq 5$, the kth-order simple path graph is a subgraph of the first-order and second-order enclosing subgraph respectively. Therefore, this study aimed to experimentally validate whether methods based on simple path graphs could match or even surpass SEAL in experimental performance when the scale of simple path graph theoretically is smaller than that of the enclosing subgraph.

Each link prediction method underwent 10 training iterations, and the average results of these 10 experiments were recorded. During the experiments, datasets PB, E.coli, and Pubmed exhibited a large number of edges. To address memory limitations, methods based on enclosing subgraph were restricted to using first-order enclosing subgraph only. Table 1 presents the experimental results on datasets without node features, while Table 2 presents the results on datasets with node features. In Tables 1 and 2, SPG4LP(kc) denotes the method based on simple path graph with restrictions on the order of simple path graph.

4.1 Limiting the Order of Simple Path Graph

From the experimental results, it can be observed that the performance of the method based on simple path graph is superior to those based on enclosing subgraph on datasets USAir, PB, and C.ele when there are no node features. Similarly, on datasets Cora and Citeseer with node features, the performance is also better, indicating competitive performance even though it may not outperform SEAL on other datasets. On datasets Power and Router, the experimental results for both methods are very close. These datasets are highly sparse, with average node degrees of only 2.67 and 2.49 respectively. As mentioned in Sect. 3, a hybrid strategy of enclosing subgraph and simple path graph is adopted when

Table 1. Comparison of AUC Values on Datasets without Node Features

Dataset	USAir	NS	PB	Yeast	C.ele	Power	Router	E.coli
CN	92.34	92.62	91.63	88.77	84.63	58.51	56.38	92.51
PA	87.68	69.91	90.14	81.70	74.29	43.85	45.56	90.32
AA	94.07	93.76	92.26	89.83	86.34	58.48	56.22	93.76
Node2Vec	91.29	90.88	84.75	93.07	83.01	74.72	64.16	89.52
LINE	83.55	82.38	75.84	86.85	68.72	53.83	65.35	80.18
VGAE	89.46	92.25	90.21	91.88	79.81	69.14	60.11	89.47
SEAL	94.73	95.34	93.72	95.71	88.28	84.52	91.28	94.53
SPG4LP(kc)	96.24	94.27	95.47	94.32	90.59	84.39	91.05	93.76
SPG4LP	96.24	97.18	96.64	96.89	90.59	86.53	91.05	95.97

dealing with sparse graphs. The datasets Power and Router fall into this situation. When k is small, there are few kth-order simple path graphs, and the subgraphs mainly consist of enclosing subgraphs. Hence, on these datasets, the results of both methods are remarkably similar.

Table 2. Comparison of AUC Values on Datasets with Node Features

Dataset	Cora	Citeseer	Pubmed
CN	62.42	58.69	52.63
PA	61.34	56.75	51.81
AA	69.25	64.67	56.54
Node2Vec	88.61	84.35	84.92
LINE	86.68	82.33	81.72
VGAE	90.13	88.72	92.24
SEAL	91.51	90.49	94.04
SPG4LP(kc)	92.43	91.58	92.89
SPG4LP	92.43	92.71	92.89

4.2 Loosening the Order of Simple Path Graph

As described in Sect. 3, our method offers a broader space for selecting subgraphs, and with the increase in the order of simple path graph, the expansion rate of the graph size is lower than that of enclosing subgraphs. Therefore, after relaxing the above limitation on the order of simple path graph, the size of simple path graph remains within a controllable experimental range. Moreover, due to the limitation on the order of simple path graph, the method based on simple path graphs did not achieve optimal performance. Thus, the experiment was extended to a larger range of k values. The last row in Tables 1 and 2, labeled SPG4LP, shows the experimental results under this circumstance.

It can be observed from the experimental results that after broadening the range of the order of simple path graph, the method based on simple path graph significantly outperforms the method based on enclosing subgraph, both on datasets with node features and without node features. While the performance of the method based on simple path graphs may be inferior to that of SEAL on some datasets when the order of simple path graph is strictly limited, it becomes noticeably superior once the range of the order of simple path graphs is expanded. Although the performance of the method based on simple path graph may slightly lag behind that of SEAL on datasets Router and Pubmed, the difference between the two methods is not significant. Therefore, considering the comprehensive experimental results, the method based on a simple path graph exhibits more advantages over SEAL after relaxing the limitation on the order of simple path graph.

However, after expanding the range of values for the order of simple path graph, does the size of simple path graph remain smaller than that of enclosing subgraph? The initial limitation on the range of k in the experiments was imposed to ensure that simple path graph is a subset of enclosing subgraph theoretically. Therefore, naturally, there arises the question of whether the size of simple path graph would remain smaller than that of enclosing subgraph after loosening this limitation. If indeed there is such a problem, then the advantage of the method based on simple path graph in terms of subgraph size will be greatly reduced.

4.3 Comparing the Sizes of Simple Path Graphs and Enclosing Subgraphs

To investigate the problem raised in the previous section, a detailed comparison and analysis of the sizes of simple path graphs and enclosing subgraph were conducted. In the experiments, the sizes of the subgraphs used by two subgraph-based methods were fully recorded. The size of a subgraph can be characterized by the number of nodes and edges it contains. Although a large number of subgraphs can be extracted from a dataset, the number of subgraphs used in the experiments is the same for both methods. Hence, the average number of nodes and edges of these subgraphs is used to measure the size of the subgraphs.

Moreover, both subgraph-based methods use subgraphs of different orders, and the performance varies with different orders of subgraphs. However, it is not necessary to record the sizes of subgraphs at each order, as the focus of the study is on comparing the sizes of subgraphs when the methods exhibit the best performance. Therefore, in the experiments, the sizes of the subgraphs used when the two methods performed the best were recorded in Table 3. Listing the sizes of subgraphs for all datasets would take up a significant amount of space. Thus, Table 3 provides the data for five datasets.

Table 3. Average Number of Nodes and Edges in Subgraphs

Dataset	Subgraph	ANN	ANE
USAir	2nd-order ESG	89.45	523.74
	4th-order SPG	26.28	157.81
NS	2nd-order ESG	20.53	42.24
	6th-order SPG	17.49	37.92
PB	1st-order ESG	51.16	697.00
	4th-order SPG	60.43	619.78
Yeast	2nd-order ESG	127.58	1465.33
	6th-order SPG	93.21	742.03
Cora	2nd-order ESG	24.52	63.45
	3rd-order SPG	5.74	11.68

Observing Table 3, it can be seen that the average number of nodes in 4th-order simple path graphs on dataset PB is slightly higher than that in 1st-order enclosing subgraphs, while the average number of edges is still higher for 1st-order enclosing subgraphs. Except for dataset PB, on other datasets, both the average number of nodes and edges in simple path graphs are lower than those in enclosing subgraphs. On datasets USAir and Cora, the simple path graphs used in the experiments are theoretically subsets of enclosing subgraphs, thus naturally resulting in smaller subgraph sizes for simple path graphs. However, on datasets NS and PB, at the point where the performance of the experiments was optimal, the simple path graphs used were no longer subsets of enclosing subgraphs, yet the sizes of simple path graphs remained smaller than those of enclosing subgraphs. In fact, except for dataset PB, simple path graphs have smaller sizes than enclosing subgraphs on other datasets.

Enclosing subgraph primarily focuses on the neighborhood information of target nodes, thus the size of enclosing subgraph is greatly influenced by hub nodes in the graph data. With the increase in the order of enclosing subgraph, the likelihood of including hub nodes also increases, resulting in a greater influence from hub nodes. In contrast, simple path graph pays more attention to the path information between target nodes, and is less affected by hub nodes. Even when simple path graph is no longer a subset of enclosing subgraph after relaxing the range of the order, the size of simple path graph remains relatively smaller. Moreover, the experimental results under this circumstance, where the method based on simple path graph performs better than that based on enclosing subgraph, further support this observation.

4.4 Relationship Between Experimental Performance and Order of Simple Path Graphs

Compared to methods based on enclosing subgraph, enclosing subgraph orders of 1 and 2 are typically chosen, while methods based on simple path graph have a wider range of choices for subgraphs. In the experiment, the order of simple path graph generally ranges from integers 2 to 7, unless the dataset is large. Within this range of orders, does the experimental performance improve or deteriorate with the increase in order? Or are there any other regularities? If there is a certain correlation between experimental performance and the order of simple path graph, then it would be possible to avoid conducting experiments on different orders of simple path graph one by one during training, thereby saving a considerable amount of training time and improving training efficiency.

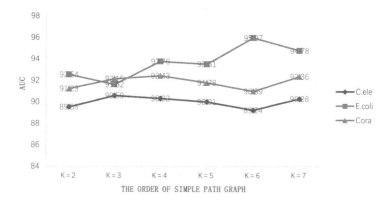

Fig. 5. Variation of AUC with the Order of Simple Path Graphs

Figure 4 presents a line graph showing how AUC varies with the order of simple path graph on datasets C.ele, E.coli, and Cora when the method based on simple path graphs is adopted. The variation trends on other datasets are similar to those shown here. In datasets C.ele, E.coli, and Cora, the best performance is achieved when using 3rd-, 6th-, and 4th-order simple path graph, respectively. However, according to the line graph in Fig. 5, there is no apparent correlation between the experimental results of the method based on simple path graph and the order of the simple path graph used. One notable point is that in all datasets used in this study, the best-performing simple path graphs are neither 2nd nor 7th order. Perhaps because the use of too small or too large subgraph sizes is detrimental to link prediction based on subgraphs. As the experimental results do not exhibit a clear correlation with the order of simple path graph, it is necessary to conduct experiments on different orders of simple path graph individually to determine the best model performance during the actual training.

4.5 Scalability of Methods Based on Simple Path Graphs

Finally, to evaluate the scalability of the method based on simple path graph and compare it with SEAL, we compared the inference times of the two link prediction methods on datasets with and without node features. A single GPU

Table 4. Inference Times on Datasets Without Node Features

	USAir	NS	PB	Yeast	C.ele	Power	Router	E.coli
SEAL(s)	28	317	142	1089	15	1593	26	324
SPG4LP(s)	19	287	98	935	10	1423	18	301

was used for the experiments, and the inference time was measured in seconds (s). The relevant results are detailed in Tables 4 and 5.

From the data in the tables, it can be observed that on datasets with different types, the method based on simple path graph generally has an advantage in inference time. There is a significant difference in inference time between different datasets, mainly due to differences in dataset size. For example, the inference time on the largest dataset Pubmed exceeds one hour. However, regardless of dataset size, the inference time for the method based on simple path graph is reduced. Nevertheless, considering that the inference time is shorter when the dataset size is smaller, the gap in inference time should widen with the increase in dataset size. However, as observed from the results in Table 5, the difference in inference time is not so obvious.

Table 5. Inference Times on Datasets With Node Features

	Cora	Citeseer	Pubmed
SEAL	862	683	5798
SPG4LP	713	653	5187

Upon detailed examination of the statistics in the tables, it can be noted that on datasets Yeast and E.coli, the inference time for methods based on simple path graph is nearly halved. In fact, compared to datasets Yeast and E.coli, Cora, Citeseer, Pubmed, and Power have lower graph densities, making these datasets sparser. As mentioned in Sect. 3, enclosing subgraph were used instead of simple path graph on relatively sparse datasets. Therefore, for these datasets, the difference in inference time between the two methods is not very significant.

5 Conclusion

The subgraph-based link prediction method plays a significant and effective role in link prediction research. In previous studies, such methods typically utilized enclosing subgraph of target node pairs as subgraphs. Enclosing subgraph focuses on the neighborhood of target nodes and their connection relationships. Compared to other link prediction methods, methods based on enclosing subgraph face challenges of large subgraph sizes and susceptibility to the influence of hub nodes in graph data. This paper proposes a method for link prediction using simple path graph between target nodes. Firstly, it is theoretically proven that under certain order limitations, simple path graph is a subset of enclosing subgraph. Therefore, simple path graph has smaller sizes compared to enclosing subgraph. Additionally, simple path graph provides a finer partition of subgraphs, offering more choices for subgraphs in subgraph-based methods. Secondly, after relaxing the range of the order of simple path graph, experimental results validate that the subgraph sizes used in the method based on simple path graph are still smaller than those used in the method based on enclosing subgraph, and also demonstrate superior experimental performance. Lastly, this paper compares the inference times of the two methods to evaluate the scalability of the methods. Experimental results also demonstrate the advantage of the method based on simple path graph in inference time.

However, when dealing with sparse graphs, the method proposed in this paper employed a simple hybrid strategy, resulting in a negligible difference in performance compared to the method based on enclosing subgraph on sparse graphs. Future research should further investigate how to adopt better strategies to address issues on sparse graphs. Additionally, after extracting simple path graphs, the subsequent neural network models used GCN and multilayer perceptron, which are relatively simple. Future efforts should explore more network models and design more effective network architectures to fully leverage the performance of the method based on simple path graphs for link prediction.

References

1. Adamic, L.A., Adar, E.: Friends and neighbors on the web. Soc. Netw. **25**(3), 211–230 (2003)
2. Barabási, A.L., Albert, R.: Emergence of scaling in random networks. Science **286**(5439), 509–512 (1999)
3. Cai, Y., Liu, S., Zheng, W., Lin, X.: Towards generating hop-constrained ST simple path graphs. Proc. ACM Manage. Data **1**(1), 1–26 (2023)
4. Chamberlain, B.P., et al.: Graph neural networks for link prediction with subgraph sketching. In: The Eleventh International Conference on Learning Representations (2022)
5. Grover, A., Leskovec, J.: node2vec: scalable feature learning for networks. In: Proceedings of the 22nd ACM SIGKDD International Conference on Knowledge Discovery and Data Mining, pp. 855–864 (2016)
6. Guimerà, R., Sales-Pardo, M.: Missing and spurious interactions and the reconstruction of complex networks. Proc. Natl. Acad. Sci. **106**(52), 22073–22078 (2009)
7. Hamilton, W., Ying, Z., Leskovec, J.: Inductive representation learning on large graphs. In: Advances in Neural Information Processing Systems, vol. 30 (2017)
8. Katz, L.: A new status index derived from sociometric analysis. Psychometrika **18**(1), 39–43 (1953)
9. Kipf, T.N., Welling, M.: Semi-supervised classification with graph convolutional networks. arXiv preprint arXiv:1609.02907 (2016)
10. Kipf, T.N., Welling, M.: Variational graph auto-encoders. arXiv preprint arXiv:1611.07308 (2016)
11. Newman, M.E.: Clustering and preferential attachment in growing networks. Phys. Rev. E **64**(2), 025102 (2001)

12. Sun, Z., Deng, Z.H., Nie, J.Y., Tang, J.: RotatE: knowledge graph embedding by relational rotation in complex space. arXiv preprint arXiv:1902.10197 (2019)
13. Tang, J., Qu, M., Wang, M., Zhang, M., Yan, J., Mei, Q.: LINE: large-scale information network embedding. In: Proceedings of the 24th International Conference on World Wide Web, pp. 1067–1077 (2015)
14. Velickovic, P., et al.: Graph attention networks. Stat **1050**(20), 10–48550 (2017)
15. Wang, C., Satuluri, V., Parthasarathy, S.: Local probabilistic models for link prediction. In: Seventh IEEE International Conference on Data Mining (ICDM 2007), pp. 322–331. IEEE (2007)
16. Xian, X., et al.: Generative graph neural networks for link prediction. arXiv preprint arXiv:2301.00169 (2022)
17. Zhang, M., Chen, Y.: Weisfeiler-Lehman neural machine for link prediction. In: Proceedings of the 23rd ACM SIGKDD International Conference on Knowledge Discovery and Data Mining, pp. 575–583 (2017)
18. Zhang, M., Chen, Y.: Link prediction based on graph neural networks. In: Advances in Neural Information Processing Systems, vol. 31 (2018)

Construction of EMU Fault Knowledge Graph Based on Large Language Model

Ziwei Han[1,2], Hui Wang[2], Yaxin Li[3(✉)], and Fangzhou Xu[3]

[1] Postgraduate Department, China Academy of Railway Sciences, Beiing 100081, China
[2] Institute of Computing Technology, China Academy of Railway Sciences Corporation Limited, Beijing 100081, China
[3] Tianjin University, Tianjin 300072, China
liyaxin2000@tju.edu.cn

Abstract. During the operation of the EMU, a large number of fault data will be generated, which record the fault overview, fault cause, treatment measures and other related fault information. Construction of knowledge graph of these data can provide knowledge for relevant staff in the maintenance process and assist them in decision-making, which is of great practical significance. Most of the existing EMU fault knowledge graph construction methods adopt supervised named entity recognition algorithms and relation extraction algorithms, which require a large amount of labeled data for training. To solve the above problems, a method for constructing EMU fault knowledge graph based on large language model is proposed. Firstly, the EMU fault ontology model is constructed based on the existing EMU fault data and drawing on the existing fault knowledge graph. Then the EMU fault triple is extracted by means of small samples using the in-context learning capability of the large language model. Finally, entity linking is performed based on the constructed fault ontology model to construct the EMU fault knowledge graph, and the graph is displayed in a visual way to guide the maintenance and diagnosis of EMU faults.

Keywords: large language model · knowledge graph · EMU fault · ontology model

1 Introduction

After more than ten years of development, China's high-speed railway EMU ranks first in the world in terms of speed, total operating mileage and scale. In the process of the operation of the EMU, some faults of the EMU will inevitably occur. Most of these fault data are stored as semi-structured and unstructured data, which makes it difficult to analyze and utilize these fault data effectively, and does not fully play the substantive role of the data.

Knowledge graph describes entities and inter-entity relations in the objective world in the form of triples, and this technological advantage is exactly in line with the demand for fault data analysis and mining in the railway safety field. Through named entity recognition, relation extraction and other technologies, the semi-structured and unstructured fault data generated in the process of EMU operation are processed, the knowledge contained therein is extracted, and the scattered EMU fault data are standardized and normalized, so that the EMU fault data can be better managed and utilised. In the process of EMU train fault detection and maintenance, it provides corresponding fault knowledge for the staff, helps the railway staff to quickly and comprehensively grasp the key information of the relevant faults, and assists them in decision-making.

Most of the current Knowledge Graph Construction (KGC) in railway safety field adopts deep learning methods to identify entities from unstructured railway fault data and perform relation extraction, and then construct a knowledge graph. Han et al. [1] explore the hidden information between railway electrical accidents, use the Bi-LSTM+CRF model for entity recognition of accident time, train, location and type content, and construct railway electrical knowledge graph by combining railway electrical accident ontology. Li [2] defines the named entity and entity relation knowledge structure of fault diagnosis and processing based on the text data of equipment faults, and proposes the named entity extraction method of multi-dimensional character feature representation + Bi-LSTM + CRF, as well as the entity relation extraction method of multi-dimensional participle feature representation + Transformer. Transformer's entity-relation extraction method, which realises the automatic extraction of key information in the textual data of equipment faults, and then constructs the knowledge graph of high-speed railway faults. Zuo [3] constructed a BERT+Bi-LSTM+CRF model for named entity recognition of railway fault text data, used a matching and rule-based method for relation extraction of railway fault knowledge entities, and constructed a knowledge graph. Dong [4], based on the fault data of railway safety accidents, proposed a railway safety entity recognition based on the ELMO-Bi-LSTM+CRF model and an entity relation extraction method incorporating Transformer's multi-head self-attention mechanism to construct a knowledge graph for intelligent high-speed railway safety and security.

Most of the above knowledge graph construction methods use deep learning. To improve the effectiveness of named entity recognition model and relationship extraction model, a large amount of well-labelled railway fault data is needed to train the models. However, most of the existing railway fault data are unlabelled, and labelling them is time-consuming and laborious, which greatly affects the construction of high-speed railway fault knowledge graph.

Recently, Large Language Models (LLMs) have achieved amazing performance on many Natural Language Processing (NLP) tasks, without the need to train or fine-tune the models, through contextual learning, and with zero and few samples, LLMs have excellent Information Extraction (IE) capabilities [5]. LLMs can be used in the main stages of knowledge graph construction such as entity identification, relation extraction, and by constructing appropriate prompts, LLMs can effectively process structured and unstructured texts and convert them into triples. To this end, this paper proposes a knowledge graph construction method based on a large model for high speed train faults, and the contributions of this paper are as follows:

1. Based on the existing EMU fault data, drawing on various existing railway knowledge graphs for ontology construction, the EMU ontology is supplemented to form a perfect conceptual ontology of the EMU fault domain. 2. According to the fault data, the fault ontology is used to construct prompts, and the LLM is used to extract triples from the text of EMU faults. 3. EMU fault knowledge graph construction based on the constructed EMU fault domain ontology.

2 Related Work

2.1 Knowledge Graph Construction

Knowledge graph construction typically employs named entity recognition [6], relation extraction [7] and entity parsing [8] techniques in order to transform unstructured text into a structured representation containing entities, relations between entities and related attributes. The approach of [9] builds a financial news knowledge graph by combining various techniques such as denotational disambiguation, named entity recognition and semantic role labelling. Luan et al. [10] developed a multi-task model for identifying entities, relations and co-referential mentions in scientific articles that can support the creation of a scientific knowledge graph. Mehta et al. [11] proposed an end-to-end knowledge graph construction system and a new deep learning-based predicate mapping model. Their system identifies and extracts entities and relations from the first text and then maps them to DBpedia namespaces. While these approaches can achieve satisfactory results and produce high-quality knowledge graphs, almost all of them utilize supervised methods and require extensive manual annotation.

To address this problem, some studies introduced Pre-Trained Language Models (PLMs) to explore the use of knowledge stored in pre-trained models to construct knowledge graphs. Wang et al. [12] devised an unsupervised method called MAMA, which works without any fine-tuning on corpora using PLMs to construct knowledge graphs. Hao et al. [13] proposed a new approach to extract knowledge graphs from any PLMs using minimal user input. By inputting a minimal definition of relations and some example entity pairs as initial hints, a new set of hints expressing the target relations in different ways is automatically generated based on the initial hints. The prompts are then weighted using confidence scores and PLM is used to search for a large number of candidate entity pairs, the candidates are ranked and the highest scoring entity pair is used as the output knowledge.

2.2 LLM For Knowledge Graph Construction

With the increase in data availability and the advancement of technology, LLMs have evolved rapidly and performed well on many NLP tasks. Experiments by

Agrawal et al. [5] and Wei et al. [14] show that new LLMs such as GPT-3.5 show significant information extraction ability with zero and small samples. Wan et al. [15] improve the relation extraction capability by retrieving examples relevant to the problem as small-sample examples to bridge the gap between LLMs and supervised models. Ashok and Lipton [16] proposed a method PromptNER for small-sample named entity recognition using LLMs. By inputting entity definitions and a small number of standard examples, PromptNER prompts the LLM to generate a list of candidate entities and the corresponding explanation that proves that it belongs to a certain entity type. Trajanoska et al. [17] proposed a method for automatic creation of knowledge graphs from raw text. This method first uses ChatGPT to jointly extract the entities and relations and then performs entity linking. And experiments were conducted using texts related to sustainable development as a use case.

3 Approach

Inspired by the literature [18], we use a large language model for the construction of the knowledge graph of EMU faults. The text of EMU faults contains a large number of concepts and is highly specialized. In order to standardize the extracted entities as well as relations, we first construct an ontology of EMU faults based on the existing EMU fault data and drawing on the existing railway knowledge graph. Then use the constructed ontology model to assist the LLM in the named entity recognition and relation extraction, and then construct the knowledge graph of EMU faults.

3.1 Task Definition

The goal of the knowledge graph construction task is to extract the knowledge contained in the text and represent it as a graph. Given a collection of texts $S = \{S_1, S_2, S_3, ..., S_n\}$, it is converted into a graph $G = \{(s_1, r_1, o_1), (s_2, r_2, o_2), (s_3, r_3, o_3), ..., (s_i, r_j, o_i) | s_i, o_i \in E, r_i \in R\}$, where E and R are the set of entities and the set of relations, respectively.

3.2 Ontology Construction

This paper constrains EMU fault entities, attributes and relationships by introducing EMU fault ontology. While standardising EMU knowledge, it establishes the upper layer data schema foundation for subsequent EMU fault entity identification and relationship extraction. In order to combine the EMU fault full-domain knowledge ontology with the downstream EMU fault entity identification and relationship extraction, a bottom-up ontology design and hierarchical construction of the five major aspects of EMU faults, namely, subject, discovery, processing, analysis and resolution, are carried out from the fault data itself. Firstly, the key concepts therein are extracted and attributes and features are defined, and then the ontology structure is designed to define the relationships between classes. Finally, drawing on the railway knowledge graph designed in the literature [2, 3, 4] for ontology fusion, the EMU ontology is supplemented, and finally a perfect conceptual ontology of the EMU fault domain is formed (Fig. 1).

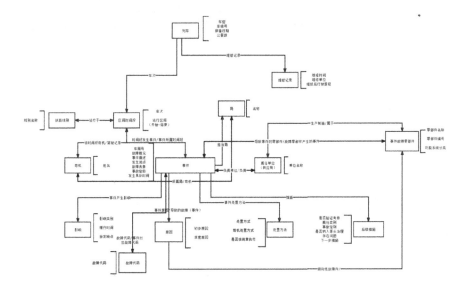

Fig. 1. EMU fault ontology model

3.3 Knowledge Graph Construction

After the construction of the ontology of the EMU fault domain is completed, the LLM is used for the construction of the knowledge graph of EMU faults. Since the EMU fault data is semi-structured data, not all fields need to be subjected to entity extraction and relation extraction, so it is necessary to firstly screen the fields; and then, according to the conceptual ontology of the EMU fault domain, the text of the EMU faults is subjected to the joint extraction of entities and relations, so as to get the triples of EMU faults. In the entity identification and relation extraction stage, the LLM we use is BaiChuan-13B-chat.

Field Filtering. The original EMU failure data has a complete semi-structured failure maintenance log for each failure, which contains several fields, such as failure overview, symptom, disposal method, disposal results, type of impact, reason, culpability, accident level, functional system classification, faulty components, manufacturing, and the last advanced maintenance. By analyzing the data, the fields of failure overview, reason, manufacturing, and last advanced maintenance are all described using short text, which contains multiple entities and relations, and the text of these fields needs to be subjected to entity recognition and relation extraction.

Table 1. Example data of EMU faults.

field name	value
Fault overview	At X:XX on XXXX-XX-XX, the driver of GXXX train, operated by XX Railway Bureau, departing from Station A to Station B on the GXXX high-speed rail line (CR400AF-B-2110 trainset, XX EMU Depot, driver: XX), reported to the control center that while the train was passing between XX North Station and XX South Station on the up line, the main circuit breaker of the 11th car failed to close (Fault code: XXXX) after the disconnection of units 734 and 732 of the XX high-speed train. The on-board technician informed the driver to manually close the main circuit breaker. Normal operation was restored at 18:29, with a delay of 5 min
Symptom	The main circuit breaker failed to close
Disposal method	Non-stop - Non-stop handling
Disposal results	no effect
Type of impact	unloaded
Reason	Upon inspection, the cause of this incident was an abnormal restart of the SMC board in the TCU cabinet of the 15th car, resulting in a hang-up of the MVB vital signal and reporting traction converter communication failure. Due to the traction converter failure in the 15th car, the main circuit breaker of the 11th car in this unit failed to close after phase separation. After replacing the TCU cabinet in the 15th car, the power supply test was successful
Culpability	Vehicle liability
Accident level	non-accident
Functional System Classification	Traction
Faulty components	TCU
Manufacturing	The XX was newly manufactured in XXXX by XX. It has accumulated a total mileage of XX thousand kilometers
Last advanced maintenance	The XX underwent its Level 5 maintenance at X in XXXX. Following the major overhaul, it has accumulated a total mileage of XX thousand kilometers

Triple Extraction. In order to construct the knowledge graph of EMU faults, it is necessary to extract the fault triples of faults in the fault text, as shown in the fault data in Table 1, the fields that need to be extracted from the triples include the four fields of fault overview, reason, manufacturing, and last advanced maintenance, which contain different types of entities and relations. Therefore, it is necessary to design prompts according to the contents of the fields and the constructed fault ontology to assist the LLM in fault entity recognition. In order to give full play to the context learning capability of the LLM [20], the designed prompt contains the following parts in addition to the input text:

Task Description: describes the operation to be performed to the LLM in the form of instruction, i.e., extracting the triples present in the given text.

List of Candidate Entity Types and Relations: According to the constructed ontology model of EMU fault, the types of entities to be extracted include event, time, name of the authority, name of the railway line, operating interval, train number, model, driver name, cause of the failure, slowdown time, maintenance time, etc. The relations to be extracted include time of occurrence, impact of the event, cause of the event, running on, bureau, train number, etc.

Demonstration Sample: In order to take advantage of the contextual learning capability of the LLM, the prompt is enriched by providing small sample data examples to guide the LLM to generate more accurate triples.

Output Format: In order to standardise the output of the LLM, the output format is defined as (Entity1, relation, Entity2) or (Entity, Attribute, Attribute Value).

Entity Linking. After the above steps, the triples are extracted from the unstructured fault text, and these triples can not constitute the final knowledge graph, because our fault data is semi-structured data, in addition to the extracted fault triples, there are also entities and attributes in the fault data itself, so it is necessary to carry out entity linking, and to add the entities, relations, and attributes into the knowledge base according to the constructed ontology model of the faults of the moving train set. In turn, the knowledge graph of EMU faults is constructed.

Each fault data records the specifics of a moving train fault event, so in this paper, entity linking is centered on the fault event ontology in the ontology model, firstly, entity nodes are constructed based on (Entity, Attribute, Attribute Value) as well as Entities and Attributes in the fault data, and then the entities are linked based on (Entity1, relation, Entity2) as well as predefined relations in the ontology model.

4 Experiment

4.1 Evaluation Metrics

In order to assess the quality of the extracted triples from the large model, we manually evaluate the extracted triples by labeling the triples with correct relation types and entities and evaluating the extracted triples using the precision (P), recall (R) and F_1. The formula is as follows:

$$Recall = \frac{TP}{TP+FN} \tag{1}$$

$$Precision = \frac{TP}{TP+FP} \tag{2}$$

$$F1-score = \frac{2*Precision*Recall}{Precision+Recall} \tag{3}$$

4.2 Experimental Environment and Setup

The experiments in this paper were run on a server Ubuntu system with an Intel(R) Xeon(R) Gold 5218R @ 2.10 GHz processor and an NVIDIA Quadro RTX 6000 graphics card. The macromodel used was Baichuan-13B-Chat, and INT8 was used for inference quantification.

4.3 Results of Triple Extraction Experiments

In this paper, experiments were conducted using 100 pieces of EMU fault data, and 790 entities and 1418 relations were obtained by extracting triples from the LLM and linking entities according to the ontology model. And the extraction results were evaluated manually. Table 2 shows the experimental results of the LLM in extracting triples on the four fields of fault overview, fault cause, new construction, and the Last advanced maintenance.

Table 2. Experimental results.

file name	Recall (%)	Precision (%)	F1-score (%)
Fault overview	82.6	64.0	72.1
Reason	63.0	68.0	65.6
Manufacturing	96.8	100	98.4
Last advanced maintenance	100	99	99.3
average	85.57	82.26	83.76

According to the experimental results in Table 2, the extraction results of the manufacturing and the last advanced maintenance are the best, with F_1 of 98.4% and 99.3%, respectively, and the extraction results of the fault overview and the reason are relatively poor, with F_1 of 72.1% and 65.6%, respectively. Combined with the data examples analysed in Table 1, the texts of manufacturing and last advanced maintenance are relatively simple and clearly structured. In the case of manufacturing, for example, there are only three types of entities contained therein, i.e., the date of new construction, the manufacturing company and the mileage, and the extraction effect is better. While the text of fault overview contains more types of entities and relationships, the text of reason, although it only contains the initial cause, the depth of the cause and the fault code, it needs to summarise the descriptions of the relevant causes and make clear the hierarchical relationships between the causes, and the extraction effect is relatively poor.

4.4 Ablation Experiments

Table 3. Ablation experimental results.

method	Recall (%)	Precision (%)	F1-score (%)
w/o demonstration sample	43.8	18.9	26.4
w/o List of Candidate Entity Types and relations	72.9	55.8	63.2
complete Prompt	82.6	64.0	72.1

In order to explore the effect of demonstration sample and candidate entity types and relations on the extraction results in the constructed prompt, this paper conducts ablation experiments on the data in the field of fault overview, and the experimental results are shown in Table. Comparing the experimental results, it can be found that adding demonstration sample and candidate entity types and relations in prompt can improve the extraction effect of the large model, F_1 by 45.7% and 8.9% respectively. The improvement of model extraction effect by adding demonstration sample in prompt is especially obvious, which indicates that by providing demonstration sample in prompt, the context learning ability of the LLM can be effectively utilized, which in turn improves its triple extraction effect (Table 3).

4.5 Display and Application of Knowledge Graph of EMU Faults

In this paper, we use 100 pieces of EMU fault data for experiments, and after triple extraction as well as entity linking, a total of 790 entities and 1418 relations are obtained out, and neo4j is used to store and display the data. Figure 2 shows some of the entities and relations of the graph. Figure 3 demonstrates the relevant information of a fault event, and it can be seen through the search that the initial cause of the D2701 fault is the automatic excision of the traction transformer, and the depth cause is the failure of the unipolar high-voltage contactor, and the faulty parts are the high-voltage contactor and so on.

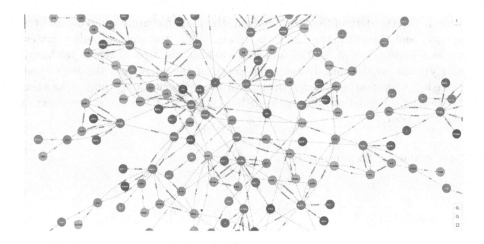

Fig. 2. EMU fault ontology model

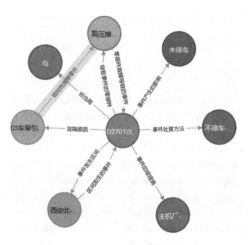

Fig. 3. EMU fault ontology model

The fault knowledge graph constructed can be used for fault cause tracing and faulty parts judgement. Take Fig. 4 as an example, when there are phenomena such as gateway failure, main disconnection automatic disconnection, and transformer excision, the parts that lead to these faults are the main control unit, and the fault causes can be traced according to different fault phenomena, such as transformer excision is caused by the faults of a certain board card.

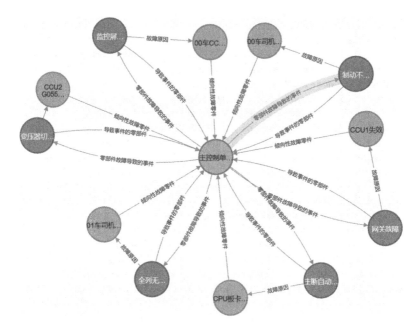

Fig. 4. EMU fault ontology model

5 Conclusion

Aiming at the problem of low structured and insufficient annotation data of EMU fault data, this paper proposes a method of constructing EMU fault knowledge graph based on LLM. Firstly, we draw on the existing railway fault knowledge graph and combine the existing EMU fault data to carry out the EMU fault ontology model. Then we use the LLM to extract the triples of unstructured fault text. Finally, according to the constructed ontology model, we carry out entity linking on the existing structured data and the extracted fault triples to form the EMU fault knowledge graph, which can provide data technical support for the subsequent research on the maintenance and querying of EMU faults.

Acknowledgement. This work is supported by the Science and Technology R&D Program of China Railway (No: N2023J044).

References

1. Han, J.: Construction and application of ontology framework for railroad electrical accidents based on knowledge graph. Master's thesis, Hebei University of Science and Technology (2020)
2. Li, X.: Research on big data analysis method of high-speed railroad safety text. Ph.D. thesis, China Academy of Railway Science (2020)

3. Zuo, J.: Research on knowledge graph modeling method based on railroad fault text data. Master's thesis, Beijing Jiaotong University (2022)
4. Dong, X.: Research on key technology of knowledge graph construction and application for intelligent high-speed rail safety and security. Ph.D. thesis, China Academy of Railway Science (2022)
5. Agrawal, M., Hegselmann, S., Lang, H., Kim, Y., Sontag, D.: Large language models are few-shot clinical information extractors. In: Goldberg, Y., Kozareva, Z., Zhang, Y. (eds.) Proceedings of the 2022 Conference on Empirical Methods in Natural Language Processing, Abu Dhabi, United Arab Emirates, pp. 1998–2022. Association for Computational Linguistics (2022). https://doi.org/10.18653/v1/2022.emnlp-main.130, https://aclanthology.org/2022.emnlp-main.130
6. Chiu, J.P., Nichols, E.: Named entity recognition with bidirectional LSTM-CNNs. Trans. Assoc. Comput. Linguist. 4, 357–370 (2016). https://doi.org/10.1162/tacl_a_00104, https://aclanthology.org/Q16-1026
7. Nguyen, T.H., Grishman, R.: Relation extraction: perspective from convolutional neural networks. In: Blunsom, P., Cohen, S., Dhillon, P., Liang, P. (eds.) Proceedings of the 1st Workshop on Vector Space Modeling for Natural Language Processing, pp. 39–48. Association for Computational Linguistics, Denver (2015). https://doi.org/10.3115/v1/W15-1506, https://aclanthology.org/W15-1506
8. Ebraheem, M., Thirumuruganathan, S., Joty, S., Ouzzani, M., Tang, N.: Distributed representations of tuples for entity resolution. Proc. VLDB Endow. 11(11), 1454–1467 (2018). https://doi.org/10.14778/3236187.3236198
9. Elhammadi, S., et al.: A high precision pipeline for financial knowledge graph construction. In: Scott, D., Bel, N., Zong, C. (eds.) Proceedings of the 28th International Conference on Computational Linguistics, Barcelona, Spain, pp. 967–977. International Committee on Computational Linguistics (2020). https://doi.org/10.18653/v1/2020.coling-main.84, https://aclanthology.org/2020.coling-main.84
10. Luan, Y., He, L., Ostendorf, M., Hajishirzi, H.: Multi-task identification of entities, relations, and coreference for scientific knowledge graph construction. In: Riloff, E., Chiang, D., Hockenmaier, J., Tsujii, J. (eds.) Proceedings of the 2018 Conference on Empirical Methods in Natural Language Processing, Brussels, Belgium, pp. 3219–3232. Association for Computational Linguistics (2018). https://doi.org/10.18653/v1/D18-1360, https://aclanthology.org/D18-1360
11. Mehta, A., Singhal, A., Karlapalem, K.: Scalable knowledge graph construction over text using deep learning based predicate mapping. In: Companion Proceedings of The 2019 World Wide Web Conference, WWW 2019, New York, NY, USA, pp. 705–713. Association for Computing Machinery (2019). https://doi.org/10.1145/3308560.3317708
12. Wang, C., Liu, X., Song, D.: Language models are open knowledge graphs. CoRR abs/2010.11967 (2020). https://arxiv.org/abs/2010.11967
13. Hao, S., et al.: BertNet: harvesting knowledge graphs with arbitrary relations from pretrained language models. In: Rogers, A., Boyd-Graber, J., Okazaki, N. (eds.) Findings of the Association for Computational Linguistics: ACL 2023, Toronto, Canada, pp. 5000–5015. Association for Computational Linguistics (2023). https://doi.org/10.18653/v1/2023.findings-acl.309, https://aclanthology.org/2023.findings-acl.309
14. Wei, X., et al.: Zero-shot information extraction via chatting with ChatGPT (2023)

15. Wan, Z., et al.: GPT-RE: in-context learning for relation extraction using large language models. In: Bouamor, H., Pino, J., Bali, K. (eds.) Proceedings of the 2023 Conference on Empirical Methods in Natural Language Processing, pp. 3534–3547. Association for Computational Linguistics, Singapore (2023). https://doi.org/10.18653/v1/2023.emnlp-main.214, https://aclanthology.org/2023.emnlp-main.214
16. Ashok, D., Lipton, Z.C.: PromptNER: prompting for named entity recognition (2023)
17. Trajanoska, M., Stojanov, R., Trajanov, D.: Enhancing knowledge graph construction using large language models. arXiv preprint arXiv:2305.04676 (2023)

Author Index

A
Amagasa, Toshiyuki 147
Ashizawa, Arisa 252

B
Bang, L. K. 77

C
Cai, Yuzheng 319
Cao, Yang 178
Chen, Xu 27

F
Feng, Hongwei 319
Fujii, Marino 293

H
Han, Ziwei 335
Hirano, Tsukasa 211
Huang, Jintao 58

I
Ishikawa, Yoshiharu 236

J
Jiang, Yuqing 110

K
Kawada, Yuka 228, 293
Khalique, Vijdan 147
Kim, Jinhyuk 302
Kitagawa, Hiroyuki 147
Kitamura, Tatsuya 187
Kiyomitsu, Hidenari 203, 277
Koh, Jia-Ling 163
Kuge, Takayuki 220

L
Lai, Ruyi 27
Lai, Wei 163

Lei, Shuya 110
Li, Jiyi 178
Li, Sheng 178
Li, Waner 27
Li, Yaxin 335
Li, Yuchen 42
Li, Yuehua 129
Li, Zhiren 319
Liang, Zibo 27
Lu, Kejing 236

M
Mibayashi, Ryota 252

N
Nadamoto, Akiyo 187
Nakai, Kanako 228
Nakayama, Yuki 269
Nguyen, Thanh Ha 203
Ni, Tianjia 236
Nishida, Takeshi 203, 277

O
Oda, Naoya 302
Ohara, Kouzou 211
Ohshima, Hiroaki 220, 228, 252, 269
Ohtsuki, Kazuhiro 203, 277

P
Pu, Jiaxi 42

S
Sasaki, Taiga 220
Sato, Shin-ya 94
Shao, Yongcun 3, 15
Shimozaki, Amon 187
Shoji, Yoshiyuki 211, 220, 269, 302
Su, Han 27
Su, Qinyuan 27
Su, Yancong 277

Sugiura, Kento 236
Sun, Yi 203, 277
Sun, Zhiqing 110

T
Tachioka, Yuuki 261
Trung, P. H. T. 77
Tsuda, Yuya 269
Tsuge, Yousuke 187

U
Umetani, Tomohiro 187

W
Wan, Zhiguo 129
Wang, Hui 335
Wang, Xiaonan 203, 277
Wang, Yanhao 42
Wang, Yishen 110

Wen, Po-Jen 163
Wu, Shaozhi 27

X
Xu, Fangzhou 335

Y
Yamamoto, Takehiro 211, 220, 228, 293
Yamamoto, Yusuke 302
Yao, Tiechui 110
Ye, Runfan 27
Yu, Fei 129
Yue, Siying 27
Yumoto, Takayuki 293

Z
Zhang, Wensi 110
Zheng, Kai 27
Zhou, Xuan 42

Printed in the United States
by Baker & Taylor Publisher Services